A Comparative Political Ecology of Exurbia

Laura E. Taylor • Patrick T. Hurley
Editors

A Comparative Political Ecology of Exurbia

Planning, Environmental Management, and Landscape Change

Springer

Editors
Laura E. Taylor
Faculty of Environmental Studies
York University
Toronto, ON, Canada

Patrick T. Hurley
Department of Environmental Studies
Ursinus College
Collegeville, PA, USA

ISBN 978-3-319-29460-5 ISBN 978-3-319-29462-9 (eBook)
DOI 10.1007/978-3-319-29462-9

Library of Congress Control Number: 2016936660

Printed on acid-free paper

This Springer imprint is published by Springer Nature
The registered company is Springer International Publishing AG Switzerland

Preface

An ideological and economic battle is under way at the fringe of cities and in rural areas. Landscapes traditionally in productive use for ranching, agriculture, timber, and mining are being reimagined for their scenic value for amenity real estate development and are being transformed into exurbia. While research into the transition from productive to "consumptive" uses is not new, the perspective of our book is new, as it suggests exurbia as an area of study in its own right, deserving of its own theoretical and empirical research frameworks. In short, understanding exurbia has been hampered too often by thinking about it only in terms of the urbanization of the rural. We take a bold step forward in explaining how and why changes in landscape ideology, deeply embedded in society's imagination of nature, have such far-reaching effects on emerging real estate markets, local planning, and people's very livelihoods in the communities affected. Far from vilifying developers for producing exurban sprawl, we delve into the reasons for home-buyers to choose areas of high natural beauty and, drawing on the field of political ecology, provide a "chain of explanation" discussing the role of environmental scientists, land-use planners, property owners, and land managers in propelling and resisting these changes.

The idea for this book emerged out of a series of sessions at the 2011 meeting of the American Association of Geographers entitled "Landscapes out of place? Political ecologies of exurbia." Three sessions were organized around three themes: land-use/land cover, managing environmental impacts, and regions and cultures. At the time, our goal was twofold: (1) promote comparative study of exurbia through dialogue between and among local case studies in order to make more apparent underlying parallels and connections between political economies that create regional variations in these landscapes and (2) engage the analytic frameworks of political ecology to increase our understanding of the uneven distribution of qualities of life and environmental impacts, which are the consequence of the pursuit of a particular way of life. We wanted to encourage discussion around the power that cultural representations (especially of nature) have while also attending to the ways in which aspirational ways of life materially alter landscapes and create uneven ecological impacts (both in place and through ecologies of consumption). Encouraged by the level of interest in exurban research and the sustained discussion we have had about the parallels and connections between exurban change and exurban places, we set to work on this book, which includes those studies we feel best exemplify how political ecology as an analytical approach helps to make sense of the kind of conflicts we see in exurbia.

Drawing on case studies from diverse parts of the United States and one from Australia, this book demonstrates how exurbanization entrenches environmental injustice through shifting access to resources from traditional producers to elite exurbanites. The process of "reterritorialization" occurs because of changing societal trends in lifestyle and the ways in which rural resources are valued. Exurbanization occurs because of shifting land markets away from productive rural economies to amenity-based economies. The process is aided by institutional changes made possible by land-use planning policy, land management decisions, and local decision-making around investment and development. In spite of recent financial corrections in the real estate market, especially in the United States, valuing scenic amenities and rural landscape values over productive resource values is transforming global countrysides. Not just a concern at the edge of eastern American cities and ranchland in the American West (although this is the subject of much of the current literature on exurbia and amenity migration), rural and peri-urban landscapes are being transformed globally by exurbanization.

This book will be of interest to those who are involved in land-use planning, conservation, and land development issues, especially in areas of high natural amenity or environmental value. While we have had the opportunity to describe many of our own experiences with exurbia and exurban change, we would like to hear from our readers about the similarities and differences they have witnessed. At this point in our research careers, we "know exurbia when we see it," as the landscape gives clues about its inhabitants: the appearance of houses whose owners are not depending upon the land for their incomes, such as gentrified farmhouses, ranchettes, and countryside homes hidden from the road with glass walls facing a scenic view at the back; and often the presence of a fair trade coffee shop, art gallery, and yoga classes also tends to be a good clue. At the local town hall, a proposal for something to do with water (such as a new well, rebuilding a septic system, or altering a stream) or new road infrastructure often heralds exurban change. The urban and regional planning literature, while concerned with issues related to growth management, does not focus rigorously on exurbia, and therefore our book can make an important contribution to the readings for planning schools as well. It is intended that the book be geared toward a senior undergraduate and graduate academic audience and the general educated public, especially those practitioners who deal with exurban change on a daily basis.

The journey from the idea of putting together a book to its completion is always filled with twists and turns, and the preparation of this book is no exception. We worked to overcome some common challenges with edited book collections, including lack of coherence and scholarly rigor, in that we guided authors in their theoretical approach from the beginning and then drew upon their findings to create a truly comparative research volume, which now should read as a coherent whole. In addition to the peer review of the book's detailed proposal, each chapter was initially and separately reviewed by each editor, then a revised version subject to a double-blind peer review and review by one of the other contributing authors, then subject to at least one additional review by an editor, and finally corrected by a copy editor. We would like to thank the chapter authors for their sustained enthusiasm throughout the process—while we did not set out to make such good friends of the chapter authors, we cherish the friends we have made along the way!

We are especially grateful for the assistance of Jacob McLean, who became a graduate of the Master in Environmental Studies program at York while working with us to critique and copy edit the entire volume. We also thank Kirsten Valentine Cadieux and Hannah Gosnell for their early contributions to discussions about comparative exurbia.

Special thanks to our respective spouses Mike and Sibel for their patience, encouragement, and moral support.

Toronto, ON, Canada Laura E. Taylor
Collegeville, PA, USA Patrick T. Hurley

Contents

Contributors

Jesse Abrams Institute for a Sustainable Environment, University of Oregon, Eugene, OR, USA

Jessica McCallum Breen Geography Department, University of Kentucky, Lexington, KY, USA

Lee K. Cerveny Pacific Northwest Research Station, U.S. Forest Service, Seattle, WA, USA

Michael H. Finewood Environmental Studies and Sciences, Pace University, Pleasantville, NY, USA

Nicholas Gill School of Geography and Sustainable Communities, University of Wollongong, Wollongong, NSW, Australia

Colleen C. Hiner Department of Geography, Texas State University, San Marcos, TX, USA

Patrick T. Hurley Department of Environmental Studies, Ursinus College, Collegeville, PA, USA

Peter Klepeis Department of Geography, Colgate University, Hamilton, NY, USA

Lou Martin Department of History, Political Science, & International Studies, Chatham University, Pittsburgh, PA, USA

Innisfree McKinnon Department of Social Sciences, University of Wisconsin, Stout, Menominee, WI, USA

Jacob McLean Faculty of Environmental Studies, York University, Toronto, ON, Canada

Brent Olson Environmental Studies, Westminster College, Salt Lake City, UT, USA

Kate Skaggs Coastal Communities Planner, Chesapeake & Coastal Program, Maryland Department of Natural Resources, Annapolis, MD, USA

Laura E. Taylor Faculty of Environmental Studies, Planning Program, York University, Toronto, ON, Canada

Jenna H. Tilt Geography, Environmental Sciences, and Marine Resource Management, College of Earth, Ocean, and Atmospheric Sciences, Oregon State University, Corvallis, OR, USA

Annette Watson Department of Political Science, Faculty in the Master of Environmental Studies Program, College of Charleston, Charleston, SC, USA

About the Contributors

Jesse Abrams is a research associate with the Institute for a Sustainable Environment at the University of Oregon. His research focuses on environmental governance, rural community change, and community-based natural resource management in the context of both public and private forests and rangelands.

Jessica McCallum Breen is a doctoral student in the Geography Department at the University of Kentucky whose research uses social media and collaborative mapping techniques to understand the role of art and creativity in urban place-making. Additionally, she is a research assistant with the newly created Mapshop, where she works with students, faculty, and community groups in the production of geographic representations to raise the critical mapping capacities of the community.

Lee K. Cerveny is a research social scientist at the Pacific Northwest Research Station, which is a research division of the US Forest Service. She is an anthropologist whose work focuses on understanding human interactions with national forests and other public lands along the urban-to-wildland continuum. She recently has been developing socio-spatial tools for capturing human values and land-uses for use in national forest planning.

Michael H. Finewood is an assistant professor of environmental studies and science at Pace University. He is a human geographer and political ecologist with research and teaching interests in water governance, climate change, and urban sustainability, with explicit attention to critical geographies and justice. His current work focuses on watershed-based approaches to stormwater management.

Nicholas Gill is an associate professor in environmental geography at the University of Wollongong in New South Wales. His research interests lie in environmental management and rural cultures, particularly cultural and social aspects of land management and land-use change, and household sustainability. His current projects focus on bushfire and invasive plant management in high-amenity rural areas.

Colleen C. Hiner is an assistant professor in the Department of Geography at Texas State University. She is a rural geographer and political ecologist with specialties in land and environmental management, (ex)urbanization, (rural) sustainability and tourism, wine and beer geographies, and agriculture (urban, peri-urban, and sustainable).

Patrick T. Hurley is an associate professor of environmental studies at Ursinus College. He is an environmental social scientist trained in the methods of human geography. His teaching and research interests center on human-environment interactions in cities and their surroundings. His current work focuses on the political ecology of land-use change and conservation in exurban southeastern Pennsylvania as well as the role that plant foraging plays in the everyday lives of people in eastern US cities.

Peter Klepeis is a professor of geography at Colgate University. His research and teaching focus on nature-society relationships and themes in sustainability science. Drawing on perspectives from political ecology and environmental history, he explores multiple forms of land system change. Currently, he is studying church forest management in Ethiopia.

Lou Martin is an associate professor of history at Chatham University and is the author of *Smokestacks in the Hills: Rural-Industrial Workers in West Virginia*. He is also a founding member of the West Virginia Mine Wars Museum, and his latest project focuses on various understandings of place in the coal industry in the 1950s and 1960s.

Innisfree McKinnon is an assistant professor of geography at the University of Wisconsin, Stout. Her research focus is political ecology and cultural geography. Specifically, she is interested in how sustainable development strategies relate to social justice, including the study of land-use governance and access to environmental amenities (e.g., parks, open spaces, and picturesque rural landscapes) in Oregon and Wisconsin. She also teaches geographic information systems with a focus on critical and participatory GIS.

Jacob McLean is a PhD student in York University's Master in Environmental Studies program. He also works as a freelance editor.

Brent Olson is an assistant professor of environmental studies at Westminster College and is a director of the Institute of Mountain Research. His research and teaching interests center on how we use, manage, and value natural and cultural landscapes. His current work focuses on the political economy of natural resources.

Kate Skaggs is a coastal planner who manages the CoastSmart Communities Program with Maryland's Coastal Zone Management program (MCZP), funded by the National Oceanic and Atmospheric Administration. The MCZP provides tools, resources, training, and assistance to local governments for addressing coastal hazards including sea level rise.

Laura E. Taylor is an associate professor of urban ecologies and environmental planning and coordinator of the graduate planning program at York University. Her research interests are the politics of nature in land-use planning, especially in exurbia, and she is the coeditor of *Landscape and the Ideology of Nature in Exurbia: Green Sprawl*. Her current research includes exurbia in the Toronto region and political ecology and land-use planning in Ontario, as well as climate change and land-use planning.

Jenna H. Tilt is an assistant professor (senior research) in the Geography program at Oregon State University. Her research focus lies at the intersection between natural environments, land-use planning, and human behavior. Current research projects include residential migration along the wildland-urban interface and outdoor recreation patterns, urban forest environments and human health and wellness, and adapting urban green infrastructure practices for rural communities.

Annette Watson is an associate professor in the Department of Political Science at the College of Charleston, where she teaches critical scholarship in geography. She develops inductive and deductive research designs, which are community-based and informed by local knowledges, to understand human-environment relationships at multiple scales. Her current projects are collaborative efforts with Southeastern Lowcountry communities and indigenous Alaskan Arctic communities that examine the sustainable practices of subsistence and commercial fishing economies.

Chapter 1
Introduction: The Broad Contours of Exurban Landscape Change

Laura E. Taylor and Patrick T. Hurley

Despite the common perception that the United States has become a "suburban nation" (see, e.g., Duany et al. 2010; Jackson 1985), exurbia has emerged as the dominant settlement pattern across the country (Berube et al. 2006), characterized by different patterns of development and different lifestyle expectations from cities, towns, and suburbs, with houses in scenic, natural areas on relatively large acreages (often with one house per 10, 20, or 40 acres or more). In fact, the total amount of land under development in exurbia is expected to increase—not only in the United States, but also Canada, Australia, and New Zealand, and increasingly in other countries—in spite of the collapse of the housing market in the late 2000s (Brown et al. 2008; Cromartie and Nelson 2009).

In this book, we approach the study of exurbia from a political ecology perspective, where exurbia is a landscape created by the fusion of urban and rural ideas, processes, and materialities. Studying exurbia is motivated by our interest as human geographers in the study of the shared destinies of people and nature in the environment. As a place, "exurbia" is often thought of and characterized by very low-density rural residential development, which is home to people who have left the city and its suburbs in search of a more rural lifestyle closer to nature. In exurbia, people "are able to live surrounded by fields and forest, birds and wildlife, seemingly far from the stresses and cares of work and society, but with all the modern conveniences, access to services, and connections to the city that modern life seems to require"

L.E. Taylor (✉)
Faculty of Environmental Studies, Planning Program, York University,
4700 Keele Street, Toronto, ON, Canada, M3J 1P3
e-mail: taylorL9@yorku.ca

P.T. Hurley
Department of Environmental Studies, Ursinus College,
P.O. Box 1000, Collegeville, PA 19426, USA
e-mail: phurley@ursinus.edu

L.E. Taylor, P.T. Hurley (eds.), *A Comparative Political Ecology of Exurbia*,
DOI 10.1007/978-3-319-29462-9_1

(Cadieux and Taylor 2013, p. 3). As more people migrate to and settle in rural areas, local communities undergo sometimes-cataclysmic transformations as previous ways of life are disrupted and new ones emerge. We like the term exurbia to describe the result of these rural transformations where exurbia emerges as a landscape that is no longer rural but is not urban either, but rather a particular kind of place where the urban and rural are "intermingled" (Taylor 2011, p. 324) or "fused" (Woods 2009, p. 853). Exurban places are still rural in some ways but economic and social changes mean that how people think about and value their environment have undergone a major shift. In exurbia, the environment is valued for its enduring natural beauty in contrast to the suburbs where the environment has been urbanized to the extent that nature is usually present only where it has been able to regenerate following development. This book explores how rural areas are transformed into exurban areas, describing the social–ecological changes through which the material and symbolic landscape of exurbia is formed. Specifically, we discuss how exurbia is shaped by land-use decision-making. Our goal is to explore a diversity of cases from around the United States, plus one case from Australia, in order to study the dimensions of exurban change, which are surprisingly similar irrespective of the distinctive landscapes and places where they play out.

By identifying and comparing the dimensions of exurban change in several disparate places, we seek to provide land-use planners and decision-makers with insights about exurban change and seek to foster a broader discussion about achieving more sustainable futures in these transitional places. In doing so, we highlight three distinct ways in which the standard literatures on exurbia, including those examining peri-urban issues or the urban–rural fringe or interface, fail to address certain aspects of exurbia's political dynamics: first, they frequently oversimplify political conflicts into useful, but not entirely accurate binaries (e.g., long-time locals vs. newcomers, exurbanites vs. developers, and environmentalists vs. everyone else); second, they privilege particular actors and knowledge sets (e.g., they privilege Western ideologies and conservation science over local and traditional ecological knowledge, and scientific study over lived experience of place); and third, they remain relatively blind to key actors and their motivations, disregarding the potential for coalitions and cooperation among actors often seen as hostile to one another, such as land developers and livelihood users such as ranchers and farmers. In the 11 chapters that follow, we explore the dynamics of exurban change in a variety of places and contexts. Our perspective as editors derives from our positions as faculty members in departments of Environmental Studies in the United States and Canada, where our scholarship on human–environment relationships is motivated by a desire not only to know more about the dynamic political ecological processes shaping the world, but also to apply these insights toward better and fairer forms of intervention. We hope these cases will help encourage readers, including other scholars and practitioners, to consider the diverse ways that the dynamics of exurbanization unfold, while teasing out implications for landscape, land-use, and environmental management and change.

Studying environmental issues related to ecological and social change in exurbia has been a challenge especially because "urban" and "rural" have proven to be ubiquitous, enduring, and ultimately limiting categories for undertaking exurban research. In many disciplines, like geography, sociology, and economics, "urban" refers to geography of cities, suburbs, and towns while "rural" basically refers to everything else (Lichter and Brown 2011; Scott et al. 2013; Woods 2011). Both are powerful analytic categories for the study of society and space, but neither has provided the conceptual depth to make sense of exurbia, this decidedly non-urban, non-rural landscape where so many people live. Exurbanization has blurred the distinctions between urban and rural land uses across the United States and in many other countries, which is why we believe greater attention should be paid to how development is remaking the landscape in exurban areas.

In the chapters that comprise this book, the authors describe changes taking place in landscapes characteristic of this fusion of urban and rural. These include parts of the American West, including California, Washington, and Oregon but also the Carolinas and Pennsylvania on the east coast, and Sydney's hinterland in Australia (Fig. 1.1). Each chapter describes the role of competing rural capitalisms in their particular case studies, highlighting how and where working landscapes such as ranchlands, orchards, vineyards, and forests have come to be intermingled with homes built to capture and satisfy the imaginations of amenity migrants. The authors consider who is participating in the transition (or not), detailing who is organizing to welcome or contest the processes at play, and specifically where these actors are situated in relation to the changing ideologies, economies, and social relations of that given place. The authors focus on the political economies of change and especially on the local political, planning, and regulatory processes through which landscapes of exurbia are made. The concluding chapter reflects on the insights these

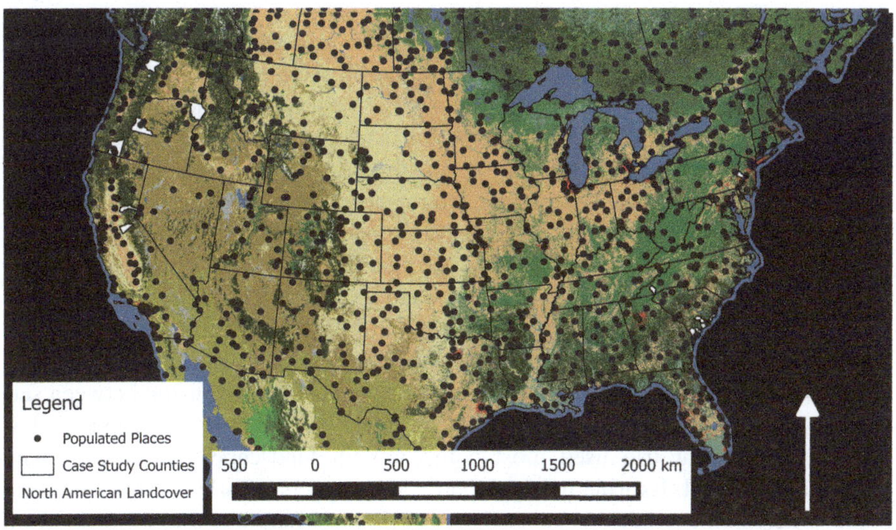

Fig. 1.1 Map of exurban counties in U.S. case study chapters. *Source*: Landcover and Populated Places courtesy of the Commission for Environmental Cooperation 2014. U.S. Counties courtesy of U.S. Census Bureau 2015. Produced by Patrick Hurley using QGIS 2.10.1 software

cases have for thinking about the question of exurban political ecological analysis, including insights for exurban places outside of the U.S. and Australia.

Next in this Introduction, we more fully introduce the idea of exurbia before turning our attention to the theoretical framework underpinning the book's approach to its study. This framework, what we call a comparative political ecology of landscape, is then discussed at length, including two concepts integral to this framework: (1) the concept of competing rural capitalisms, which we believe is a major contribution to future studies of exurbia because of its ability to help researchers see how both production and consumption continue over the long term in exurbia, sometimes in political economic competition with each other, but often symbiotically; and (2) uneven environmental management, which is a concept that captures the ways in which decision-making about land use is often highly institutionalized, includes similar types of actors in a variety of exurban situations, and responds to multi-scalar economic, social, and political shifts, but yet is highly contingent upon local landscapes and environmental histories. This introductory chapter concludes by offering a series of traits that emerged from our analysis of the similarities between case studies in this comparative political ecological research on landscape change. To this end, we detail "seven markers," or widely shared characteristics that partially define the broad contours of exurban landscape change. We see these markers as a useful diagnostic for future discussions within what we hope will be even more systematically comparative studies of exurbia.

1.1 Understanding Exurban Change

The term "exurbia" is attributed to A. C. Spectorsky, a journalist who wrote a book entitled *The Exurbanites* in 1955 satirizing the lifestyle of those who commuted from their country homes to work in Manhattan. He explored the ideal and reality of those who saw the countryside as an escape from the chaos of the modern city but for whom the escape came at a high personal price (and, as we see today, a high ecological and social price, too). While geographers had long realized that there was a distinctive form of residential settlement beyond the suburbs, it was arguably not until the 1970s that scholarship began to describe exurbia as the furthest edge of the commuting zone beyond the edge of cities and to delineate areas characterized by very low-density homes in areas of scenic natural beauty (Fava 1975; Lamb 1983; see also Taylor 2011). Today, we use the term to describe more than a geographic zone; the term has come to denote a landscape sustained by a tension between settlement and conservation. As such, exurbia provides an important conceptual addition to the familiar trio used most commonly in landscape taxonomy: urban, suburban, and rural. Exurbia neatly captures the concept of urban-connected rural living, where some residents choose to live as close to nature and as far away from the city as they can, despite the connection in their daily lives to economic and social spheres in city and suburban centers (Taylor 2011; Taylor and Cadieux 2013). Based on our studies, however, we have concluded that exurbia is not just another form of urban living or of rural living; it is an urban–rural hybrid purposefully created and maintained as the antithesis of both urban and rural life. Moreover, exurbia is "recognizable as a landscape of housing set in a matrix of vegetation—'nature'—that is different from urban greenspace, which is usually vegetation and other 'nature' set in a matrix of the built environment" (Taylor and Cadieux 2013, p. 2). While the landscape may look rural, the dynamics producing the exurban landscape are based in amenity rather than resource economies and related landscape values. While these dynamics may seem urban, however, the imperative of conserving nature as a permanent part of the landscape motivates very different social, cultural, and political relationships with the land than is the case in most central city and suburban areas. Conceptually, therefore, exurban landscapes are neither urban, suburban, nor truly rural because the processes that shape exurban landscape change are so very distinct from urban and rural contexts.

Exurbanization is not, as some observers would have it, just a step in the process of urbanization: that is, a step on a continuum of transforming rural places into urban ones (e.g., Lang 2003). Although sometimes exurban places, over time, will be overwhelmed by the suburbs of a growing metropolitan region and will be encompassed by a city, in this book we are interested in exploring the idea of exurbia as a stable, long-lasting condition. We think that this may be a revelation for some readers who see exurbia as a spatial zone where the rural countryside is transitioning to city or suburb. Instead, we focus on exurbia as a permanent landscape where urbanization is actively opposed.

How and why does exurban change happen? What are the dimensions and dynamics of exurban change? Exurban landscapes are the result of shifting global economic conditions and capital flows, which compel change in almost all aspects of exurban life. In "developed" countries, economies have largely—but not entirely—shifted from predominantly resource production and manufacturing to knowledge-based sectors (Winkler et al. 2012). As the case studies in this book demonstrate, rural places undergo transformative change as economic restructuring on a global scale is seen and experienced at the local landscape level. So, too, the effects of nearby cities and their expansion may serve to drive some aspects of these rural transformations (Sandberg et al. 2013; Walker and Hurley 2011). Places once valued for their raw timber, ranch lands, and mineral and energy resources in earlier days of industrial capitalism have seen extraction and production move to other countries in the "global south." The trees, water, and rocks that were the raw material for the industrial economy have *largely* lost their industrial commodity value. The same trees, water, and rocks are instead often revalued as amenities for exurban lifestyles and provide the natural resource foundation for new real estate economies. These economic revaluations are reflected in the visible landscape. Landscapes undergoing a rural to exurban shift continue to have a distinctive rural character but are increasingly seen as places of high natural beauty connected to urban centers by road, air, and telecommunication technologies. These landscapes are characterized by increased real estate values and a corresponding increase in contentious local politics related to land-use planning and environmental management practices, all of which shape landscape change and produce the settlement patterns we recognize as exurban.

Through the process of exurbanization, resource economies once related to the extraction of objects of nature (trees, metal, coal, etc.) are incompletely replaced by exurban economies where profit is maximized by keeping the forests and mountains (mostly) intact. Whereas timber, minerals, and water are resources for exploitation in a resource economy, these same natural elements become objects of amenity in an exurban economy, most often expressed through real estate values. Over the long run, however, exurbanization can be just as environmentally destructive as extraction. For example, new home building and landscaping interfere with the movement of wildlife and the flow of streams and stormwater, disrupt indigenous and other local traditions of hunting and fishing, and introduce exotic plant species that interfere with established native species. To address destruction and degradation caused by exurban settlement, state intervention in exurban economies, especially through land-use planning and decision-making, is often sought by local communities. For example, increased nature conservation and land stewardship are often seen as legitimate measures for limiting the negative impacts of exurbanization, often with the intervention itself upsetting long-standing social relations and cultural practices for people in exurbanizing communities.

How people live in the landscape is, of course, deeply related to their livelihoods, where "livelihood" is meant in its broadest sense: that is, people's means of social and economic support. Today, people's livelihoods no longer have to be economically connected by their labor to the land on which they live: for example, a ranch may be home to a corporate executive who commutes from home to a company based in a distant city via the internet or by air. Or an exurbanite living on a farm may commute by car to a nearby city and work for a multi-national company with a global reach. In both cases, living on a ranch or a farm is a lifestyle choice. Although neither makes their living from the land, and both typically rely on paid labor to maintain the landscape, the land remains an important part of an exurbanite's livelihood strategy, as real estate investment is, for many, a key form of capital investment and long-term income (Marcouiller et al. 2011). Nonetheless, the landscape level is the spatial scale where people's lifestyles and livelihoods influence their engagement with the land, their sense of place, and their identities. It is through analysis of political ecology at the landscape level that we can comparatively study the use of the land and related impacts by, say, the ranch-owning corporate executive and a cattle rancher/landowner (see e.g., Gosnell et al. 2006) as people's livelihoods transform landscapes through individual and community land-use choices and environmental management approaches. In any given place, a landscape's transformation from rural to exurban may be highly contested by those whose lives and livelihoods depend on the valuation of the landscape's objects of nature as either amenities for conservation or resources for industrial commodification, or both. As will become clear through comparison of the case studies in this book, the interaction of global and local processes influences livelihoods, land use, and landscape change and these dynamics are an important part of understanding exurbia.

Exurbanization is driven by the in-migration of urban people and their capital to rural areas. As such, exurbia can be seen as derived from the arrival of so-called "amenity migrants," or individuals seeking an idealized rural lifestyle in areas of high natural and scenic beauty (Gosnell and Abrams 2011; Taylor and Cadieux 2013). Amenity migrants, for our purposes here, are people in the process of transitioning from urban to exurban places (for a review, see Abrams et al. 2012), while the term "exurbanites" refers to those who have settled and who often then become engaged with the politics of exurbanization. With the influx of amenity migrants, local places undergo social and ecological upheaval and are no longer held together by economic and social ties associated with economies of resource production. Instead, these places are increasingly interlinked with multiple cities through markets related to the knowledge economy, real estate, and tourism, and are digitally linked to communities extending well beyond the local. There is a substantial body of study on amenity migrants and the push/pull factors behind their moves (Abrams et al. 2012) and although we extensively discuss the participation of exurban "actors" in the negotiation of landscape change, we acknowledge that there is still much to be learned in future studies about the experience of people living in exurban areas, especially in terms of gender, race, and class dimensions.

Exurbanization often also means the arrival of new ways of thinking about land-use and environmental management. By environmental management, we refer to approaches taken to manage change in a given environment, like tree planting and clearing of invasive species, where those approaches are influenced by the ideology of nature and ideas about the social, cultural, and ecological dimensions of a place that constitute the environment in question. The ideology of nature refers to common sense understandings of "what is culturally acceptable, or potentially even possible, to say about a topic such as nature" (Taylor and Cadieux 2013, p. 10). Importantly, environmental management practices related to resource economies are very different from those related to exurban economies. That is, exurban environmental management approaches are the result of tensions between (at least) two types of economies: "'traditional' natural resource-based production and 'new' economies and cultures of aesthetic landscape 'consumption'" (Walker and Fortmann 2003, p. 470). For Walker and Fortmann, these tensions are encompassed by the term "competing forms of rural capitalism," and we have adapted the conceptual framework of competing rural capitalisms as an important aspect of making sense of exurban landscape change, including specific environmental management dimensions, as discussed below and throughout the chapters of the book.

1.1.1 Competing Rural Capitalisms

Exurban landscape change has been studied from many angles by human geographers and political ecologists over the years, including by us, the editors (see the special issue of *GeoJournal* edited by Cadieux and Hurley 2011; see also Cadieux and Taylor 2013). Geographers have studied landscape change outside of cities using diverse methods and theoretical perspectives, including those grounded in demography, social and political geography, cultural landscapes studies, and environmental studies (see Taylor 2011 for a review). Our comparative political ecology of landscape approach is in contrast to studies that focus solely on the social drivers of exurban change (Benson and Osbaldiston 2014), or on changes in social relations and cultural practices in rural areas from a sociological perspective (e.g., Schewe et al. 2012), or on the implications of habitat fragmentation from a conservation biology perspective (Brown et al. 2008; Theobald 2005), or on the economics of exurbanization (Marcouiller et al. 2011). We focus more comprehensively on how shifts in global capitalism generate similar local outcomes across places experiencing rural to exurban landscape change.

At its most basic, capitalism is understood as the dominant economic system of our time, where the means of production of the goods and services that people need and want are privately owned (see Williams 1976, pp. 50–52). As the dominant economic system, capitalism is seen as the foundation for the social system because of the influence of the economy on social relations. Under capitalism, ownership of the means of production is limited to relatively few people who are mostly looking to increase their own wealth. Those who own the means of production depend upon others—the "working class"—through whose labor goods and services are produced and who purchase most of those goods and services. The goal of the owners is to find new ways to increase their profits and the goal of the working class is to increase the value of their labor so they can afford more goods and services (see Watts 2009 for a succinct overview, including the influence of Karl Marx and others on the present day understanding of capitalism as a social and economic system).

Exurbanization is a process whereby one form of capitalism (based on industrial resource extraction) comes into competition with another form of capitalism based on real estate and other non-extractive measures of value (Gosnell and Abrams 2011; Travis 2007). The transition from productive to consumptive capitalism is described as the process of "reterritorialization" by Brogden and Greenberg in which an "interest group redefines commodity values and achieves the power to rearrange access rights to a natural *system* so that earlier commodity values become obsolete and disprivileged" (2003, p. 291, emphasis in original). In their article, Brogden and Greenberg do not acknowledge Deleuze and Guattari's writing on deterritorialization/reterritorialization (1987), which is foundational to our discussion of competing rural capitalisms. Deleuze and Guattari (1987) explore the destabilization of cultural meaning and help us to consider how landscape values that dominate in a particular place can become destabilized and break apart, making room for alternative understandings of a place. Reterritorialization makes way for new ideologies and meanings, a process exemplified by Brogden and Greenberg in their case study of Arizona, where ranchers lost access to grazing lands on both public and private lands. When their neighbors' private ranches were sold off to amenity migrants, grazing leases on public lands were withdrawn by public land managers in response to the exurbanites' desire for the conservation of "wild" natural areas. Such outcomes highlight the ways exurbanites use power and money to change the cultural valuation of the exurban landscape, thereby challenging natural resource use in host communities and dramatically changing environmental management practices and ultimately the landscape itself.

The idea of reterritorialization has been useful in understanding the power of shifting landscape values in exurban change, but existing scholarship does not systematically acknowledge the role that land-use planning and decision-making at various levels of government play in facilitating reassignment of access to or control of resources from one group to another. As we will discuss, scholars in the U.S. tend to focus on the role of the federal government in regulating access to public lands. However, in a place where the value of land and its appropriate uses is determined through local land-use planning processes and not through government-centered decision-making processes, this lack of engagement by scholars with planning and territory represents a gap in research.

The process of transferring resource access from one group to another is also described in Robbins' (2006) study of elk management in the Northern Yellowstone region of Montana. There, many long-time Montanans are losing access to customary areas to hunt, as wealthier hunters from out-of-state increasingly own and control private land historically used for access. In his study, Robbins (2006) examines how the environmental knowledge of local hunters differs from other stakeholders involved in the formation of elk management policy. Seizing upon scientific knowledge to legitimize their goals, environmentalists and state wildlife managers claim greater legitimacy as a means to combat the environmental knowledge (and supposed self-interest) of local hunters. By impugning the environmental knowledge of locals, environmentalists, state wildlife managers, and other stakeholders prevent that knowledge from being translated into policy and thus contribute to wresting resource control away from local hunters and long-term rural residents. Often in the reterritorialization process, the conflict is framed as a moral conflict, which is especially poignant where locals see hunting as a cultural tradition and livelihood practice in contrast to environmentalists who see hunting as murder. Each side attempts to vilify the other in spite of the fact that all sides are seeking outcomes that they imagine will benefit their community (Brogden and Greenberg 2003).

In the Arizona and Montana cases, amenity migration is a part of the process of reterritorialization as places undergo the transition from rural to exurban. Gosnell and Abrams (2011) define amenity migration as a distinct pattern of human migration characterized by the seasonal or permanent movement of largely affluent urban or suburban populations to scenic/nature-rich and/or culturally rich rural areas. Additionally, in their conceptualization of amenity migration drivers, Gosnell and Abrams stress the globalized aspects of the phenomenon. Traditional natural resource extraction industries in developed nations, including the U.S., are disinvesting in many areas, in favor of timber and mining in less-developed regions of the globe with lower wages and weaker state oversight. This departure of rural industries in the U.S. opens up the opportunity for in-migrants who, because of wealth, age, or advances in technology that allow them to be increasingly mobile, begin investing in these increasingly post-industrial (economically depressed and depopulated, yet amenity-rich) areas. Over time, local rural economies shift as tourism and second home markets begin to dominate (Gosnell and Abrams 2011). As the Arizona and Montana cases demonstrate, the dominance of tourism, second home, and exurban housing markets accelerates the economic shifts involved, as both tourists and in-migrants compel the redefinition of environmental imaginaries[1] and perceive many types of traditional extractive land uses in a negative light.

Beyond these global-scale rural restructuring forces, there are also forces at the regional, state, and local levels that enable or even encourage nonmetropolitan growth associated with amenity migration. Government investment in infrastructure, both transportation and communication, in support of rural development in depressed areas, helps to make rural areas more accessible to amenity migrants as well as locals. Pro-growth policies, including government subsidies such as tax breaks and other public investments, further encourage businesses to relocate to rural areas (Travis 2007). Local industries may take advantage of these dynamics to transform themselves within this context. The internal social, political, and institutional structures once supporting traditional extraction may shift to encouraging exurbanization, thereby enabling residents to benefit economically from the natural amenity values of the landscape (Colladoes and Duane 1999; see also Olson, Chap. 6). The concept of reterritorialization is useful as it captures the power shift in the political economy as the rural-to-exurban commodity-to-amenity valuation of the landscape takes place. However, reterritorialization in the existing literature oversimplifies the political ecological changes taking place by giving the impression that the shift from productive to amenity use is absolute.

Therefore, we suggest competing rural capitalisms as a means to see how both production and consumption continue in exurbia. As discussed by Breen, Hurley, and Taylor (Chap. 9), "rural capitalisms" are economic but also ideological, in that people compete with each other to make sure their ideas about nature shape the landscape: "The concept of competing rural capitalisms considers the shift in ideologies of nature as necessary to understanding the shift in political economic power relations that is the focus of reterritorialization" (p. 200). But while people can sit on their barstool and argue about the kind of landscape they wish to have in their community, it is land-use regulation that inscribes the more politically/economically powerful person or group's vision into that landscape.

Changes in state regulation of land use are characteristic of the shift in rural economies. Land-use planning regimes are often almost non-existent in pre-exurban rural areas, as has been the case in many areas of the American West (Ghose 2004; Robbins et al. 2012, with the notable exception of Oregon; see Chap. 6, where Olson traces the history of change in Oregon's support from agriculture to exurban-style development). But even in places with land-use planning in place, exurban residential development triggers increased government intervention in the property market through an often-cataclysmic change in local power relations and politics.

[1]The concept of environmental imaginaries (Peet and Watts 1996) is useful in political ecology for studying the relative influence of particular ideologies of nature in landscape politics, where ideologies have the power to define the nature of nature in a particular place (e.g., "endangered species habitat," "natural heritage system") and to determine the corresponding land uses permitted. (For a detailed discussion of the use of the concept of environmental imaginaries in exurban political ecology, see Chap. 2).

Representing the rural to exurban change as a purely urban/rural or newcomer/long-time resident culture clash and power shift does not tell the whole story. Walker and Fortmann (2003; see also Hurley and Walker 2004) observed that amenity migrants often find political supporters among long-time locals, who object to particular types of natural resource extraction and wish to take control of the growth and land management agenda. Together this coalition of actors with interests around landscape values seeks to influence decision-making about regulatory practices and priorities, including challenging traditional *and* emerging land-use practices through land-use planning processes (Hurley and Walker 2004; Walker and Fortmann 2003). For example, Abrams (Chap. 4) found that conflicts over exurban development in Wallowa County, Oregon did not fall neatly along old-timer/newcomer lines but rather represented coalitions of strange bedfellows. Further, real estate developers may influence the creation of more standardized forms of development approval within land-use decision-making processes, in part because they seek to protect natural amenity in the landscape as a form of resource commons (Robbins et al. 2012). In so doing, new types of land-use planning approaches or decisions in exurban contexts are not just intended to shape the built environment.

Landscape values rooted in historic forms of extraction, such as ranching, logging, and mining, are the basis for resistance to those new uses of the land, which are set out in new planning regulations, for example open space conservation and residential development. Land-use regulatory interventions are seen as necessary to protect a continuing tradition of extracting economic value from local landscapes. Whereas extraction once harvested the productive powers of the land, residential development takes advantage of the land for building sites with spectacular amenities. Determining whether there is too much development and whether that development harms the amenity values is of much less concern. Indeed, most forms of regulation that would reduce the creation of building entitlements—i.e., in the form of newly created parcels—are seen as anathema to property rights. Likewise, restrictions on where and how to build are seen as reducing the profit-making potential of the land, just as environmental protections put in place on ranching, logging, or mining would be.

Competing rural capitalisms captures the idea that there is more than a simple dialectic between historic extractive use on the one hand and amenity use on the other. As we have discussed, the literature characterizes the exurban shift as one from productive to consumptive uses but fails to problematize the complexity and contradictions within the shift and then the hybrid landscapes that emerge. The transition from *Gemeinschaft* to *Gesellschaft* (the idea that the rural world is being lost to the vagaries of modernization as described in 1887 by Tönnies; see Bunce 1994, p. 14) is never complete in exurban areas. Where rural lands are urbanized or suburbanized there is a clearer change; once (sub)urbanized, these lands are no longer places of industrial commodity production (timber, minerals, crops) and instead become places designed for the consumption of the landscape through urban uses (such as fully serviced subdivisions or shopping centers). Exurban landscapes, however, are in flux with many kinds of values and ideologies of nature in competition. Exurban places are neither dominated by extractive rural nor amenity uses, nor is the shift in community relations simply from old to new; the value of nature is always in tension, thus the concept of competing rural capitalisms makes sense as more than the temporary and teleological process of reterritorialization.

The new exurban planning framework secures the natural landscape value dimensions of newly emerging economies and influences how socio-cultural values such as quality of life develop. We encouraged all chapter authors to draw out the land-use planning processes in their case study areas. For example in Chap. 3, Hiner describes planning as the political arena in which residents came together to articulate their contested ecologies or landscape visions for Calaveras County in California. In Chap. 9, Breen, Hurley, and Taylor describe the introduction of land-use zoning in the form of steep slope ordinances to limit the construction of exurban homes in the mountains of North Carolina. Similarly, in both Chaps. 7 and 11, authors Watson and Skaggs, and Tilt and Cerveny describe the introduction of complex land-use regulation to manage landscape change resulting from the transformation of resource industries into real estate development companies.

The complexity of the competing rural capitalisms framework is directly discussed by several chapter authors. In Chap. 9, Breen, Hurley, and Taylor examine "the ways in which different industries and associated economic actors are tied to different ways of extracting value from landscapes and landscape features" (p. 198). Likewise, in Chap. 5, McKinnon's case study of Jackson County, Oregon provides an excellent example of an exurban landscape maintained over the very long term by *complementary* rural capitalisms: orchards and residential real estate investment. But while complementary at the local level, at the state-level agriculture is seen as being threatened by or competing with sprawl and the recent focus on land-use regulation runs the risk of upsetting the balance in the future.

1.1.2 Uneven Environmental Management

The complexity of the exurban change may be shaped not only by competing rural capitalisms, but also by the differing strategies of environmental management that these actors may embrace. In research on the emergence of conservation interventions in western Canada, Reed (2007) suggests that the changing drivers of amenity migration, when mixed with local politics, lead to "uneven environmental management" outcomes. That is, different combinations of civil society come together to protect iconic landscapes and vulnerable species in biosphere reserves through different mechanisms. These protection efforts create new forms of public and private conservation territories, with consequences for how these landscapes can be used (i.e., whether or not the landscapes can be used for continued extraction, for passive recreation, or whether local people can even access them for so-called non-consumptive uses like foraging or hiking) (Sandberg et al. 2013). For example, in coastal South Carolina, private residential developments planned and designed to conserve environmentally significant landscapes as a permanent part of the community contribute to a loss of traditional access to natural resource supplies for African American livelihood users (Hurley and Halfacre 2011). In both cases, the research illustrates aspects of a landscape protection-oriented rural form of capitalism. The notion of uneven environmental management, in a conceptual form applied to exurban land-use decision-making, is helpful for thinking about the coalitions of people and interests and the acceptance of some rules and the outright rejections of others within relatively similar contexts. Thus, even when similar globally influenced ideas about landscape protection and ecological restoration are at play, local social and political relations mediate how those processes shape local landscapes.

The rural to exurban transition is at its heart a change in the types of environmental management that once characterized rural economies. Reed sees the governance of the complex interconnections in these places as the intersection of property exchange, reterritorialization, valuation, and planning. For her, environmental management regimes are "formal and informal institutional arrangements through which public, private, and civic interests work simultaneously (together or apart, in synch or at odds with one another) and within different sets of power relations to influence, make, and/or carry out governing decisions about environmental and resource management" (2007, pp. 321–322). An example of the reshaping of a landscape through the public process of reterritorialization is King County in Washington State, as described in Chap. 11 by Tilt and Cerveny. They describe the actors involved in the process of shifting from forestry to residential development where the future of the natural environment in the County was at the heart of land-use planning and politics. Theirs is an excellent example of a transformation in environmental management with a very uneven outcome across three local communities in the same County, with some areas set aside for nature conservation, some for conservation but with permitted activities related to recreation, and others designated for residential development.

Such efforts to navigate this resource transition are not novel, however, as Olson's case study of Deschutes County, Oregon demonstrates in Chap. 6. Olson's study illustrates the economic conditions and power dynamics through which the agricultural to amenity recreation transition occurs. In his research on the creation of Black Butte Ranch and other exurban residential settlements in Central Oregon, he follows individuals in government who changed state regulations to accommodate real estate development and corporations that changed their companies from the forest industry to benefit from more profitable real estate development, replacing timber. In the Wallowa County case, also in Oregon, Abrams (Chap. 4) describes how different actors had access to different institutional tools related to land-use planning, including state-wide policy, local county authority under that policy, legislative and voter-approved exceptions to the policy, historic preservation laws, rights of appeal, etc. Abrams shows how actors used their political, economic, and discursive powers to bring these tools to bear to achieve their desired goals for the landscape.

Finewood and Martin's case study of Beaufort County in South Carolina (Chap. 8) provides a useful examination of an environmental management regime that has been in existence now for several decades. In the context of this book, this research is thought provoking because it demonstrates how the contemporary landscape—highly valued for its low country tidal beauty—is actually the site of an epic challenge between a proposed "productive" industrial use and amenity-based use values. The success of the amenity valuation of the natural landscape is so complete today that the memory is lost of both the political choice between the proposed industrial plant and the displacement of previous informal resource-based uses of the landscape. The conflict between rural environmental imaginaries is complete in Beaufort County as the possibility of a rural landscape dominated by heavy resource extraction has been defeated; the exurban landscape prevails, with a mix of small-scale fishing and amenity development.

Klepeis and Gill (Chap. 10) provide an excellent detailed discussion of the challenges that communities living in hybrid exurban landscapes face. Their research on invasive species in New South Wales, Australia is a story at the heart of an emerging new management regime in which ranchers and exurbanites are considering alternative ways of approaching ecological management, including forming alliances to improve knowledge exchange, as well as environmental regulation. In contrast to other chapters in the volume, Klepeis and Gill are not focused on explaining the reasons for the unevenness of access to resources. Rather, they examine the problems that emerge as a result of that unevenness and detail the problems associated with this new environmental regime, where both productive and consumptive rural capitalisms compete in shaping the exurban landscape.

1.2 A (Comparative) Political Ecology Approach to Understanding Exurban Change

In this book we employ a comparative political ecology to study exurban change. By comparative we mean an approach employing key tenets of political ecology to study multiple cases from distinct places. For some readers, the idea of a comparative political ecology may seem novel. To some extent this is true, but as we show below, there is a subset of the political ecology literature that has engaged with this idea and approach (e.g., Reed 2007). For other readers, the field of political ecology may be entirely new and so we first offer an introduction to the field.

As an interdisciplinary critical social science, political ecology has several goals, which are summarized by Robbins (2011, pp. 15–16). These goals include studying the connections between global economic forces and local environmental management choices, and seeking to understand the politics of access and control behind environmental injustices. Robbins (2011, p. 99) describes the methodological approach of political ecology as a "hatchet" and a "seed," two tools for focusing on "equity and sustainability" where the hatchet offers a way to "take apart flawed, dangerous, and politically problematic accounts," and where the seed promises "grow[th] into new socio-ecologies." The goal of our book is to take a hatchet to oversimplified accounts of urban–rural relationships and to offer competing rural capitalisms and uneven environmental management as the conceptual seeds for an understanding of exurbia leading to more sustainable land-use planning.

A detailed discussion of political ecology's history as an approach to doing research is set out in Paulson et al. (2003), where political ecology is described as an integrative approach emphasizing the dialectical relationship between society and the environment. According to them, political ecologists generally study the biophysical and ecological environment "together with human knowledge and practice" (Paulson et al.

2003, p. 205). In this way, nature and the environment are not merely something "out there" and separate from humans, but rather are something of which humans are inherently a part and about which humans accrue knowledge through different modes of interaction. In commenting on the many transitions facing rural places worldwide, Woods (2011) discusses the difficulty in conceptualizing the phenomenon of exurban change and identifies the benefits of considering exurbia a hybrid, relational space, which can only be fully understood by considering both natural and social relationships together. The idea of hybridity (Latour 2004)—studying nature and society together—is foundational to political ecology and is essential to make sense of places like exurbia. While many scholars and practitioners have a tendency to focus on either ecological processes, hydrological processes, demographic change, or the number of development permits, political ecology insists that an integrative approach (i.e., seeing all of those as mutually constitutive) is necessary to understanding the larger picture. In other words, political ecologists do not view environmental science and management as external to social relations, but rather as culturally and politically situated. As such, environmental management practices are understood by political ecologists to be inherently political, where politics are defined as "the practices and processes through which power, in its multiple forms, is wielded and negotiated" (Paulson et al. 2003, p. 209). This political ecological understanding leads to the recognition that science, far from achieving so-called "objectivity," often has the effect of "legitimiz[ing] the interests of certain groups over others" (Paulson et al. 2003, p. 209).

A political ecological analysis, then, understands exurbia as produced by the interrelationship of economic, societal, cultural, political, institutional, and ecological processes, especially those related to land-use planning and decision-making (Walker and Fortmann 2003). As the economy shifts from more resource extraction to more amenity real estate, all other aspects of life shift with it and the landscape is where these processes converge. These shifts may be characterized by extreme conflict (Hurley and Walker 2004), or a mixture of conflict and cooperation coined as "coopetition" by Larsen and Hutton (2012), or they may even be complementary (see McKinnon, Chap. 5). Therefore, relational theory (Murdoch 2005) is also important to a political ecological approach to understanding exurbia. Relational theory offers the post-structural perspective that space and society are co-constituted, meaning that "space is made not by (underlying) structures but by diverse (physical, biological, social, cultural) processes; in turn, these processes are made by the relations established between [human and nonhuman] entities of various kinds" (Murdoch 2005, p. 19). Using a relational approach, we realize that it is at times difficult to pinpoint exurbia as a spatial zone because exurbia is constituted by intersecting processes and ideas that are often invisible—that is, not clearly indicated by visible, physical markers on the landscape. For example, speculative real estate investment is difficult to see but often underpins scenic near-urban farmland. Likewise, projects undertaken to inventory and map areas of environmental significance are not visible, but constitute the "paper landscape" discussed in county meeting halls. In our work, we make visible what might otherwise be invisible political processes, especially regarding land-use and resource management. For example, in Chap. 2, Hurley and Taylor discuss the conflict over the future of Nevada County,

California where, in spite of decades of political negotiation over the extent of natural landscape protection, no homes have been built and the visible landscape still looks much the same as it did in 1995 when the first plan was proposed. Similarly, in Chap. 8, Finewood and Martin describe the historiography of race and class conflicts in Hilton Head's tourist landscape, where these conflicts are no longer visible due to the success of the processes that fashioned the contemporary amenity landscape.

Another key theme to emerge within the political ecology literature is the focus on the evolution and deployment of particular geographies associated with conservation practice. Political ecologists point out the ways in which conservation practices often rely on both the "top-down" logics that enforce particular spatialities and the efforts to privatize the types of nature or ecologies identified especially through the use of supposedly apolitcal environmental science (Zimmerer 2006). In particular, Zimmerer (2000, 2006) describes a proliferation of new environmental management schemes, including a pervasive use of land-use zones that literally "contain in space" the practices of humans. For example, it is common in North America for wildlife habitats to be identified as zones for conservation by the international non-profit organization Ducks Unlimited through high-level, top-down negotiations with government as well as sometimes through privately negotiated schemes. Similarly, but on a regional scale, the Heritage Conservancy, as described by Hurley and Taylor (Chap. 2) in the Quakertown, PA case study, depended on environmental science to make the case for conserving large areas of the Quakertown Swamp. Wildlife habitat conservation zoning severely limits most public uses. For political ecology scholars, these interventions raise critical questions about access and control.

In fact, Zimmerer's work highlights the clash between historical property/resource regimes and emerging amenity-based property regimes associated with new conservation areas, raising critical questions about the ability of communities—and particular individuals within these communities—to legitimate their environmental imaginaries in the spaces of conservation that result. For example, Sandberg et al. (2013), in work documenting exurban struggles along the Oak Ridges Moraine in southern Ontario, discuss the distinctive ways that natural resource industries engage with conservation discourses. They trace the strategies used by aggregate companies to moderate the effect, and even implementation, of new conservation schemes intended to reduce their industries' negative impacts on the landscape. While conservation of large areas of the Moraine has been successful through regional planning, aggregate resource operations are still permitted in all but the areas of highest biophysical vulnerability because of the political success of the aggregate companies in controlling the natural science and landscape discourse. Sandberg et al. illustrate a key point in exurban study: the emergence of conservation landscapes of all kinds is "for the management of biogeophysical impacts and *the expansion of markets*" (Zimmerer 2000, p. 359, emphasis added; see also Zimmerer 2006).

Concern over the relationship between conservation efforts and expansion of markets is a prevalent theme within political ecology, including research that examines neoliberalism's correlation with environmental issues and ecological governance. Neoliberalism is "a process of market-driven social and spatial transformation" (Brenner and Theodore 2005, p. 102) and is useful in thinking about exurban development because it helps to conceptualize how state intervention in land-use regulation makes room for economic change. Although a fixed definition of neoliberalism is hard to pin down—in part due to the fact that neoliberalism "stands for a complex assemblage of ideological commitments, discursive representations, and institutional practices, all propagated by highly specific class alliances and organized at multiple geographical scales" (McCarthy and Prudham 2004, p. 276)—for our study of exurbia we identify certain basic elements of the concept. These common elements include a seemingly unrelenting faith in the market to generate the most socially acceptable outcomes where this reliance on the market has led to the privatization and commodification of nearly everything across all sectors of society, including nature. Accompanying this faith in the market is a restructuring of state authority to create and protect these privatization and commodification dynamics (McCarthy 2006), such as state intervention through exurban land-use planning and regulation (Sandberg and Wekerle 2010). Spaces of conservation created in parallel with spaces for development are the subject of Chap. 7 by Watson and Skaggs, who compare for-profit and not-for-profit approaches to neoliberal environmental planning in the rural to exurban transition. Similarly, Tilt and Cerveny (Chap. 11) describe Washington State's role in creating large areas of conservation lands as part of their approach to managing the shift in the economy from timber to residential development.

In discussing the politics of forestry alternatives, McCarthy (2006, p. 98) highlights neoliberalism's "faith in civil society and its components, including NGOs and communities (evident in their increasing role in governance, and in the explosive growth of NGOs), and a profound focus on individuals as economic and social actors." We have found that environmental non-governmental organizations (ENGOs) often play an important role in the process of the exurban transition (see also Sandberg et al. 2013), as we discuss in the example of the Quakertown Swamp in Chap. 2 where the efforts of ENGOs resulted in large protected areas.

Definitions of neoliberalism encompass a broad spectrum ranging from a narrow conceptualization that emphasizes global trade policy and implications for national economies, to a wider definition that describes the reconfiguration of "economic and political governance" according to the principles of liberal theory (McCarthy 2005, p. 997). Because this wider definition implies that neoliberalism is one of the most pervasive influences shaping society and the environment, we wonder to what extent theorists of neoliberalism are overemphasizing the effect of the economy in shaping social relations.[2] Indeed, we sometimes feel that too much of a focus on neoliberalization may have the power to blind us to our involvement in capitalism, with the effect that our agency in the current neoliberal order is diminished. We take the stance that neoliberalism is "us," meaning that collectively we *are* the market: we created the institutions of governance and shaped the planning process, albeit with individuals having differing levels of influence, and therefore we have the abil-

[2] For an introduction to neoliberalism in geography scholarship, see Harvey (2007); also Gibson-Graham 2006 for a critique of scholarship

ity to make change. We are intending to understand the rural to urban transition in order to do something about it.

Throughout the book, both editors and chapter authors draw upon the foundational work on political ecology, which some readers new to political ecology may find useful. Blaikie and Brookfield's *Land Degradation and Society* (1987), which was reviewed in a special issue of *Geoforum* (see Muldavin 2008), is an early example of the political ecology approach. Comprehensive surveys of political ecology are provided by Robbins (2011), Neumann (2005), and Forsyth (2003) and most recently, two extensive contributed volumes edited by Bryant (2015) and Perreault et al. (2015). Likewise, Cadieux and Taylor (2013) describe the addition of landscape studies to the political ecology approach. Yet, each chapter author includes their own descriptions of the way their political ecology of landscape approach draws from these and other foundational scholars in the field. These works remind us that, what may first appear as very local land-use and environmental management issues, are instead often the result of multi-scalar global economic changes and very similar processes occurring in places around the world.

Comparative work in political ecology, by which we mean work undertaken explicitly across a number of different sites chosen for their value in examining a target phenomenon or process, is relatively rare. Instead, research regularly rests on individual case studies. But a persistent critique of the case study approach common to political ecological research has suggested a lack of scientific rigor (i.e., undertaking studies with a very low *n* or sample size). This critique potentially undermines the important insights political ecology offers in understanding nature–society relationships. For example, Büscher (2008, p. 230), in his editorial in the journal *Conservation Biology*, suggests that natural scientists should do what they "do best," which is to "provide the data and analyses that allow us to see what we as humanity are doing to the planet," but we think he fails to make the connection to the value of political ecological analysis, which is to do what we do best: to *also* explore the "how" and "why" of what we are doing to the planet. For conservation scientists interested in exurbia, this book may seem to have a narrow case study focus. They may have a point when they suggest that the researchers in this book are not undertaking empirical analysis of past or potential ecological change in their study areas, but rather only derive their understandings of environmental impacts from secondary sources. Nonetheless, all authors in this book are "ecologically engaged" and include analysis from environmental science in their work.

Across 11 chapters, this book includes comparative political ecological research on 16 case studies of exurban change. We demonstrate how similar political economic processes, constellations of similar actors, and common contemporary land-use and environmental governance techniques interact with very different landscapes, each with unique histories of settlement, natural resource use, extraction, and management, forms of governance, and approaches to land-use planning. Based on these cases, we suggest exurbia includes a number of key markers of change, which reflect the social–ecological transitions at play. In what follows we elaborate upon these key markers and then turn to a discussion of the importance of an attention to environmental justice when studying exurbia.

1.3 Locating Places Experiencing Exurban Change

One of the most common sentiments of people in exurbanizing areas that we have encountered during research in different parts of the United States, Canada, and Australia is a feeling that their local experience with exurban landscape change is unique. While we acknowledge the great variety of exurban experience and that particular places do exhibit distinctive outcomes, we have learned from undertaking a comparative political ecology of case studies that experiences of particular places are quite similar. Just within this volume, there is an enormous variety in the landscapes discussed, from tidal estuaries and mountains of the Carolinas, to rolling Pennsylvania farmland outside of Philadelphia, to the Sierra Nevada Mountains of California, and the tablelands on the plateau outside of Sydney in southeastern Australia. The communities and local cultures associated with these landscapes are similarly diverse. Because of the focus within local communities on local landscape changes—how people value the landscape, what kind of knowledge about the land is valued—the commonalities between different exurban experiences are often difficult to see. Local case studies of exurban landscape change, such as the ones in this book, are highly regionalized and there is a tendency to think that a particular region is experiencing a unique transformation, precisely because exurban change is so closely tied to the ecologies of local landscapes. How are 10-acre ranchettes in the foothills of California related to suburban-style homes on three acres in the Pennsylvania countryside? We believe the dynamics of these exurban landscapes are quite similar.

The literature on exurbia, based on our reading, reinforces this sense of local or regional uniqueness, with the American West as the focus of perhaps the most extensive range of studies about exurban environmental conflicts, especially by political ecologists (beginning with Walker and Fortmann 2003, but extending through to the present day, e.g., Gosnell and Abrams 2011). Despite a trend toward political ecological explorations of exurbia in diverse parts of the world—North America especially—much of the published work remains focused on the American West (Hurley and Carr 2010). In the book, we broaden the geography of exurban studies through comparison of local experiences from the West (Oregon, California) with cases from other less-studied places (Pennsylvania, North and South Carolina, Washington State, not a typical Western study area, and Australia).

Although the causes of exurbanization might be similar from place to place, the local experience of change differs. For example, the landscapes of North Carolina and Washington State are both mountainous but look and feel very different and so the reaction to exurbanization is different, and so is the landscape outcome. In North Carolina, as described by Breen, Hurley, and Taylor in Chap. 9, one county introduced zoning to mitigate slope stabilization issues with the goal of limiting the construction of new exurban homes. Support for the proposed zoning ordinance came from a seemingly unlikely coalition of individuals, for reasons far beyond concern for natural hazards. The result has been a continuous, although perhaps less dense, landscape of scattered homes along valley slopes. Whereas in Chap. 11, Tilt and Cerveny discuss how one county in Washington State zoned large tracts of land for nature conservation, thereby displacing new exurban development to lands defined as less ecologically significant, with the result that exurban homes at much higher densities are clustered next to large areas of natural "wilderness" (see images in Chap. 12). In both cases, the reimagination of nature from its productive mining and timber use to amenity use set the planning process in motion where local politicians, land-use officials, and other actors acted to protect local landscape values and to control landscape change without getting in the way of economic investment generally in those areas.

To better assist with diagnosing similarities across exurban cases, we propose "seven markers" of exurban landscape change, which we think will help observers better recognize the exurban transition (Table 1.1). We discuss each marker in turn below.

1.3.1 Rural Character

The first marker of an exurbanizing area is the persistence of rural landscape character. By rural landscape character, we mean land cover commonly associated with traditional rural land uses: agriculture (e.g., farms, orchards, fruit-producing land covers, and/or horticultural operations); forestry (e.g., timber plantations, woodlots, Christmas tree farms); ranchlands (with or without livestock); and other forms of natural resource extraction (e.g., sand and gravel pits, mining, etc.), where such land cover persists over a relatively large area. An exurbanizing area is distinguishable from a suburban area in that the rural character is maintained because some aspect of the previous

Table 1.1 Markers of exurban change

1	*Rural character* •The persistence of traditionally rural land uses and land cover in the visible landscape
2	*Access* •The ease of accessibility from exurbanizing areas to urban (or suburban) centers
3	*Changes in local economy* •The perspective from within the community that changes are highly localized and unique
4	*Ideologies of nature* •The reconfiguration of how people value nature in the area: nature-as-amenity and nature-as-commodity
5	*New perspectives on land management* •The revolution in land management approaches toward conservation, especially for lands seen to belong to the sense of place and culture of the local community
6	*Coalition-building in the community* •The rise of coalitions of people in the community to secure future land uses and property rights
7	*Land-use planning* •The emergence of planning as an arena to mediate conflicts caused by the shift in land use from industrial extraction to amenity real estate

resource economy persists or new resource-based uses have emerged or are emerging. For example, in Chap. 4, Abrams discusses the persistence of ranching in Wallowa County, Oregon, but where ranchers have sold their land to amenity migrants, who hire the ranchers back to maintain the rural landscape character. In other cases, ecological conservation may emerge as the dominant use of natural resources, creating a landscape that looks more like wilderness than a settled rural area. Although many social and economic changes may be occurring in the background, the visible landscape maintains an overwhelming rural appearance or character.

1.3.2 Access

An exurban area is accessible to urban centers through diverse transportation and telecommunication networks. Commuting distance is one aspect of "access," where exurbanites may be within the commutershed of an urban or suburban employment center and may commute by car to work most days. Demographers use the limit of the commutershed as one way of defining the extent of metropolitan areas (Berube et al. 2006). For example, Tilt and Cerveny (Chap. 11) use commuting as a marker of exurbia in their Washington State case study. In addition to accessibility of the wider world by residents, the market accessibility of the exurban area may be changing, for instance where a farmer's market for locally produced food and craft items in the city center provides exurban farmers with a source of income or where a local airfield provides access for tourists and for the more distant export of locally produced artisanal food and other goods.

1.3.3 Changes in Local Economy

With the withdrawal of large-scale rural resource industries and the declining profitability of smaller-scale resource holders, conflicts develop in an exurbanizing area over what the "new" economy will be and what constitutes appropriate use of land and natural resources. Each party in these conflicts has a particular vision about how nature should be valued and what constitutes appropriate land use in their area; with each political skirmish, a shift or change in permitted use occurs, potentially with new areas open for development and others shielded from the bulldozer (e.g., see Chap. 6 in which Olson documents the shift from agriculture to tourism in Deschutes County, central Oregon; or Chap. 7 in which Watson and Skaggs discuss competing landscape visions for new exurban development in South Carolina's Lowcountry). Indeed, the hallmark of exurban transformation is a partitioning of the landscape into areas for development (places of capital accumulation through real estate development) and areas with some level of conservation or protection (publicly or privately held protected areas unavailable for real estate profit-making). The outcome of exurban landscape change is a fragmented natural landscape of competing (yet often interdependent) rural capitalisms as discussed above.

1.3.4 Ideology of Nature

As discussed by Cadieux and Taylor (2013), there is a shift in landscape ideology that accompanies exurban change. The conventional view that natural landscapes have value for their resource commodity use undergoes a shift in an exurbanizing area, so that those same landscapes come to be seen by some as "natural," "wild," "untouched," and worthy of conservation. Where land values may have been depressed because of the withdrawal of interest in the landscape for its resource commodity value, appreciation of scenic and natural beauty may create new economic value for the land-as-amenity within the real estate sector. Some inhabitants see natural resource extraction as ecologically damaging and environmentally unsustainable and pressure industry to withdraw even further. However, productive uses that maintain scenic landscapes (e.g., most kinds of farming and ranching), or small-scale craftsmanship using natural materials in sustainable ways, may be reinvigorated because of their role in protecting or creating the amenity value. As discussed in detail in Chap. 2 by Hurley and Taylor, securing an "environmental imaginary" is a highly political move. The power to inscribe a particular ideology of nature into planning and zoning decision-making processes has far-reaching effects on resource-based livelihoods, the community's sense of place, and the potential for future forms of landscape change.

1.3.5 Land Management

In exurbanizing areas, the use of land may change along with the kind of activities and practices sanctioned by individuals and the community. Rural lands in productive use for timber, ranching, or mining under the ownership of a large resource corporation or even individual owners are often treated principally as aesthetic commons in practice, with locals experiencing few restrictions on hunting, foraging, informal recreation, and other activities. In contrast, the use of land in the exurban context is highly regulated and policed. Exurban homeowners surveil their properties with an eye to enforcing their right to privacy and securing their environmental imaginary of living in nature away from the lights, sounds, and strangers associated with modern city life. Conservation land managers in exurban areas may limit public use in defense of ecological restoration—especially in land trusts but also in public state and national parks—and thereby reinforce the idea that nature is at its best when tucked away from human use, as discussed by Olson in Chap. 6 where the Sierra Club used the *National Environmental Protection Act* to try to block development in Deschutes County, Oregon. In some cases, however, productive land uses are able to co-exist with the exurban environmental imaginary of land-as-amenity, a point exemplified in Chap. 7 wherein Watson and Skaggs show how the Savannah River Preserve was conceived as an area where traditional subsistence uses, like hunting and fishing, would be protected from residential development.

1.3.6 Coalition-Building

The sixth marker is the rise of sometimes surprising coalitions of people in the community to secure future property rights as the transition to an amenity exurban landscape is highly contested by those with a personal stake in the transformation. Generally, exurban political dynamics are assumed to be a contest between longtime locals (e.g., commodity producers and laborers) and newcomers (e.g., developers and homebuyers). However, this perspective, while representing typical power dynamics and aspects of control, fails to encompass the multiplicity of variables at work in the rural to exurban change. Indeed, newcomers may see abundant opportunities to benefit from the new real estate economy, including engaging in development themselves, while their neighbors who have farmed the land for generations work diligently to align their politics to ensure their agricultural livelihoods—and the local agricultural economy—survive long into the future. In contrast, other newcomers may see their capital stake in the landscape tied to preserving the rural character of the landscape around them. The alliances that form between individuals and groups to ensure the dominance of particular environmental imaginaries may vary from place to place but the following types of actors are common to almost every exurban situation: new amenity migrants, established exurbanites, local producers including farmers and ranchers, real estate developers, local businesses, global companies, planners, politicians, state and federal government bureaucrats, and non-profit environmental groups. As many of the chapters in this book highlight, there is a surprising variability in the ways in which the actors build coalitions to embrace or protest transformations of the local economy. For example, in Chap. 3, Hiner discusses the power of local actors in supporting or frustrating the landscape changes desired by landowners, where social relations play out in the politics of the planning process. In Chap. 8, Finewood and Martin present an historiography of the South Carolina Lowlands where white exurban residents joined African American fisherman in blocking a proposal for a chemical plant. Finally, Chap. 9 by Breen, Hurley, and Taylor discusses in detail the complex dynamics of coalitions of actors in the exurban transition in Jackson County, North Carolina.

1.3.7 Land-Use Planning

Landscape preservation, natural resource conservation, and ecological sustainability—or similar discourses—become increasingly important in an exurban context and state intervention may commonly involve different approaches that seek to bar destructive practices. The role of land-use planning becomes very important as the mechanism by which a community seeks to control the use of land; indeed, it is through the planning process that the dominant environmental imaginary is assigned. Through the planning process, lands are delineated and labeled according to the natural value conferred to those lands. Planning interventions are typically referred to in planning literature as legislation, policy, practice, public consultation, and decision-making, where planning is codified in vision documents, plans, and development agreements. In an area with very little state intervention in land-use planning, the rise of highly politicized processes around property use and environmental management is disconcerting to many in the community. Instead, an exurbanizing community should anticipate the emergence of planning as an arena to mediate conflicts caused by the transformation of land use from industrial extraction to amenity real estate. Even in an area with land-use planning experience, the level of detail and oversight increases so that those who succeed in having their land-use aspirations codified in comprehensive plans and zoning are those with the most political sophistication and who have succeeded in drawing together coalitions of people willing to speak out (and vote!) in support of their vision.

As we discuss in the chapters in the book, local land-use planning processes—at the intersection of local interests and larger processes of change—produce very different plans, government and non-profit interventions, and landscape outcomes. This is true even in areas of the same state with very similar economic, social, and political histories, in areas experiencing the same extractive/productive-to-amenity/consumptive economic shifts, in areas characterized by the same material landscape, and in areas subject to the exact same state-level planning regime. For example, Watson and Skaggs (Chap. 7) describe two very different approaches to managing exurbanization in the South Carolina Lowcountry: one approach by a for-profit developer resulting in plans to conserve three-quarters of the land area and the second approach by a not-for-profit initiative to use conservation easements to protect lands from residential development.

Based on the collective research presented in this book, we believe all seven of these markers will be present when rural landscapes are undergoing exurban transformation. While one or two markers may be easy to see, others may be less visible. But it is our belief that, considered together, this collection of markers offers a comprehensive framework for understanding the broad environmental changes in exurbia.

1.4 Conclusion: Exurban Political Ecology and Sustainability

As book editors, we are motivated by a desire to manage natural landscapes over the long term in ways that equitably meet the needs and desires of diverse human and nonhuman constituencies. This concern derives from a very real fear of the loss of ecological services to support future generations and from an environmental ethic of stewardship of the planet. Given that so many areas within the United States, Canada, Australia, and elsewhere are experiencing the exurbanization process, and given our belief that political ecology has critical tools for examining the social–political and ecological changes that occur, we think undertaking a comparative political ecology examination is insightful. The case studies in this book were chosen with a deliberate focus on developing material for a comparative work. As such, we asked each of the authors to consider how the exurban landscape transition played out in their local area. In doing so, they draw on Reed's conceptual framework to think about the intersection of regional environments, regional economies, changes in institutions, and the change in or continuity of social relations and local cultures. In their chapters, authors focus on competing rural capitalisms and uneven environmental management with particular attention given to the institutional structures of the state and engagement with diverse land-use planning regimes.

The authors seek to understand the role of the ideology of nature among people participating (or not) in the transition. They consider who is participating and how in the transition, and specifically where these actors are situated in relation to the changing economic and social relations in a given place, including whether these actors might gain or lose political power as a result of the transition. This includes especially how racialized and class differences play out in relation to land-use and environmental change and the approaches to regulation that arise to contend with these changes. Moreover, chapter authors avoid apolitical examinations of the technical and policy interventions increasingly common to exurbanization. In cases where conservation science is enrolled in the politics of landscape change, authors question the power of conservation science to tell the "truth" about local landscapes, especially where it is in contrast to ways in which people experience those landscapes.

In keeping with the rich tradition of qualitative research methods at the heart of political ecology, our authors use diverse, grounded approaches to find out what people think about exurban change in their area. They do so through the use of actors' own words and perspectives to capture the lived experiences of those participating in and living through the exurban transition. Specific methods that are deployed by contributing authors include participant observation, focus groups, interviews (open-ended, semi-structured), surveys, and reviews of discourse in popular media (including print and digital sources). Further, visual research methods, including mapping and photography, are used to gather data and bring to life the material conditions and aesthetic differences of the landscapes compared. Finally, some authors deploy mixed methods, where qualitative data is analyzed using quantitative approaches that include coding analysis (e.g., Q method [Chap. 9] and NVivo [Chap. 11]) to provide insight into the political fault lines within and between communities experiencing exurbia.

By focusing on land-use planning and its role in dealing with competing rural capitalisms and in producing uneven environmental management regimes, we see opportunities to intercede in processes that would otherwise reproduce existing elite-nature relationships, marginalizing the less powerful. A key part of our motivation for studying the emergence of uneven environmental management is a consideration of environmental justice. As such, we ask the following: are the landscapes produced through the rural to exurban transition environmentally just? Our consideration of environmental justice follows Walker's (2012) three concepts: distributive justice, procedural justice, and justice as recognition. Therefore, we take into account distributive justice—that is, the distribution of environmental benefits and risks—by asking if exurbia is produced at the expense of others, and if so, what might be opportunities for intervention. Procedural justice considers the fairness of decision-making procedures and processes, and is especially important given our consideration of land-use planning in mediating exurban landscape conflicts and the power dynamics influencing those processes. And finally, "recognition" of multiple ways of knowing is an important aspect of environmental justice, which in studying exurbia refers especially to the multiplicity of environmental imaginaries that may exist in a particular landscape.

Unfortunately, the trend in response to exurban development pressure seems to be to reactively create places of nature (conservation areas) and places for development (zoned for large-lot residences), which does not result in a just distribution of conservation and development as lands catering to amenity migrants tend to border conservation areas of all kinds. Thus, nature provides the setting for exurbanites' quality of life, a quality of life not shared equally among all social groups. Creating conservation areas is a popular response to exurbanization because these areas can be delineated by seemingly objective scientific studies, which express the natural heritage and ecosystem values that underpin new amenity values (Forsyth 2003). Communities might have more success in creating meaningful landscapes where local culture and natural processes may continue over the long term when they proactively identify and respond to the markers of local exurban change. In other words, other local public issues, in addition to nature conservation, could be considered if policies and regulations are established ahead of time. For example, the new environmental management regime could be based in more collaborative decision-making, consider more equitable distribution of access to natural spaces and ecosystem services, and consider the provision of affordable housing.

The cases in this book examine changes in state intervention and environmental management with the hope that readers may be better equipped to deal with their own community's transition. Readers should have a better appreciation of what is going on in their own communities through the comparisons we offer with places in the eastern and western United States and Australia. If real change for a more sustainable future is to take place, readers should consider their local place in the broadest context, which means to "look up" to consider state, national, and global institutions and organizations as the "centers of knowledge and power with which" the local finds itself "entwined" (Robbins 2002, pp. 1510–1511). Being forewarned about coming change is better than being reactive and attempting to mediate and mitigate negative externalities after "the horse has left the barn" (Walker 2006).

References

Abrams, J. B., Gill, N., Gosnell, H., & Klepeis, P. (2012). Re-creating the rural, reconstructing nature: An international literature review of the environmental implications of amenity migration. *Conservation and Society, 10*(3), 270–284. doi:10.4103/0972-4923.101837.

Benson, M., & Osbaldiston, N. (Eds.). (2014). *Understanding lifestyle migration: Theoretical approaches to migration and the quest for a better way of life*. Basingstoke, England: Palgrave MacMillan.

Berube, A., Singer, A., Wilson, J. H., & Frey, W. H. (2006). *Finding exurbia: America's fast-growing communities at the metropolitan fringe*. The Brookings Institution. Retrieved from

http://www.brookings.edu/~/media/research/files/reports/2006/10/metropolitanpolicy-berube/20061017_exurbia.pdf.
Blaikie, P., & Brookfield, H. (1987). *Land degradation and society*. New York, NY: Methuen.
Brenner, N., & Theodore, N. (2005). Neoliberalism and the urban condition. *City, 9*(1), 101–107. doi:10.1080/13604810500092106.
Brogden, M. J., & Greenberg, J. B. (2003). The fight for the west: A political ecology of land use conflicts in Arizona. *Human Organization, 62*(3), 289–298. doi:10.17730/humo.62.3.0jdr5nte0b8vf7aw.
Brown, D. G., Robinson, D. T., An, L., Nassauer, J. I., Zellner, M., Rand, W., et al. (2008). Exurbia from the bottom-up: Confronting empirical challenges to characterizing a complex system. *Geoforum, 39*(2), 805–818. doi:10.1016/j.geoforum.2007.02.010.
Bryant, R. L. (Ed.). (2015). *The international handbook of political ecology*. Cheltenham: Edward Elgar Publishing Limited.
Bunce, M. F. (1994). *The countryside ideal: Anglo-American images of landscape*. London: Routledge.
Büscher, B. E. (2008). Conservation, neoliberalism, and social science: A critical reflection on the SCB 2007 Annual Meeting in South Africa. *Conservation Biology: The Journal of the Society for Conservation Biology, 22*(2), 229–231. doi:10.1111/j.1523-1739.2008.00894.x.
Cadieux, K. V., & Hurley, P. T. (2011). Amenity migration, exurbia, and emerging rural landscapes: Global natural amenity as place and as process. *GeoJournal, 76*(4), 297–302. doi:10.1007/s10708-009-9335-0.
Cadieux, K. V., & Taylor, L. E. (Eds.). (2013). *Landscape and the ideology of nature in exurbia: Green sprawl*. New York: Routledge.
Colladoes, C., & Duane, T. P. (1999). Natural capital and quality of life: A model for evaluating the sustainability of alternative regional development paths. *Ecological Economics, 30*(3), 441–460. doi:10.1016/S0921-8009(99)00020-8.
Cromartie, J., & Nelson, P. B. (2009). Baby boom migration tilts toward rural America. *Amber Waves*, 1–7. Retrieved from http://www.ers.usda.gov/amber-waves/2009-september/baby-boom-migration-tilts-toward-rural-america.aspx#.Vgl87Xg1P5g.
Deleuze, G., & Guattari, F. (1987). *A thousand plateaus: Capitalism and schizophrenia*. Minneapolis: University of Minnesota.
Duany, A., Plater-Zyberk, E., & Speck, J. (2010). *Suburban nation: The rise of sprawl and the decline of the American Dream (10th anniv.)*. New York: North Point Press.
Fava, S. (1975). Beyond suburbia. *Annals of the American Academy of Political and Social Science, 422*(1), 10–24. doi:10.1177/000271627542200103.
Forsyth, T. (2003). *Critical political ecology: The politics of environmental science*. New York: Routledge.
Ghose, R. (2004). Big sky or big sprawl? Rural gentrification and the changing cultural landscape of Missoula, Montana. *Urban Geography, 25*(6), 528–549. doi:10.2747/0272-3638.25.6.528.
Gibson-Graham, J. K. (2006). *The end of capitalism (as we knew it): A feminist critique of political economy*. Minneapolis: University of Minnesota Press.
Gosnell, H., & Abrams, J. B. (2011). Amenity migration: Diverse conceptualizations of drivers, socioeconomic dimensions, and emerging challenges. *GeoJournal, 76*(4), 303–322. doi:10.1007/s10708-009-9295-4.
Gosnell, H., Haggerty, J. H., & Travis, W. R. (2006). Ranchland ownership change in the Greater Yellowstone Ecosystem, 1990–2001: Implications for conservation. *Society & Natural Resources, 19*(8), 743–758. doi:10.1080/08941920600801181.
Harvey, D. (2007). *A brief history of neoliberalism* [ProQuest ebrary version].
Hurley, P. T., & Carr, E. R. (2010). Forum on political ecology of the US South. *Southeastern Geographer, 50*(1), 99–109. doi:10.1353/sgo.0.0063.
Hurley, P. T., & Halfacre, A. C. (2011). Dodging alligators, rattlesnakes, and backyard docks: A political ecology of sweetgrass basket-making and conservation in the South Carolina Lowcountry, USA. *GeoJournal, 76*(4), 383–399. doi:10.1007/s10708-009-9276-7.

Hurley, P. T., & Walker, P. A. (2004). Whose vision? Conspiracy theory and land-use planning in Nevada County, California. *Environment and Planning A, 36*(9), 1529–1547. doi:10.1068/a36186.

Jackson, K. T. (1985). *Crabgrass frontier: The suburbanization of the United States*. New York: Oxford University Press.

Lamb, R. (1983). The extent and form of urban sprawl. *Growth and Change, 14*(1), 40–47. doi:10.1111/j.1468-2257.1983.tb00395.x.

Lang, R. E. (2003). *Edgeless cities: Exploring the elusive metropolis*. Washington, DC: The Brookings Institution.

Larsen, S., & Hutton, C. (2012). Community discourse and the emerging amenity landscapes of the rural American West. *GeoJournal, 77*(5), 651–665. doi:10.1007/s10708-011-9410-1.

Latour, B. (2004). *Politics of nature: How to bring the sciences into democracy* (Trans. C. Porter). Cambridge, MA: Harvard University Press.

Lichter, D. T., & Brown, D. L. (2011). Rural America in an urban society: Changing spatial and social boundaries. *Annual Review of Sociology, 37*(1), 565–592. doi:10.1146/annurev-soc-081309-150208.

Marcouiller, D. W., Lapping, M. B., & Furuseth, O. (Eds.). (2011). *Rural housing, exurbanization, and amenity-driven development: Contrasting the "haves" and the "have nots."* Surrey: Ashgate.

McCarthy, J. (2005). Devolution in the woods: Community forestry as hybrid neoliberalism. *Environment and Planning A, 37*(6), 995–1014. doi:10.1068/a36266.

McCarthy, J. (2006). Neoliberalism and the politics of alternatives: Community forestry in British Columbia and the United States. *Annals of the Association of American Geographers, 96*(1), 84–104. doi:10.1111/j.1467-8306.2006.00500.x.

McCarthy, J., & Prudham, S. (2004). Neoliberal nature and the nature of neoliberalism. *Geoforum, 35*(3), 275–283. doi:10.1016/j.geoforum.2003.07.003.

Muldavin, J. (2008). The time and place for political ecology: An introduction to the articles honoring the life-work of Piers Blaikie. *Geoforum, 39*(2), 687–697. doi:10.1016/j.geoforum.2007.07.003.

Murdoch, J. (2005). *Post-structuralist geography: A guide to relational space* [ProQuest ebrary version]. doi:10.4135/9781446221426.

Neumann, R. P. (2005). *Making political ecology*. London: Hodder Arnold.

Paulson, S., Gezon, L. L., & Watts, M. (2003). Locating the political in political ecology: An introduction. *Human Organization, 62*(3), 205–218. doi: 10.17730/humo.62.3.e5xcjnd6y8v09n6b.

Peet, R., & Watts, M. (1996). Liberation ecology: Development, sustainability, and environment in an age of market triumphalism. In R. Peet & M. Watts (Eds.), *Liberation ecologies: Environment, development, social movements* (pp. 1–45). London: Routledge.

Perreault, T., Bridge, G., & McCarthy, J. (Eds.). (2015). *The Routledge handbook of political ecology*. Abingdon/New York: Routledge.

Reed, M. G. (2007). Uneven environmental management: A Canadian comparative political ecology. *Environment and Planning A, 39*(2), 320–338. doi:10.1068/a38217.

Robbins, P. (2002). Letter to the editor. *Environment and Planning A, 34*(08), 1509–1513. doi:10.1068/a34217.

Robbins, P. (2006). The politics of barstool biology: Environmental knowledge and power in greater Northern Yellowstone. *Geoforum, 37*(2), 185–199. doi:10.1016/j.geoforum.2004.11.011.

Robbins, P. (2011). *Critical introductions to geography: Political ecology: A critical introduction* (2nd ed.) [ProQuest ebrary version].

Robbins, P., Martin, S., & Gilbertz, S. (2012). Developing the commons: The contradictions of growth in exurban Montana. *The Professional Geographer, 64*(3), 317–331. doi:10.1080/00330124.2011.601193.

Sandberg, L. A., & Wekerle, G. R. (2010). Reaping nature's dividends: Neoliberalization and gentrification of nature. *Journal of Environmental Policy & Planning, 12*(1), 41–57. doi:10.1080/15239080903371915.

Sandberg, L. A., Wekerle, G. R., & Gilbert, L. (2013). *The Oak Ridges Moraine battles: Development, sprawl, and nature conservation in the Toronto region*. Toronto: University of Toronto Press.

Schewe, R., Field, D. R., Frosch, D. J., Clendenning, G., & Jensen, D. (2012). *Condos in the woods: The growth of seasonal and retirement homes in Northern Wisconsin*. Madison, WI: University of Wisconsin Press.

Scott, A. J., Carter, C., Reed, M. R., Larkham, P., Adams, D., Morton, N., et al. (2013). Disintegrated development at the rural–urban fringe: Re-connecting spatial planning theory and practice. *Progress in Planning, 83*, 1–52. doi:10.1016/j.progress.2012.09.001.

Spectorsky, A. C. (1955). *The exurbanites*. Philadelphia: Lippincott.

Taylor, L. E. (2011). No boundaries: Exurbia and the study of contemporary urban dispersion. *GeoJournal, 76*(4), 323–339. doi:10.1007/s10708-009-9300-y.

Taylor, L. E., & Cadieux, K. V. (2013). Sprawl and the ideology of nature. In K. V. Cadieux & L. E. Taylor (Eds.), *Landscape and the ideology of nature in exurbia: Green sprawl* (pp. 1–30). New York: Routledge.

Theobald, D. M. (2005). Landscape patterns of exurban growth in the USA from 1980 to 2020. *Ecology and Society, 10*(1), 32. Retrieved from http://www.ecologyandsociety.org/vol10/iss1/art32/.

Travis, W. R. (2007). *New geographies of the American West: Land use and the changing patterns of place*. Washington, DC: Island Press.

U.S. Census Bureau. (2015). *Cartographic boundary shapefiles—Counties*. Retrieved from https://www.census.gov/geo/maps-data/data/cbf/cbf_counties.html.

Walker, P. A. (2006, April). Panelist. *Amenity migration, exurbia, and emerging rural landscapes*. Session conducted at the Association of American Geographers Annual Meeting, Chicago.

Walker, G. (2012). *Environmental justice: Concepts, evidence and politics*. New York: Routledge.

Walker, P. A., & Fortmann, L. (2003). Whose landscape? A political ecology of the "exurban" Sierra. *Cultural Geographies, 10*(4), 469–491. doi:10.1191/1474474003eu285oa.

Walker, P. A., & Hurley, P. T. (2011). *Planning paradise: Politics and visioning of land use in Oregon*. Tuscon: University of Arizona Press.

Watts, M. (2009). Capitalism. In D. Gregory, R. Johnston, and G. Pratt (Eds.), *Dictionary of human geography* (5th ed., pp. 59–64). Hoboken, NJ: Wiley-Blackwell [ProQuest ebrary version].

Williams, R. (1976). *Keywords: A vocabulary of culture and society*. Glasgow: Fontana/Croon Helm.

Winkler, R., Cheng, C., & Golding, S. (2012). Boom or bust? Population dynamics in natural resource-dependent counties. In L. J. Kulcsár & K. J. Curtis (Eds.), *International handbook of rural demography* (pp. 349–367). Dodrecht: Springer. doi:10.1007/978-94-007-1842-5.

Woods, M. (2009). Rural geography: Blurring boundaries and making connections. *Progress in Human Geography, 33*(6), 849–858. doi:10.1177/0309132508105001.

Woods, M. (2011). *Rural*. Abingdon, Oxon: Routledge.

Zimmerer, K. S. (2000). The reworking of conservation geographies: Nonequilibrium landscapes and nature-society hybrids. *Annals of the Association of American Geographers, 90*(2), 356–369. doi:10.1111/0004-5608.00199.

Zimmerer, K. S. (2006). Cultural ecology: At the interface with political ecology—The new geographies of environmental conservation and globalization. *Progress in Human Geography, 30*(1), 63–78. doi:10.1191/0309132506ph591pr.

Chapter 2
From Swamp to Ridgeline: Exploring Exurbia in Southeastern Pennsylvania and the Sierra Nevada Foothills of California

Patrick T. Hurley and Laura E. Taylor

In the outskirts of Philadelphia, along the eastern seaboard of the United States, the scenic rolling countryside is an attractive choice for homebuyers seeking an escape from the crush and noise of the city and the monotony of the suburbs. In this part of southeastern Pennsylvania, historic farmhouses, acres of green farmland and hardwood forests, and gentle ridgelines provide the scenic beauty, closeness to nature, and serenity that these in-migrants seek. In short, this rural feel and associated quality of life provide the foundations for a strong real estate market. Three thousand miles away, in the Sierra Nevada Mountains of the American West, similar opportunities attract new residents, but in a very different cultural and ecological landscape. In Nevada County, California, new arrivals are struck by the beauty of extensive oak woodlands, pine forests, historic irrigation ditches, and quaint towns established during the Gold Rush, and are attracted by recreation opportunities in the nearby national forest. Southeastern Pennsylvania and the Sierra Nevada Foothills feature exurbs, where people are looking for a place in the country, close to natural amenities. Yet, as more and more people choose to live there, the very amenities of the countryside are threatened by over-development for exurban housing. Indeed, as with many rural areas experiencing relatively rapid in-migration during the 1990s and early 2000s, in both southeastern Pennsylvania and Nevada County the resulting exurban residential development was not welcomed by everyone and certain areas in the landscape were seen as too important to be "sacrificed" to residential uses and new homes.

P.T. Hurley (✉)
Department of Environmental Studies, Ursinus College,
P.O. Box 1000, Collegeville, PA 19426, USA
e-mail: phurley@ursinus.edu

L.E. Taylor
Faculty of Environmental Studies, Planning Program, York University,
4700 Keele Street, Toronto, ON M3J 1P3, Canada
e-mail: taylorL9@yorku.ca

L.E. Taylor, P.T. Hurley (eds.), *A Comparative Political Ecology of Exurbia*,
DOI 10.1007/978-3-319-29462-9_2

Philadelphia, Pennsylvania's metropolitan outskirts and the slopes of Nevada County, California are typical of amenity landscapes outside of many large cities around the world where natural beauty and a rural lifestyle draw people to live. But how do communities deal with this in-migration and the low-density development it often fosters? What types of landscape features and associated longstanding natural resource uses do people value enough—whether through institutional mechanisms or individual commitments—to "protect" during this type of exurban transition?

From the perspective of political ecology, the future of the Quakertown Swamp and Nevada County foothills of the Sierra Nevada is being determined by dominant environmental imaginaries championed by powerful political and economic actors, often drawing on natural science or technical discourses and privileged access to the land-use planning process to cement new visions of appropriate land use and the landscape patterns they produce.

The landscapes of southeastern Pennsylvania and California's Sierra Nevada have been shaped by political and natural histories that make landscape conservation and large-scale coordinated ecological management challenging. We turn to these two particular case studies to illustrate some typical dynamics associated with exurbanization. Moreover, we introduce important analytical questions for examining exurbia and point toward the ways that political ecology can help to better understand the ways exurbanization transforms communities and landscapes. This chapter does not specifically apply political ecological analysis to the cases under consideration, rather it is intended to set the tone for what follows in the rest of the book: a comparative analysis of the experiences of several communities in the United States and one in Australia. In each of these cases, as the cases in this chapter illustrate, communities are dealing with exurban change and their experiences highlight how the pressures of exurban change are so similar, despite the fact that local outcomes are or can be so different.

2.1 Protecting the Swamp or Conserving Natural Amenity? Land-Use Planning in Exurban Pennsylvania

When entering the Philadelphia Metropolitan area, an hour's drive northwest of the city, observers encounter exurbia: low-density residential housing spread throughout a fragmented landscape of fields and forests. The ongoing exurban transition of the Pennsylvania countryside is controversial to some, but observant visitors will also notice the diverse ways in which actors are responding to this exurban change. Located in western Bucks County on an elevated ridgeline, the Quakertown Swamp (Fig. 2.1) exemplifies key tensions of the rural transition that is exurbanization. At the same time, Bucks County is indicative of many metropolitan areas in the United States, where although the county may not be widely recognized for its natural amenities (McGranahan 1999), the area's beauty offers clear natural benefits for those living in the region.

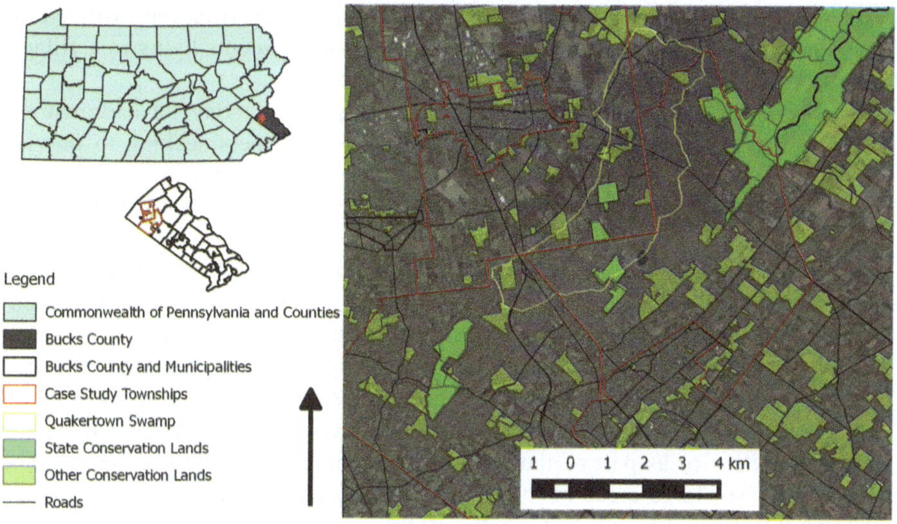

Fig 2.1 Map of the Quakertown Swamp in Bucks County, Pennsylvania (*Source*: Commonwealth of Pennsylvania and Bucks County as well as State Conservation Lands shapefiles courtesy of the Pennsylvania Geospatial Data Clearinghouse 2015. Bucks County landcover and Other Conservation Lands courtesy of Delaware River Valley Planning Commission, 2010 and 2011. Produced by Patrick Hurley using QGIS 2.10.1 software)

Known as the Great Swamp since the early 1700s, the Quakertown Swamp features many areas unsuitable (too rocky, too wet, or areas simply too small) for modern farming. In recent decades, the area has returned to a heavily wooded swamp and has become more ecologically diverse. Much of the area is underlain by impermeable rock and includes a small creek and extensive wetland system, which provides habitat to several important bird and plant species. Adjacent areas, also characterized by boulder fields, have similarly deterred farming efforts, and have aided the persistence and resurgence of forests in the area. Only farther afield, and moving outside these geologically resistant areas, did farming continue as a characteristic land use in the area during the late twentieth century.[1]

People and nature have shaped the contemporary Quakertown landscape over many, many years. Especially through the most recent era of colonial settlement, natural processes have competed with agriculture, industry, and many other human uses of the land. The Swamp proper is part of the larger Tohickon Creek watershed; the Tohickon flows across this landscape and empties into the nearby Lake Nockamixon. The lake is actually a reservoir that was created as part of a new state park developed in the late 1960s to provide recreational opportunities to nearby

[1] The analysis of landscape change discussed here is the result of using grounded visualization techniques (see Hurley et al. 2008; Knigge and Cope 2006), including air photo interpretation and coupled interviews with long-time locals. Fieldwork and qualitative interviews with Quakertown municipal officials were undertaken and completed by Hurley and Ursinus College student Max Lehner as part of an independent research project.

residents and visitors from Philadelphia. The creation of the reservoir involved the flooding of the Tohickon Run valley and relocation of residents living in a village, now flooded.

The ecological diversity and rural character present today have drawn the attention of home buyers seeking urban escapes, developers seeking to provide these opportunities, and environmental qualities are recognized by local townships, regional conservation organizations, and even federal agencies. Like many ridgelines with rugged geology in this part of the state, the rocky elevations of the Quakertown Swamp area are densely forested, particularly when compared to nearby lower elevation areas and valleys long ago converted to and largely maintained in row-crop and grain agriculture. As such, the Swamp and the surrounding area have emerged historically as an island of forest in a sea of farmland. The area is one of a series of ridges stretching from Central Pennsylvania through northwest New Jersey, all the way to Connecticut, which are together recognized by the U.S. Forest Service as part of the ecologically important Highlands Region. The undesirability of large areas of the Swamp for farming helped to create forested landscapes, which although historically useful as woodland and timber resources are highly valued for ecological conservation in the present day. The ecological diversity and rural character of the landscape sets the stage for a quintessentially exurban competition over rural landscape values: that is, a competition between different conservation and development approaches.

Securing a particular ecological value for transitional landscapes such as Quakertown Swamp is a highly political move. The resulting land-use politics become particularly complicated when different economic sectors and actors make competing demands on the landscape. In the case of the Swamp, conservation has been an active feature of the area's historical development. Lands purchased by the state of Pennsylvania in the early twentieth century are now part of the state game lands system and are areas of wildlife conservation. Nearby Nockamixon State Park offers ample recreational opportunities, marketing itself as "close enough to Philadelphia for a day trip, but far enough away for a vacation" (PA DCNR 2015). This proximity was enhanced by the construction of Interstate 476, which has created easy access for visitors from and commuters to Philadelphia. As a result, these townships and their landscapes have become sought-after areas for rural residential, or exurban living and these Pennsylvania townships have all of the markers of exurbia, as discussed in Chap. 1.

The Swamp and adjoining portions of the wider rocky landscape are characterized by fragmented land-use planning and decision-making; that is, planning decisions concerning the protection of agricultural heritage, nature conservation, and exurban development vary between the townships. The protection of agricultural heritage is related to Pennsylvania's constitutional organization as a commonwealth. Unlike in other states, counties have relatively little influence on land-use decision-making. Instead, townships and boroughs are responsible for planning and zoning. Despite this degree of local land-use control, Pennsylvania courts have interpreted constitutional provisions in ways that protect the interests of economic development. Importantly, once a municipality has zoned an area for a future land use, developers' entitlements for projects are often well-protected within state law (Walker and Hurley 2011). The courts have indicated that municipalities must accommodate all reasonable uses within their jurisdictions, meaning zoning cannot be used to exclude unwanted uses. To this end, the courts have placed limitations on the use of minimum parcel sizes typically used in other exurban jurisdictions to exclude particular types of residential development, so that in Pennsylvania two acre minimums are seen as the largest legally defensible minimum lot size (compared to 40 acres or more in other states). Thus, individual jurisdictions are limited in their ability to stop unwanted residential development in expanding urban and rural areas.

Reflecting this fragmented political territoriality, the governance of land use and development in the Swamp is split across three separate townships—East Rockhill, West Rockhill, and Richland (the ridgeline continues through other townships) and their respective land areas. By the mid-2000s, Richland Township, because of its proximity to the Borough of Quakertown, was experiencing suburban expansion into the southern end of the township where the Swamp is located. By contrast, East Rockhill is characterized by large expanses of exurban development and conserved areas. Likewise, much of West Rockhill is characterized by exurban patterns of development. But as all three townships have wrestled with the impacts of ongoing exurban development during recent decades, including finding planning and policy mechanisms to minimize development, their efforts to stop development in the Swamp are constrained by court precedents. Nevertheless, these municipalities are actively concerned about ongoing residential development. Their responses, as we shall see, raise questions not just about policy approaches, but the landscape features or characteristics that have become the focus of their efforts.

This fragmentation of land-use decision-making mirrors not only the political boundaries of local townships, but also the social contours of the area. Attitudes among residents toward both conservation and development in these Bucks County townships are divergent, in that approaches to transforming or protecting the landscape differ among individuals and groups. While development is a contentious and often unwelcome prospect in places where many people would rather see less traffic, fewer crowded schools, more farmland, and more forests (but also nice shopping opportunities and other urban services), targeted efforts to protect landscapes from development by designating specific areas for ecological conservation may not always gain popular support. Instead, other goals and priorities, such as aesthetics, may gain traction. For many observers, these cases and the effort to stop development or to "slow it"—through the so-called "slow growth" policy responses—are seen as a Not-in-My-Backyard (NIMBY) response of new exurbanites to further development. These new residents, attracted by the scenic countryside, are seen as obstructing local property owners from seeking to benefit from potential land sales.

Such views, however, often overlook the critical role that competing ideas about desired landscape aesthetics, in situ place-based economic transformations, and political constraints on land-use decision-making at higher levels of government play in the emergence of specific development- and conservation-oriented land-use politics. The need for long-term planning is often reactive and urgent, including finding ways to balance competing visions of the landscape. Yet planning decisions are often mired in political, social, economic, and ecological sands, shifting and changing the ways that any one particular jurisdiction can slow some forms of development. Meanwhile development can be easier in another jurisdiction through zoning that provides new opportunities for developers, even if and when contentious politics dominate the process.

Responding to development pressure, Bucks County (and other southeastern Pennsylvanian counties) have undertaken significant land conservation efforts and supported township efforts to shape patterns of development. Bucks County sought approval for, and voters approved, open space conservation funding during the 2000s, including bond measures to finance land acquisition for public open space and conservation easements on private lands. For example, in 1997 Bucks County voters passed a measure to fund the existing parks program, but also approved the funds necessary for agricultural land preservation and open space protection. To further these efforts, townships were then invited to develop open space plans that identified potential parkland, natural areas, and/or farmland worthy of protection. Using these plans, the townships could draw on the County funds (while providing matching monies from their own budgets) to acquire land outright or purchase conservation easements. During this process, both the County and area land trusts created their own landscape analyses, which were intended to provide guidance on exactly which areas of the county were most valuable according to conservation science frameworks (see Hurley and Walker 2004; Sandberg et al. 2013). The county worked with local botanists and ecological scientists to develop a natural areas inventory, which identified numerous sites with important ecological or wildlife values. Meanwhile, the County's' land trusts identified key areas, too, based on a combination of natural values, ecological criteria, and historic characteristics. Quakertown Swamp and adjacent portions of the wider ridgeline were recognized as an important geographical area upon which these parallel but separate analyses converged. Encouraged by these efforts and funding streams, regional land trusts stepped in to assist some townships with land acquisition.

Still, development and conservation approaches toward the Quakertown Swamp landscape—and the Tohickon Creek watershed of which it is a part—have taken on different forms, depending on which of the three townships one considers. As described above, the Swamp is split among three townships; the Tohickon Creek begins in West Rockhill, and then flows east into East Rockhill, before turning north into Richland Township. Finally, the creek turns back to the east and reenters East Rockhill. Throughout its journey, the creek encompasses areas of adjacent wetlands and smaller sections of open water. As such, collectively the Swamp and areas of adjacent forest represent both an alluring natural feature for developers and a feature of ecological concern. Getting land conserved on the ground has required a mixture of willing landowners and efforts by local governments to either work with non-profits to acquire land or steer development away from particular areas within the constraints of state law. In doing so, some conservation efforts have prioritized areas of the Swamp acknowledged as particularly valuable for their ecological importance, while *most* efforts have focused more broadly on less ecologically specific (but natural amenity oriented) open space.

Importantly, not all of the townships wished to directly embrace the activities of conservation organizations. Lands acquired by a regional land trust—in contrast to earlier properties acquired by the state—have all been within the Richland Township area. However, these conservation efforts derive in part from conflicts over land-use decision-making within the township about where and how development should proceed but conservation has been secured in ways that could be seen as separate from the use of the township's formal, institutional land-use powers (e.g., zoning). For example, the non-profit Heritage Conservancy specifically identified environmentally significant lands in key parts of the Swamp to purchase for conservation purposes. Disappointingly for conservationists, some long-time landowners were unwilling to let go of potential development entitlements and the income development from future land sales might generate. In many cases, however, lands acquired by the organization have been placed permanently into conservation and are now part of a community park system. These efforts have been enhanced by the use of conservation easements in other portions of the watershed outside the study area linking local site-level efforts to regional-scale natural heritage systems.

By contrast, much of the land conservation in the West and East Rockhill townships has come through either use of conservation easements or zoning, respectively. But, in contrast to Richland, the conservation process in West and East Rockhill largely disregarded the recommendations of the county's natural inventories. Specifically recognizing that the Swamp's "natural, historical, and cultural legacy [was] at risk," West Rockhill moved in the late 1990s to create its own conservation committee and, according to a township official, "leveraged County and State funding to help meet the landowner's financial needs and save... critical piece[s] of open space." In the process, the township focused its efforts on acquiring forested areas and farmland, which, according to one interviewee, were identified as representative of the "total breadth of West Rockhill." The result of such acquisitions of forest and farmland has been the modest progress toward conservation of the Swamp, namely through private conservation easements, even as many other areas of the township have been actively targeted. Meanwhile, using yet a different tack, East Rockhill largely decided minimum-lot zoning of 1 unit per 2 acres would suffice to ensure that residential development would not radically alter the area's rural qualities. Conservation of land through active means has centered on the acquisition of open farmland, both to create recreational park opportunities and conserve the agricultural resource base. Whether the use of conservation science in the natural inventories results in an ideal conservation outcome is highly indeterminate. In the end, all three townships set out to limit exurban development, but each township took a different approach. In spite of experiencing the same transition from a relatively rural area to one characterized by a more exurban economy, and in spite of having the same general ecological features (including the ridgeline) associated with the Quakertown Swamp, not to mention having the same county jurisdiction and funding mechanisms, the unique local politics and land-use planning approaches of each township resulted in uneven conservation attention to the Swamp.

The case of Quakertown Swamp demonstrates how one swamp and its wider watershed are transformed into three distinct landscapes through institutional

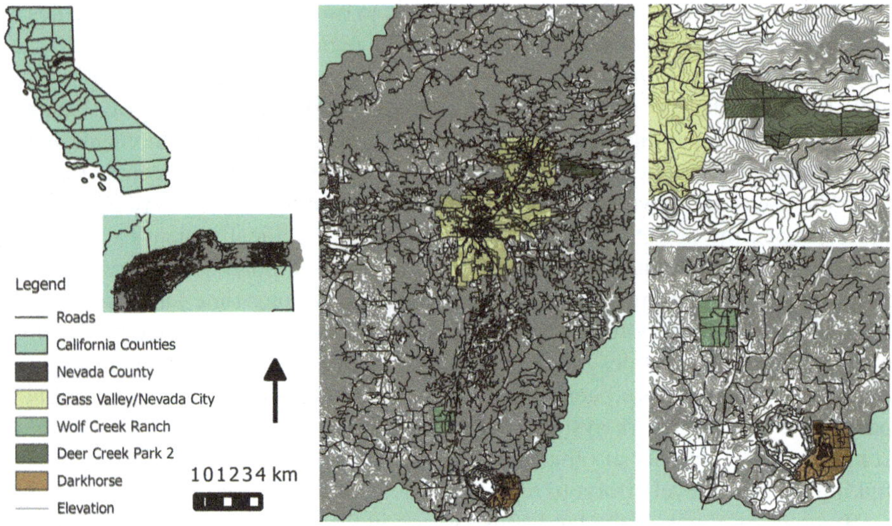

Fig. 2.2 Map of the Proposals featuring clustered development approaches in Nevada County, California (*Source*: Roads, cities, parcels, and elevation data courtesy of Nevada County 2015. Produced by Patrick Hurley using QGIS 2.10.1 software)

dynamics and formal processes of land-use planning, influence, and decision-making. The Quakertown example is interesting because it shows how sense of place and natural landscape values are enshrined at a very local level, albeit responding to, interacting with, and relying on dynamics and processes that are anything but local. The particular culture and environmental imaginary of each of the three townships have resulted in conservation of large ecologically significant areas of the Swamp in some places and the continued fragmentation of the Swamp by residential development in others. Thus, fragmented conservation is the outcome of political negotiation over environmental imaginaries within the land-use planning regime. Indeed, it is clear that the environmental imaginary of the Quakertown swamp is less about rational and integrated ecological conservation efforts to protect nature, and more of a growth management tactic used by locals.

2.2 Getting the Design Right or Getting the Right Design? Appropriately Clustering Residential Development in Nevada County, CA

Located in the Central Sierra Nevada Mountains of California, Nevada County has been a hotbed of exurban land-use conflict (Duane 1999; Hurley and Walker 2004; Walker and Fortmann 2003) (Fig. 2.2). With a colonial settlement history dating back to California's Gold Rush days in the mid-1800s, Nevada County's landscapes have long been used for the production of cattle, orchard fruit, timber, and valuable ores. But beginning in the 1960s, this part of California saw the arrival of ever increasing numbers of individuals seeking recreational and second home opportunities. The county's experience with exurbanization is typical of changes occurring in many counties with high natural amenity values (especially the American West) (McGranahan 1999; see USDA 2012 for a visualization). Indeed, Nevada County has experienced a tripling of the county's population since 1970, the rise of a real estate economy based primarily on low-density exurban development, and other changes to the area's historical economy and culture of mining, ranching, and farming. With this transition, exurbanization has ensued through the in-migration of high-tech workers and industries, equity and investment fund migration, and the resettlement of large numbers of urban retirees (Duane 1999; Walker and Fortmann 2003). At the same time, the county undertook efforts to implement new ideas based in a more conservation-oriented approach to land-use planning in response to land-use pressures (Duane 1999, Hurley and Walker 2003).

Community conflict over the implementation of new land-use planning ideas and associated land-use governance has dominated politics in Nevada County for at least the past two decades. Conflicts have included battles during the mid-1990s over the creation of a new General Plan for the county, battles during the early 2000s over efforts to implement countywide conservation measures that resulted from adopted policy in the 1995 General Plan, and more recently over efforts to reopen a gold mine outside of the Town of Grass Valley, which has been closed since the 1950s. While most of this scholarly attention has focused on battles of land conservation and development practice at the county scale, numerous battles were fought at the development project- or site-scale during the late 1990s and early 2000s.

Like land-use conflicts at the county level discussed elsewhere (Duane 1999; Hurley and Walker 2004; Walker and Fortmann 2003), these conflicts help to illustrate how different landscape values and ideas about appropriate uses of open space intersect to shape the very design elements of residential projects that emerge in these transitional rural places.

Among the battles over individual development projects in Nevada County generally, the story of the Wolf Creek Ranch Estates (discussed below along with the two parallel cases of Darkhorse, and Deer Creek) is indicative of exurban land-use planning conflicts. The Wolf Creek Ranch project proposal and review process is useful in illustrating key dynamics of the exurban transition, with important implications for thinking about political ecological examinations of exurbia. Wolf Creek Ranch Estates represents the efforts of a family with a multi-generational commitment to ranching to engage in the land development process. For conservationists in the county, the site was seen as an important undeveloped area with ecological value, given the site's extensive oak woodlands and the eponymous creek that runs through the site. But contention over the proposal was also mired in the history of land-use decision-making as applied to this portion of the southern county.[2]

[2] The analysis of land-use conflict discussed here is the result of extensive ethnographic fieldwork conducted in the early 2000s, including interviews with county officials, developers, and residents. Fieldwork was undertaken as part of dissertation research (Hurley 2004).

From the start, Wolf Creek was likely to be contentious. First, the zoning that laid the groundwork for the project resulted from the county's 1995 General Plan process, which had introduced a higher level of allowable units into areas that had been previously zoned for and were generally characterized by low-density residential parcels of 5 and 10 acres. In doing so, this process resulted in increased development rights for particular landowners, including the family owners of the Wolf Creek Ranch parcel; moreover, the relative locations of the projects largely ensured design proposals divergent from surrounding development patterns. The 1995 General Plan process also introduced into policy and codified the use of "clustering" for project proposals. Clustering allows developers to build dwelling units—typically single-family houses but also townhomes—at higher densities on a much smaller part of the property, while ensuring that larger portions of the site will be maintained as open space. As a voluntary policy, the inclusion of clustering in the Plan opened up the possibility for all future projects to maintain higher densities and numbers of buildable parcels for their projects. When merged together, these planning dynamics almost ensured that any ensuing project that employed the new kind of clustering would be controversial, given the extent to which patterns of development and aesthetics would diverge from nearby landscape norms.

Of the areas given increased development rights, nearly all of the ensuing project proposals put forth incorporated the clustering option, albeit with the remaining open space areas used for landscape conservation. The first and most well-known project, Darkhorse, emerged almost immediately after the General Plan was adopted. Proposed by the Baldwin Family and an investment group from nearby Roseville, the 1046-acre project set out to convert an area that, according to a local land-use official had largely been "used for generations as ranch land," into what one developer described as a "world class golf course and housing subdivision."[3] This situation meant that the developers were eligible for a maximum of 300 single family homes on 846 acres, with the remaining 200 acres approved for the golf course, which angered conservationists who questioned whether the project was consistent with the area's rural and pastoral character. Thus, Darkhorse set a new precedent in the county by demonstrating the ways in which the county's new policies might be used to achieve the maximum number of units in spite of the environmentally sensitive context of the site.

By contrast, the lesser-known Deer Creek Park 2 project was a subdivision proposal on land that is heavily forested, with typical forest cover of ponderosa pine, black oak, manzanita, incense cedar, and Douglas fir. In localized areas, the site is characterized by steep terrain; indeed, a longstanding criterion within county policy dictates that all slopes above 30 % are unsafe to build upon. While Darkhorse represented a rural subdivision in the rolling hills of the southern part of the county, where oak woodlands and open fields characterize a landscape shaped by an agrarian history of cattle grazing, Deer Creek Park 2 represents a test of county policy on lands historically managed for timber production. The project proposal, put forth by the family-owned Terra Alta Development Corporation, called for the maximum

[3] Lake of the Pines, 8–7–2003.

number of 193 residential units on 580 acres in an area characterized by narrow, winding, and relatively steep roads. Implementation of the project would result in 404 acres designated "open space," a small amount of land for a communal sewage disposal facility, and home sites ranging from half-acre to 2-acre lots. Given that Terra Alta was headed by long-time local Lance Amaral, who is also known for the exploits of his timber company and other contentious projects, it did not necessarily come as a surprise that this proposal would also have a twist.

Besides project opponents' reaction to the project as classic sprawl (The Union Editorial Board 2004) and fears about "increased traffic, noise and possible pollution" to nearby Deer Creek (Druzin 2003), resistance to the project also centered on the company's efforts to retain ownership of the remaining open space for continued timber production. Specifically, this design feature was intended to conserve a portion of the county's resource base, by drawing on the county's policy that clustering helps "to support grazing, forest management, and crop production coexistent with residential uses." Thus, Deer Creek Park 2's use of clustering not only would provide the developer with the opportunity to maintain the maximum number of dwelling units, but also the opportunity to own and manage the open space for continued economic gain. The proposal angered many neighbors and activists who did not see the protection of the county's natural resource base, in this case timber lands, as in keeping with the nature-as-wilderness amenity valuation they imagined for the conservation lands.

During the ensuing years, the project continued to encounter stiff resistance as it made its way through the planning process. Opponents of the project were eventually able to reform the project proposal, including a reduction in the number of total units (from 193 to 62), the inclusion of buffer parcels around the residential area intended to protect against forest fires, and the addition of an area subject to specific watershed management efforts. The proposal has continued to slowly move through subsequent phases of the planning process in the years since, incurring resistance from various neighborhood groups using appeals all along the way to either stop the project, continue to reduce the number of allowable units, and/or change the eventual design layout.

Thus, from the perspective of controversial uses of clustering in Nevada County, Wolf Creek Ranch was in good company. But in the case of Wolf Creek Ranch, the issue of how open space would be used—the persistence of the working landscape—was less of an issue. Instead, the notion of clustering itself and the patterns of development it produced were challenged, both in terms of the overall number of units and how those units would be built. Located in the southwestern portion of the county, the project was originally conceived in 1991 and was envisioned as a planned development of 5-acre lots. However, the consultant, who at the time was reviewing the county's General Plan, recommended 10-acre parcels to the family who owned the land and was interested in developing, citing the lack of public water and sewer (Nevada County Planning Department 2004a). In response, the family submitted an application for a General Plan designation change "in exchange for a commitment to bring public water to the site and a clustered" project with a comprehensive site plan. The family also promised to provide substantial open space, which resulted in them obtaining an average 3-acre density for the project site (Nevada County Planning Department 2004b, p. 1). As a result, the project proposal went from the recommended 69 units to a proposed 230 units. This 3-acre average density, plus the clustering of parcels necessary to maintain this number, has been at the core of every plan since the project was formally submitted to the planning department for consideration in 1997.

From the time the Wolf Creek Estates project was submitted for formal review, individual neighbors, neighbors' groups, and other groups in the county concerned with growth, such as a land-use advocacy group known as the Rural Quality Coalition, have fought to reduce the size and intensity of the project. Indeed, throughout the years of review, neighborhood groups and environmental activists continually questioned the willingness of the developer to produce a proposal that was consistent with the county's General Plan. For example, The South Nevada County Concerned Citizens[4] challenged the project's consistency with the General Plan, noting discrepancies in the calculation of the project's average density and the county's comprehensive plan language focusing on larger parcels. In doing so, opponents further challenged notions of what it means to be rural, with letters submitted to the planning commission emphasizing the extent to which this density and the resulting aesthetics, loss of trees, and increased traffic would harm the "agricultural nature of our home" (PMC 2000, pp. 3–78). Another person at a land-use hearing testified that the "project hasn't been clustered to maintain rural quality or protect the quality of life or resources…[This project has] too many impacts and too few benefits". The same individual, a member of the South County Neighbors Association (another organization in opposition) and a landscape architect by training, asked the planning commission "to deny the project" (Nevada County Planning Commission 2004).

Over the course of the next several months, the county's planning commission and ultimately the county's Board of Supervisors would debate the developer's proposed plans as well as the competing plans submitted by the neighbors' association. Significantly, the South County Neighbors Association was able to force the issue of the size and density of the project through an appeal of the project's initial environmental review. In lodging their challenge, the group pointed out that significant and unavoidable negative environmental impacts were being overlooked, and they insisted that the project applicants should be required by the planning committee to lower the project's density. The County's decision-making body, the Board of Supervisors, which was comprised of a "planned growth" majority at the time (see Hurley and Walker 2004), upheld the appeal and ordered the developer to consider alternative project designs, including two suggested by South County Area Neighbors.

The two alternatives suggested by South County Neighbors Association reduced the overall number of lots available for development. The first alternative argued

[4]This is a separate organization from the group, South County Area Neighbors, referred to at the beginning of this chapter, further highlighting the extent to which these issues are being politicized.

for "a total residential lot count of 69 lots" based on using a minimum lot size of 10 acres to calculate the allowable number of units, but with "lot sizes of 5 acres and greater" and where the "remaining 333 acres would be dedicated as open space and privately owned and maintained by a homeowners association" (PMC 2001, pp. 1–6). Importantly, the neighbors' association argued that, "the lot sizes under this alternative would be more equivalent to lot sizes in the surrounding area than would those under the proposed project" (PMC 2001, pp. 1–6; see Fig. 2.2). Moreover, this alternative would maintain significant open space buffers between the project and the majority of existing landowners adjacent to the project. The second project alternative offered by the neighborhood association involved the elimination of 115 lots, or half of those that had been proposed. In this case, the equation used to calculate the number of housing units was based on a 5-acre minimum lot size, but the site design provided by the group showed the resulting units on much smaller lots, covering only 287.5 acres, and keeping the remaining 403.5 acres as dedicated open space. This layout design also created extensive open space buffers between the majority of surrounding neighbors and the proposed residences.

Upon completion, the revised environmental document showed that the design put forth by the Wolf Creek developer was not the option with the least environmental impact. Interestingly, the alternative designs proposed for evaluation by the developer, which maintained 230 units for the parcel, had fewer impacts than their preferred option, but had greater impacts than the reduced intensity versions submitted by the South County Neighbors Association. The neighborhood group had shown that while project design was obviously part of the problem, the real issue was the politics that had led to awarding the developer with, what was in their opinion, the high number of possible housing starts from the outset.

Yet the finding of the environmental review has not resolved the question of what minimum lot size (under a conventional, non-clustered approach) should be used to calculate the appropriate number of homes, nor has it led to a decision about how many homes will actually be built. Instead, the developers have consistently maintained the 230 units as necessary to the success of the project, leading to a continued political fight by the neighbors against the project. At a later meeting, the developer proposed to reduce an area of dense housing clusters on part of the project. In their place, the owners of the property (or project applicants) had decided to retain three parcels (each with agricultural easements) in their personal ownership and a 10-acre part of the project, formerly slated for a park, would now be the site for 55 future condominium units. After a grueling meeting, planning commissioners decided not to rule on the proposal, but the planning commission's chair[5] implored the developer to work with the neighbors to resolve the issues involved. One opponent—also a

[5] The current chair of the planning commission was considered by many observers in Nevada County to be "in the minority" because he was appointed by a "planned growth" supervisor. Planning commissioners in Nevada County, however, have a history of demonstrating their independence, although the votes on many decisions related to development that are appealed to the Board of Supervisors often fall along ideological lines.

former planning commissioner—has suggested that regardless of what the planning commission decides, the developer is likely to appeal the decision to the Board of Supervisors where it is believed the project would be approved at the density preferred by the developer (the current proposal). Perhaps typifying the cynicism among some opponents was the appearance of condominiums in the project, and the failure of the commission to outright reject the idea. Like Deer Creek Park 2, Wolf Creek Ranch Estates has yet to physically materialize on the Nevada County landscape. These two housing projects are, in essence, hidden imprints on the landscape that have yet to materialize but reflect the politics of land use in exurbia.

The Nevada County cases demonstrate how the aesthetics of a rural landscape associated with agriculture and forestry, but more recently transformed through low-density development, are the site of intensive site-level engagements by exurban actors. The transition from rural to exurban plays out through the planning process, which is the focus of actors' negotiation over the economic, political, social, and ecological changes that come with the transition. The future exurban landscape is emerging out of the conflicting "environmental imaginaries" (discussed in the introduction and below). But how are we to make sense of the development outcomes in the cases discussed above and the forces that produce these divergent outcomes, both in southeastern Pennsylvania and in Nevada County, California? Are these land-use outcomes simply the result of newcomers and their preferred ideas about what the landscape should look like, including which landscapes should be privileged and, thus, which natural resource activities should no longer be permitted by regulatory code or socially tolerated in the landscape? Or are these outcomes the result of complex, emergent forces that are not as easily characterized by the newcomer/long-time local binary?

2.3 Implications of the Quakertown and Nevada County Cases When Viewed from a Political Ecology Perspective

Returning to considerations of how these cases inform our study of exurbia, we draw on Maureen Reed's (2007a) exploration of environmental governance and the so-called "environmental management regimes" to posit that "land-use planning regimes" within particular areas explain a lot about the differences we observe in patterns of exurban development and the resultant conflicts. Reed defines "regional environmental management regimes" as the "formal and informal institutional arrangements in which public, private, and civic interests work simultaneously (together or apart, in synch or at odds with one another) to influence, make, and/or carry out governing decisions about environmental and resource management" (pp. 321–322). Reed's use of "environmental management regimes" intersects well with what Calhoun (2002) describes as the "broad framework of rules and norms" that together comprise planning. For example, these rules and norms are typically referred to in planning literature as legislation, policy, practice, public consultation, and decision-making. When deployed together, Reed and Calhoun's concepts provide a mechanism for examining how ideas about nature and landscape are negotiated within the dynamics of land-use planning, including the environmental management approaches and efforts to conserve particular places that result. Moreover, we draw attention to the ways these regimes affect trajectories of land-use change within an area. To be clear, comprehensive plans, zoning, subdivision control, and other "tools" of planning are limited in their effect on shaping landscapes; as planning tools, they are only as successful as the rules and norms that guide them and those who wield them. Rather, it is the politics in real time (constantly unfolding, never static) of comprehensive plan-making, zoning code writing, and the approval of subdivision plans and site plans where political economic values become embedded in the material landscape and compel new approaches to environmental management.

As with environmental management regimes (see Chap. 1), land-use planning regimes do not spontaneously arise. Rather, they emerge from the intersection of changing social, cultural, political, economic, and ecological dynamics and the ways new knowledges about the environment, technological capacities, and design options interact with these dynamics. In choosing how to deploy new knowledges, technologies, and designs, though, ideas about nature, its fragility, and appropriate uses—what political ecologists refer to as "environmental imaginaries"—come to influence decision-making moving forward. Indeed, as political ecology scholars have suggested, control over the dominant environmental imaginary shaping policy is a powerful political tool. Watts and Peet (1996, p. 263) suggest the concept of "environmental imaginary" as "a way of imagining nature, including visions of those forms of social and individual practice, which are ethically proper and morally right with regard to nature." The concept of environmental imaginaries in relation to exurbanization captures the discursive (and performative) nature of environmental knowledge and how ideas of nature are "formed, contested and practiced in the course of specific trajectories of political-economic change" (Watts and Peet 1996, p. 263). Perhaps the most important aspect of Watts and Peet's concept of environmental imaginaries is their contention that environmental imaginaries are central to environmental debates, as people come together to create social movements to push back against well entrenched and institutionalized notions of nature. Despite Peet and Watts' (1996) focus on the "developing" world, we see much to draw from their work and their questioning of the imposition of Western understandings of nature that displace traditional indigenous knowledge (and in our cases, indigenous knowledge as well as others in touch with the land (e.g., ranchers, farmers, and foragers)). In our cases, we see important parallels between studies of the "developing" world and studies of whose vision of the environment dominates in rural to exurban transitions in the "First World."

In other words, no matter where conflicts between resource and amenity landscape values are playing out, whether in the "developed" or "developing" world, environmental imaginaries are being fought over. In the technical, rational-comprehensive planning process, nature is represented through western environmental science (habitats, ecosystems and, most recently, natural heritage features and systems). But ideological nature, especially powerful in the hands of amenity migrants and those who would house them, is not data for analysis in planning. People's sense of place in relation to intimate and relational lived experience and transcendental natural beauty is not easily captured or represented in the planning process. Much of planning scholarship suggests better forms of "post-rational" collaborative and communicative processes are needed in order to draw in people's experience (Healey 1997; Innis and Booher 2010). The hope of post-rational processes is that the "common" interest will be expressed in ways that shape future land-use plans. However, that the planning process as it stands inherently favors the elite who have the education and time needed to participate effectively means that the outcomes of the planning process often inadequately reflect the environmental imaginaries of all who those outcomes will impact.

In their study of landscape change in the Appalachian Mountains in the eastern United States, Nesbitt and Weiner (2001) draw on Peet and Watt's concept of environmental imaginaries to great effect in thinking about the way everyday lived experience shapes individual frames of reference toward nature. Nesbitt and Weiner's examination demonstrates the way these everyday experiences are tied to livelihood interactions with particular environments and resources. The historical and contemporary struggles between rural capitalisms in the Appalachians—between land use for timber, agriculture, recreation, biodiversity conservation, and exurban residential development—is exactly the kind of rural to exurban transition addressed by authors in the chapters that follow. However, as much as land-use conflict and the regulation of property are at the center of the West Virginia case, the role of land-use planning is not explicitly discussed. The Quakertown case study suggests that the jurisdictional fragmentation of planning and land-use regulation as well as differences in environmental imaginaries produce different results even within the same landscape. Likewise, the Nevada County case demonstrates how these struggles take place through the complex decision-making processes that characterize planning and associated governance dynamics.

The political economic approach used by political ecologists, highlighted here by the work of Peet and Watts (1996) and Nesbitt and Weiner (2001), can be fruitfully extended to consider actors in the rural transition, especially as their roles relate to land-use decision-making. Actors, or stakeholders, are those who individually or through communities of place and other forms of alliance, debate environmental imaginaries and appropriate forms of management (Reed 2007a, b; Robbins 2006). In the case of Nevada County, central actors and discourses conflicted in a collaborative planning and visioning process for a county in the throes of the rural to exurban transition (Hurley and Walker 2004; Walker and Fortmann 2003). The vision proposed by county leaders at the time was to move toward a more conservation-based planning approach. County stakeholders (ranchers, environmentalists, exurbanites) understood how planning could "redefine where development should or should not take place" (Hurley and Walker 2004, p. 1542) and therefore redefine access and control over landscape and nature. The examination of the decision-making around the protracted conflicts over the design of Wolf Creek Ranch begins to bring these conflicts into closer focus. So, too, the extent to which particular imaginaries shape the use of different regulatory tools in the case of Quakertown help us to see how these actors engage with, use, and contest different ways of transforming landscapes.

Land-use planning has come into being as a way governments shape access to land and control resources. Historically, planning in the U.S. and Canada co-evolved with the establishment of state and local government to deal with issues of public health and safety (Hall 2014). Today, how members of the public are able to meaningfully participate in decision-making about land use and environmental management is a preoccupation of the planning literature (Quick and Feldman 2011), including not only local voices but also members of institutions and organizations (especially conservation organizations) working across regions and national boundaries to influence land-use decisions. Through land-use planning, society makes and follows rules and regulations as well as determines "rules-in-use" through accepted planning practice. "Differentiated social actors gain access to and control over resources through institutionalized practices [including planning]" (Watts and Peet 2004, p. 26). And while Watts and Peet's attention is toward understanding environmental conflicts in "developing" regions of the world, their political ecological analytical framework has much utility for the study of cases in the United States, Canada, Australia, and elsewhere where institutions are more established and time-tested and yet exhibit similar tensions in the politics of environmental imaginaries (Hurley and Walker 2004; Robbins 2002; Walker 2003; Walker and Hurley 2011).

Considering the Quakertown Swamp and the Nevada County cases from a political ecology perspective suggests the need for a more nuanced consideration of the efforts to manage the rural to exurban transition. These cases demand answering a series of questions about how key dynamics shape the decision of individuals to sell land, how development attempts to transform these parcels play out in relation to changing planning frameworks, and how these processes of change fit within planning systems and community expectations about what landscapes will look like in the future. In answering these questions, the political ecological focus on ideology and power in communities and institutions opens up new questions about the interaction of forces producing both uneven patterns of residential development and nature conservation. Which landscapes are protected and which are sacrificed? As well, environmental management is uneven; that is, which human activities are allowable in conserved areas (e.g., hunting and foraging), which species and their habitats are privileged for restoration efforts (e.g., native species, agricultural land covers), and which landscape management protocols take priority in different areas—all of these variables depend on the dynamics of specific places and regions, with the result that landscapes are managed "unevenly". Further, political ecology suggests the need to connect land-use decisions, and the ways landscape changes occur in particular places, to global political economies. Increasingly, this means seeing how decision-making is characterized by landscape-level interventions and associated environmental politics (see Hurley and Walker 2004; Sandberg et al. 2013).

By using political ecological analysis, an approach built on "ethnography and intensive case study focusing on micropolitics" (Robbins 2002, p. 1509), the following chapters in the book seek to further illustrate the diverse efforts to manage the rural to exurban transition. They document the complexity of actors and the diverse interactions among these actors in new contexts, drawing on the critical concepts briefly described above and in Chap. 1. In doing so, they highlight how these interactions influence the trajectories of landscape change in exurban places. Further, when taken together, these chapters further demonstrate the analytical strengths of political ecological analysis for understanding exurban change, while providing a model for thinking about future comparative work in the field.

References

Calhoun, C. (Ed.) (2002). Regime. In *Dictionary of the social sciences*. Retrieved from http://www.oxfordreference.com/view/10.1093/acref/9780195123715.001.0001/acref-9780195123715-e-1420.

Druzin, H. (2003, 25 August). Diaz: County could lose its charm. *The Union*. Retrieved from http://www.theunion.com/news/2332082-113/news-none.

Duane, T. (1999). *Shaping the Sierra: Nature, culture, and conflict in the changing West*. Berkeley, CA: University of California Press.

DVRPC. (2010). DVRPC 2010 digital orthoimagery—Mr. SID County Mosaics. Retrieved from http://www.dvrpc.org/Aerials/.

DVRPC. (2011). *2011 Protected open space*. Retrieved from http://www.dvrpc.org/Mapping/data.htm.

Hall, P. (2014). *Cities of tomorrow: An intellectual history of urban planning and design in the twentieth century* (4th ed.). Oxford: Blackwell.

Healey, P. (1997). *Collaborative planning: Shaping places in fragmented societies*. Vancouver, British Columbia, Canada: University of British Columbia Press.

Hurley, P. (2004). *Whose vision? The political ecology of land-use policy in Nevada County, California*. Unpublished doctoral dissertation. University of Oregon, Eugene, OR.

Hurley, P., Halfacre, A., Levine, N., & Burke, M. (2008). Finding a "disappearing" non-timber forest resource: Using grounded visualization to explore urbanization impacts on sweetgrass basket-making in greater Mt Pleasant, SC. *The Professional Geographer, 60*(4), 1–23. doi:10.1080/00330120802288941.

Hurley, P., & Walker, P. (2004). Whose vision? 'Conspiracy theory' and land-use planning in Nevada County, California. *Environment and Planning A, 36*(9), 1529–1547. doi:10.1068/a36186.

Innis, J., & Booher, D. (2010). *Planning with complexity: An introduction to collaborative rationality for public policy*. New York: Routledge.

Knigge, L. D., & Cope, M. (2006). Grounded visualization: Integrating the analysis of qualitative and quantitative data through grounded theory and visualization. *Environment and Planning A, 38*(11), 2021–2037. doi:10.1068/a37327.

McGranahan, D. (1999). Natural amenities drive rural population change. Washington, DC: USDA, Agricultural Economic Report No. 781, USDA Economic Research Service.

Nesbitt, J. T., & Weiner, D. (2001). Conflicting environmental imaginaries and the politics of nature in Central Appalachia. *Geoforum, 32*(3), 333–349. doi:10.1016/S0016-7185(00)00047-6.

Nevada County. (2015). *Digital data download*. Retrieved from http://www.mynevadacounty.com/nc/igs/gis/Pages/MetaData.aspx.

Nevada County Planning Commission. (2004). *Recorded proceedings of the Commission, 2-12-2004*. Nevada City, CA: Nevada County Planning Department.

Nevada County Planning Department. (2004a). *Nevada County Planning Commission staff report* (File# FM97-007 & Z97-007).

Nevada County Planning Department. (2004b). *Wolf Creek Ranch project findings*. Retrieved from http://docs.co.nevada.ca.us/dscgi/ds.py/Get/File-69658/Project_Facts_&_Findings.doc.
Pacific Municipal Consultants (PMC). (2000). *Wolf Creek Ranch Estates final environmental impact report* (SCH No. 1997102097). Sacramento, CA.
Pacific Municipal Consultants (PMC). (2001). *Wolf Creek Ranch Estates revised draft environmental impact report* (SCH No. 1997102097). Sacramento, CA.
Peet, R., & Watts, M. (Eds.). (1996). *Liberation ecologies: Environment, development, and social movements*. London: Routledge.
Pennsylvania Department of Conservation and Natural Resources (PA DCNR). (2015). *Nockamixon State Park*. Retrieved from http://www.dcnr.state.pa.us/stateparks/findapark/nockamixon/.
Pennsylvania Geospatial Data Clearinghouse. (2015). *Pennsylvania spatial data access*. Retrieved from http://www.pasda.psu.edu/uci/SearchPage.aspx.
Quick, K. S., & Feldman, M. S. (2011). Distinguishing participation and inclusion. *Journal of Planning Education and Research, 31*(3), 272–290. doi:10.1177/0739456X11410979.
Reed, M. (2007a). Uneven environmental management: A Canadian comparative political ecology. *Environment and Planning A, 39*(2), 320–338. doi:10.1068/a38217.
Reed, M. (2007b). Uneven environmental management: A Canadian perspective. *Environmental Management, 39*(1), 30–49. doi:10.1007/s00267-005-0159-5.
Robbins, P. (2002). Obstacles to a first world political ecology: Looking near without looking up. *Environment and Planning A, 34*(8), 1509–1513. doi:10.1068/a34217.
Robbins, P. (2006). The politics of barstool biology: Environmental knowledge and power in greater Northern Yellowstone. *Geoforum, 37*(2), 185–199. doi:10.1016/j.geoforum.2004.11.011.
Sandberg, A., Wekerle, G., & Gilbert, L. (2013). *The Oak Ridges Moraine battles: Development, sprawl, and nature conservation in the Toronto Region*. Toronto, Ontario, Canada: University of Toronto Press.
The Union Editorial Board. (2004, February 10). The Union Q & A: Olivia Diaz, District 1. *The Union*, Grass Valley, CA.
Walker, P., & Fortmann, L. (2003). Whose landscape? A political ecology of the 'exurban' Sierra. *Cultural Geographies, 10*(4), 469–491. doi:10.1191/1474474003eu285oa.
Walker, P., & Hurley, P. (2011). *Planning paradise: Politics and visioning of land use in Oregon*. Tucson, AZ: University of Arizona Press.
Watts, M., & Peet, R. (1996). Conclusion: Towards a theory of liberation ecology. In R. Peet & M. Watts (Eds.), *Liberation ecologies: Environment, development, and social movements* (pp. 260–269). London: Routledge.
Watts, M., & Peet, R. (2004). Liberating political ecology. In R. Peet & M. Watts (Eds.), *Liberation ecologies* (2nd ed.). London: Routledge.

Chapter 3
Divergent Perspectives and Contested Ecologies: Three Cases of Land-Use Change in Calaveras County, California

Colleen C. Hiner

3.1 The Exurban Context: A Place of Simultaneous Social, Economic, and Ecological Change

Exurbia occupies the edges and borders between rural and urban spaces and places. It is significant not just for its spatial patterns and implications, but also for its social, cultural, and political ones. In other words, exurbia can be defined both by its physical structure—often large-lot, low-density housing occupied by urban-oriented residents—as well as by the diverse and sometimes disputed narratives of nature and rurality found there. In this chapter, I argue that exurbia is characterized by the concept of *contested ecologies*, wherein contrasting perspectives of the meaning, values, and/or function of land or resources lead to disagreements over the future of a particular place, environment, or landscape. *Contested ecologies* amount to differing viewpoints of not just people but also environment in place. As a focus of my discussion, I use a comparative case study of three examples of rural to exurban land-use change in Calaveras County, California, to investigate the significance of divergent environmental and ideological perspectives for land-use decision-making and environmental management in exurbia.

"Exurbia" is not a new geographical concept for scholarly inquiry; indeed, the extra-urban, peri-rural, "in-between" spatial and cultural zone known as exurbia has been explored and explained by numerous scholars over time (including, but certainly not limited to: Abrams et al. 2012; Ban and Ahlqvist 2009; Brown et al. 2008; Cadieux and Hurley 2011; Cadieux and Taylor 2013; Nelson 1992; Spectorsky 1955; Taylor 2011; Walker and Fortmann 2003). This substantial body of literature identifies the impacts of exurbia as broad (Taylor 2011), including impacts which are: *ecological or environmental*, such as habitat fragmentation or destruction, which

C.C. Hiner (✉)
Department of Geography, Texas State University,
601 University Drive, San Marcos, TX 78666, USA
e-mail: cchiner@txstate.edu

L.E. Taylor, P.T. Hurley (eds.), *A Comparative Political Ecology of Exurbia*,
DOI 10.1007/978-3-319-29462-9_3

creates discontinuous management zones, and other impacts of sprawling residential development; *social*, i.e., being a source of conflict between differing interest groups, such as "new" versus "old" residents or liberals versus conservatives; and *economic*, such that economies shift from being more primary sector or resource based to more amenity-driven and/or post-productivist or consumption-based.

A number of scholars have described how amenity migration and associated exurbanization are changing the ways that rural areas are viewed, used and managed, noting, in short, how amenity migration has shifted the landscapes and social norms of places along the rural–urban edge (Nesbitt and Weiner 2001; Travis 2007; Cadieux and Hurley 2011; Gosnell and Abrams 2011; Taylor 2011). Exurbanites value the rural "in particular ways that emphasize their value as sites of landscape consumption" (Cadieux and Hurley 2011, p. 298), which influences those places sought after as amenity destinations as well as those that are not. The shifting and plural perspectives on the function and value of rural landscapes can have profound implications for the governance of rural space. Specifically, "rural governmental institutions facing exurbanization and amenity migration are often unequipped to grapple with the multiple competing interests that constitute diverse and changing rural agendas" (Cadieux and Hurley 2011, p. 297). In short, exurban land-use change "complicates rural environments, governance, and resource management" (Cadieux and Hurley 2011, p. 298) where it occurs.

These shifting dynamics alter what Reed (2007, pp. 321–322) calls "environmental-management regimes," which she defines as the following:

> …formal and informal institutional arrangements through which public, private, and civic interests work simultaneously (together or apart, in synch or at odds with one another) and within different sets of power relations to influence, make, and/or carry out governing decisions about environmental and resource management.

While such environmental management regimes are literally everywhere—whether we recognize them or not—the functional shifts occurring in environment and society in exurbia make understanding their workings both challenging and critical as the implications for land-use change in the exurban context are multiple and profound. I add to this literature by describing three cases of landscape change in a particular locale and by applying the concept of *contested ecologies*, wherein disagreements over the meaning, values, and/or function of land or resources lead to contrasting perspectives of a particular place, environment, or landscape (Hiner forthcoming).

As a political ecologist, I approach social and ecological change as mutually constitutive and power-laden and seek to understand the use of natural resources as mediated by biophysical characteristics and processes as well as socio-political ones (Zimmerer and Bassett 2003; Paulson and Gezon 2005; Blaikie and Brookfield 1987; Robbins 2011). Specifically, I consider this study a "regional" political ecology, as described by Blaikie and Brookfield (1987) and elaborated by others (Black 1990; Walker 2003; Neumann 2010), which seeks to ground theorizing in particular places while also drawing out the linkages and "chains of explanation" to help explain local circumstances and outcomes. Three cases of land use and management are presented here, each with distinct and divergent land-use outcomes, and, which elucidate the kind of conflicts identified by the exurbia and political ecology literature. At their core, these disagreements are about differing conceptions of rural space, contrasting environmental imaginaries, and diverse preferences for how society, the environment, and the government function, which I frame as *contested ecologies*, instances wherein the environment itself, let alone the "problems" associated with it, is contested.

In the following sections, I describe the site, situation, and methods of the study; "set the scene" of exurbia in general and in this particular place; present three case studies and their contested ecologies; and offer an in-depth analysis of one case given its wide-ranging political and policy implications. I mainly focus on one case in my discussion because it is the most compelling analytically, offering a clear-cut and fascinating view into the varying perspectives on land use and private property present in Calaveras County. Moreover, the case presents a site in which the ecological implications of various land-use outcomes are readily apparent. I conclude with an evaluation of the significance of such divergent perspectives for land-use decision-making and environmental management in the context of *contested ecologies* and their implications for environmental management (regimes).

3.1.1 Site, Situation, and Methods

The context of this research is Calaveras County, California (USA), an ecologically heterogeneous county located in the Sierra Nevada "foothills," stretching from the floor of the Central Valley into the alpine reaches of the Sierra Nevada mountains, mainly characterized by rolling oak woodland (Sierra Nevada Ecosystem Project 1996). With a population of approximately 45,000 (Census Bureau 2010), a generally undeveloped landscape, and only one incorporated city, the county is rural by conventional descriptions (Cromartie and Bucholtz 2008; Woods 2005). While mainly demographically homogenous (the 2010 US Census described Calaveras County as 88.9 % white (Census Bureau 2010)), socioculturally, the county has undergone shifts in recent decades away from a strong, historical emphasis on primary production and resource extraction to an economy that is more mixed, more focused on consumption-based activities (Walker and Fortmann 2003), and which features a greater diversity of interests, backgrounds, and occupations among its residents (Mintier and Associates, Environmental Science Associates and Calaveras County Community Development Agency 2008; Momsen 1996; Hiner 2014, 2015, and forthcoming).

In terms of methods, the study incorporates a detailed analysis of three cases of divergent land management strategies in three different locations in the study area, Calaveras County, California (Fig. 3.1). Using these cases, I investigate the values various rural residents embrace and how they mobilize those values and ideologies to influence and enact land-use change. Methods include in-depth, semi-structured interviews, a written demographic and political/ideological survey, participant observation, and public document analysis (DeLyser et al. 2010; Patton 2002; Sayer 2010; Tashakkori and Teddlie 1998). I interviewed the major political officials as well as representatives from various interests and perspectives using a cumulative, "snowball" (Patton 2002) or inductive (Sayer 2010) sampling method, conducting

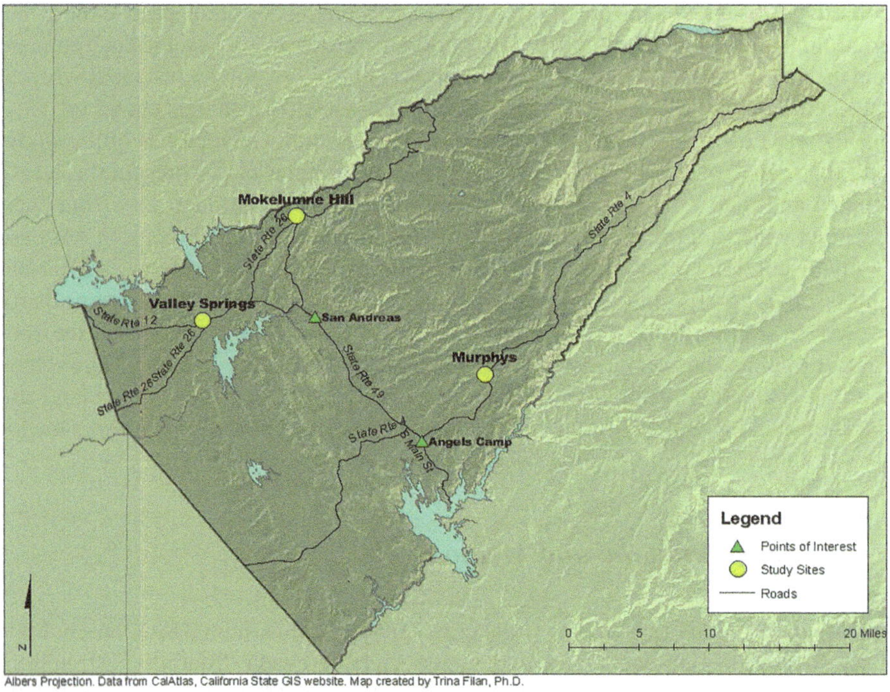

Fig. 3.1 Map of study sites. Produced by T. Filan

51 interviews total. Participation observation occurred during primary data collection between January and June 2010; however, this study was also informed by an ongoing interest in the area beginning in January 2008.

3.2 Exurbia, a Place of Contrasts

As exurban settlement patterns have spread across the landscape, one might expect a cultural homogenization to occur alongside the physical standardization. However, while exurban penetration into rural areas has pushed ever deeper, the perceived differences between urban and rural populations have persisted. Indeed, as I argue elsewhere, as the functional differences between rural and urban places have diminished, the significance of cultural and ideological differences within those areas has increased (Hiner 2015). In short, in order to cope with decreasing space-time[1] between the rural and the urban, cultural ideals and preferences between groups that

[1] There are numerous takes and discussions of how time and space are interconnected and socially mediated, making the perception of space and time vary from place to place, time to time, and from person to person (see, for example: Harvey 1996; Massey 1999; Massey 2001; Harrison et al. 2004; Merriman 2012).

perceive themselves to be different from one another have become even more entrenched. That is not to say that we should reify the rural/urban or been-here/come-here divide, but the reality of social and political conflict in places in flux cannot be denied (Hiner 2014, 2015, and forthcoming).

In order to "set the scene" for the kinds of divergence one might find in a typical exurban place, I offer two quotes from respondents. The first respondent is an elected community leader, who states:

> As people drive through the county, they really enjoy the fact that there are trees and grass and cows. Where I think that people sometimes get confused is that they say that land ought to stay that way, and they get quite up in arms when somebody needs or wants to sell a parcel. And this dreaded word of "developer" comes into the conversation, and suddenly people want to organize to keep a developer from taking away their view. [But] they don't own the view. I have to keep reminding people that, at the end of the day, if they want to preserve the view, if it's that important to them, then they should go buy the land and keep it looking like it does.

The other two respondents are a pair of in-migrants who are active in local governance processes, who contend:

In-migrant 1: I'm not sure you [can] have economic viability unless—

In-migrant 2: There's an environmental sustainability…If you put rooftops on all of these [hills] and you put so many straws in the ground that none of them get water anymore, then there's no economic viability…

In-migrant 1: I would also think there's a spiritual aesthetic element to it. One of our hydrogeologists wrote a letter to the editor. I think he was up in Twain Heart, talking about the proposed night sky ordinance where you've got to aim your light down and not let it bounce up. He very eloquently, I thought, said, "It's good for the soul to be able to look out and see the stars." If we lose that, we've lost something essential in terms of what it is to be human and that our spirit needs substance. So, I think yeah, at some level, who ultimately wants to live in a cement warren? I guess you do what you have to do, but we ought to be able to avoid that with decent planning…I don't know whether that's spiritual, good for the soul, or simply aesthetic.

Taken together these quotes demonstrate the potential for difference in perspective in regards to land use and perceptions of the value(s) of land. In one, you have the idea that open space and viewscapes are enjoyed by all but are nevertheless privately owned. Which means, if you care about those elements of the landscape, you should go out and buy some. Disregarding for a moment that this kind of sentiment ignores the economic, social, or structural issues which might inhibit someone from buying property for preservation, it is indicative of the point of view that the government's role is not to micro-manage land use. That is the role of the market, in this case through real estate, and, ultimately, private landowner stewardship decisions. In the second quote, the respondents present the idea that there are public goods (and, thus, public "bads") emanating from private property and that those public goods and bads can—and should—be accounted for in planning efforts. Moreover, the (public) natural amenities provided by such landscapes provide more than simple, tangible goods—they provide spiritual renewal.

In-migrants, particularly those from urban areas, while seen as lacking local knowledge (of traditions, culture, processes), nevertheless "often play pivotal roles in regulating local space" by participating in local governance activities (Gosnell and Abrams 2011, p. 310). This trend can have two opposite outcomes. On the one hand, as outsiders become heavily involved in politics and decision-making, "they" (perceived outsiders or "others") can end up controlling planning processes and outcomes, often facilitating more strenuous regulatory regimes of which "rural" private property interests commonly disapprove. On the other side, ironically, some newcomers can be stronger supporters of private property rights than long-time residents—even as "traditional" rural players are often characterized as (hyper) conservative. As noted by Yung and Belsky (2007), new rural land owners may be more strict in their interpretation of private property rights in regards to public access and collaborative management.

In sum, the divergent constructions of place and community found in rural areas means there is much room for conflict. Some residents seek the highest form of freedom from government intervention—no matter the cost, smell, or inconvenience—and others came to rural areas with certain idyllic expectations (Bunce 1994). When those expectations are not met or are thrown into question, stakeholders may "work to reconstruct rural space to better match" their ideal (Gosnell and Abrams 2011, p. 311). Indeed, newcomers may adopt a "close the gate behind you" or "last settler" mentality, trying to discourage or prevent further in-migration to protect the asset(s) they recently acquired (see for example Gosnell and Abrams 2011), even though they themselves came to a rural area via the mechanisms they sometimes then oppose (e.g., land development, vacation homes, tourism) (Abrams et al. 2012; Cadieux 2008). The social fabric of rural areas in transition is not smooth, but rough in places. In Calaveras County, actors representing both the "private property rights" and "public good" sides of the debate, as well as perspectives elsewhere along the political spectrum, whether "new" or "old" to the community, are present (Hiner 2015). And, significantly, respondents' political ideologies are not dependent upon whether they are "been-heres" or "come-heres" (Hiner 2014).

3.3 Three Cases and Their Contested Ecologies

In this section, I present three case studies and their *contested ecologies*; in the next, I provide a deeper analysis of one of the cases. Each of the cases selected are located within Calaveras County, California (Fig. 3.1), and present several controversies and differences in perspective. I detail the *contested ecologies* that emerged from each case, outlining how each is simultaneously politically, ecologically, and socially constructed. Further, I interrogate which political/ideological, economic, or ecological elements or factors are mobilized by various stakeholders in the ongoing process of environmental negotiation and management in rural areas.

1. *Garamendi Ranch* (Mokelumne Hill, CA): The ranch is private property and the rancher is encouraged to pursue the "highest and best use" of the land. However, the ranch also serves as a public good or common resource, providing valuable—and uncompensated—benefits.

2. *Ironstone Vineyards* (Murphys, CA): The Ironstone winery and event center is an economic engine for the region, but one that has radically changed the landscape and lifestyle/culture of the area. The change has been generally perceived as one for the better, but not according to all.
3. *The Ridge at Trinitas* (Valley Springs, CA): The Trinitas golf course, as it is informally called, is a multifunctional property used for agricultural production as well as recreational activities. It is a site of divisive conflict and contestation over conceptions of private property rights, public versus private goods, and the meaning of rurality.

In the following sections, each case will be described in detail, followed by an in-depth treatment of the Trinitas case as it serves as a focal point of analysis.

3.3.1 Garamendi Ranch

Garamendi Ranch is located near the town of Mokelumne Hill in Calaveras County. The family patriarch purchased the land during the nineteenth century Gold Rush and, through various other purchases over the years, consolidated smaller holdings into one 2000+-acre parcel. Although not the single largest land holding in the county, the Garamendi Ranch is significantly larger than most residents' parcels. Moreover, the acreage is located along Highway 49, a well-traveled highway connecting Jackson (a regional business/commercial hub in Amador County) and "Moke Hill," the town where the ranch is located. Farther south along that same road is San Andreas, the county seat, Angels Camp, the only incorporated city in the county, and, ultimately, Sonora, another regional hub located in Tuolumne County. Although located along a major thoroughfare for the area, and thus seemingly being readily physically visible, the ranch is less visible symbolically and politically in terms of residents and stakeholders recognizing the resource and thus placing value on it, as indicated by my interview data.

Nevertheless, those respondents who could speak specifically about the property mentioned that the land manager is a model steward of the land, inflicting no negative impacts, and, indeed, likely having a positive environmental impact based on his land management. The main use of the 2000+-acres is cattle grazing, as it has been for generations. There are a few other activities on the property, including: some light manufacturing (a "floating island" company fabricates their waterway-filtering "islands" on the site which are then shipped for use in degraded ponds, lakes, and other waterways), some residual mining (historically, mining was a predominant land use on the property), and the placement of utility infrastructure (a cell phone tower is now installed on the property for a lease) (see Fig. 3.2a–c). Additionally, several hundred acres have been placed into a conservation easement—a voluntary, but legally binding, agreement to restrict development rights, and thereby protect open space, in perpetuity (Quast et al. 2012).

The challenge for cow-calf cattle ranchers (those that keep a permanent herd of cattle for producing calves which are then sold annually) is that such operations are no longer especially profitable endeavors. The land manager at Garamendi Ranch

Fig. 3.2 "Floating island" water filtration pads are fabricated on site at Garamendi Ranch. The site also has some residual mining uses. *Photo credit*: C. C. Hiner, 2010

c

Fig. 3.2 (continued)

is in some sense a relic in the new service-based, import-dependent, globalized economy (Ilbery 1998; Sayre 2011; Woods 2005). However, he continues to ranch because he values the lifestyle, the environmental and cultural heritage that it conserves, and the consolidated parcel that it protects. He is not alone is this evaluation. In fact, this kind of holistic and multi-valued perspective of agriculture, referred to as "multifunctionality," is common in the European context (Huylenbroeck and Durand 2003; Wilson 2008, 2010). The challenge in applying the multifunctionality paradigm is determining how to value the multiple non-commodity benefits of land use (Haaland et al. 2011; Wiltshire et al. 2011).

The challenges aside, the land manager at Garamendi Ranch thinks it is appropriate that he be compensated for the values and services he produces above and beyond the minimal revenue he can garner from his cattle. Others, when queried about it, seem to agree, but also simultaneously demonstrate an unyielding, ideological commitment to private property rights, which can prevent some from even considering the options available for such compensation (Hiner 2015). There is a small but vocal political perspective in the area that argues environmental protections and regulation are simply a mechanism being used by subversive, socialist groups to degrade or to outright dismantle private property rights (Advocate 2012; Alcalá 2012). This kind of perspective makes policy or programs aimed at compensating land managers for environmental services provided difficult to negotiate politically.

That being said, whether compensated or not, the land manager fundamentally sees himself a steward of the land—land he and his family have managed for over 150 years. This perspective is supported by ecological research: "Private landowners are often de facto stewards of biodiversity and ecosystem services. In California's Sierra Nevada foothills, ranchers frequently present the only defense against biological invasions in private rangelands" (Aslan et al. 2009, p. 28). As such, ranchlands are not just iconic features of the rural landscape in the Sierra Nevada, they are an important part of ecosystem management (Aslan et al. 2009; Epanchin-Niell et al. 2010; Huntsinger and Hopkinson 1996; Huntsinger et al. 2010; Sayre 2006). As such, landowners play key roles in ecosystem management and sustainability. "Whether we are concerned with sustaining economic growth, conserving natural resources for future growth, or preserving natural resources for their social, aesthetic, or recreational value, the integrity of ecosystems and their viability is a major challenge facing society today" (Weltz and Dunn 2003, p. 370). As the land manager at Garamendi explained to me:

> Actually here in a relatively short period of time, environmental groups have partnered with ranchers knowing that if we keep that rancher there, then that land will remain in a condition that we want it, you know, that we would like to see. Rather than him going out, and the land being cut up into subdivisions and ranchettes or whatever.

Despite the important role land managers like the Garamendis play, my interviews generally revealed that the actions and management decisions made on large properties such as this one are relatively invisible to the "lay person" or resident. However, ranchers' commitment to stewarding the environment does not go entirely unnoticed. One county leader makes supporting ranchers a key element of his platform:

> Our ranchers and farmers provide more than just food here in Calaveras County. The rolling hills and beauty of well-kept ranches and farms help make our county one of the most scenic in the state. In fact, the fate of a number of [Calaveras County] communities hinges almost entirely on the continued inclination and financial ability of a few ranching and farming families to work their land rather than sell it for subdivisions. In light of the harsh economics facing most family farms and ranches, this amounts to a form of self-sacrifice or philanthropy. It is in our community interest to help them succeed. (Wilensky 2008)

Moreover, agriculture, as an interest group, is relatively well placed in Calaveras County, especially in comparison to the neighboring counties whose interests are more divided (Respondent 26, March 2010). The "Ag Coalition," as it is known, is a high profile public interest group in Calaveras consisting of members from each of the major agricultural groups in the county: the Farm Bureau, Calaveras Grown, the Cattlemen's Association, and Calaveras Wine Grape Alliance. Each of these groups elects two representatives to the Ag[riculture] Coalition, plus there are two ex officio members, the Farm Advisor and the Agricultural Commissioner. Essentially, the group "rather than… sort of willy-nilly representing agriculture separately," advocates for agricultural issues to the local government, as indicated in the comments from one of my respondents:

> If at some point [an issue] comes out as not going our way, then the Ag Coalition [gets] right back into the political arena…It involves a lot of time with individual supervisors, pulling a lot of favors, all of those things that you have to do. And then coming in in strength to the Board meeting. But you need to make sure you have all of your ducks in a row individually first. Fortunately the group overall hasn't lost strength…Yeah, we've had quite a few fights over the last few years.

The Ag Coalition has successfully lobbied for several major policy provisions benefiting agriculture in the county. Their first was the creation of an agricultural zoning ordinance to define, promote, and protect multifunctional activities on agricultural land in the county. Second, the Ag Coalition successfully negotiated the inclusion of an Agricultural Element in the updated General Plan; an agricultural element is not a required feature of the state-mandated general plan, but it is an option and this group argued that for an "ag county" like Calaveras it was an important addition. Third, the Board of Supervisors instated an Agricultural Dispute Resolution Committee, an extra-legal mechanism by which agricultural disputes between conflicted parties can be resolved.

As demonstrated here, in Calaveras County, agricultural interests are as integral to environmental outcomes as well as political ones. But the significance of landowners' management does not make the job any easier. Ranchers are working toward multiple goals, which can at times be contradictory. Environmental protections aimed at preserving sensitive habitat or threatened species may be well intentioned, but to a landowner managing multiple agendas, they can become overwhelming. My data indicate that some ranchers feel environmental regulations that are too intrusive are actually counterproductive toward their aims. Over half of the ranchers/farmers I spoke to (8 of 13) expressed some version of a stewardship ethic, and the same proportion of them also mentioned regulation—especially overzealous or overreaching regulation—as a detriment, or a potential detriment, to their ability to manage the land in the way they feel is most appropriate (Hiner 2014, 2015, and forthcoming).

In sum, the case of Garamendi Ranch can be characterized this way: the land manager is a model land steward in a profession with decreasing profitability (at least within current configurations of the national and global economy), but he sees ranching as a lifestyle choice. His land-use choices and management preserve agriculture, open space, and local history as well as providing ecosystem services to the wider community. Moreover, the landscape he preserves is emblematic of what rural residents expect, desire, and prefer in rural spaces. Ranches like Garamendi's preserve viewsheds, protect watersheds, and provide open space buffers between rural enclaves (Fig. 3.3). However, the provision of ecological and aesthetic public goods currently occurs with little to no actual compensation to the provider, and concerns over private property rights make valuing the common goods provided difficult (Hiner 2015).

3.3.2 *Ironstone*

The story of Ironstone is one of entrepreneurial enterprise, shifting landscapes, and socioeconomic and cultural transformation. When John Kautz married Gail Kramer, they eventually took over her family's land in Murphys, California. They bought some additional acreage over time to bring their property to a total of 1100 acres. Just under 100 of that is now planted in winegrapes and, aside from the portion of the land dedicated to the winery and event center facilities, the rest is used for cattle grazing. One of their sons, Stephen, runs cattle on the land as a side operation as well as serving as President of Ironstone Vineyards. In addition, the Ironstone property also serves as an industrial winery, processing many thousands more acres of

Fig. 3.3 A rancher, with his land spreading out "as far as the eye can see" (Warrin and Gomes 2001). Large landholdings such as this provide valuable ecosystem services as well as pleasant viewsheds. *Photo credit*: C. C. Hiner, 2010

grapes grown near Lodi, to the west in the Central Valley of California (Fig. 3.4a–c). The facility is also used as a concert and special event venue (Fig. 3.4d), which can accommodate several thousand attendees. The property is located on a two-lane road a couple of miles outside of the town of Murphys (see Fig. 3.4e).

The Kautzes have been very successful in their business ventures and are known in the community for being philanthropists and community partners. Their summer concert series and year-round festivals and performance bookings are a primary form of entertainment for local people as well as being a major tourist draw. And although there were a couple of wineries in the area prior to the genesis of Ironstone, the wine industry, inclusive of the growing, processing, and selling of wine, in Calaveras—and particularly Murphys—blossomed after the Kautzes installed the winery and built the event facility.

The Kautzes understood that cattle alone would not be able to financially support their land, given property taxes and operating costs, so they sought out a higher value product. In one insider's words:

> When we bought [the land] we knew that we had to increase the revenue off of the ranch, because otherwise the taxes and the carrying costs would eat us alive. So we planted apples and had some very nice apple orchards. But then, a number of years later, China got into the act and essentially put California apple growers out of business. So we had taken 'em out and we put in wine grapes, still trying to find higher revenue.

But more than simply seeking to increase revenues, the Kautzes had a vision for the area. In the extensive quote below, an insider describes the impetus and process of transforming Ironstone and Murphys into what it is today:

Fig. 3.4 Ironstone processing and event facilities, and the road leading to them. *Photo credit*: C. C. Hiner, 2010

Fig. 3.4 (continued)

Fig. 3.4 (continued)

> When we broke ground in 1989 for Ironstone, there was no other operation that even had a conceptual idea of doing what we're creating. And the first thing we did was actually try to utilize one of the shafts and tunnels the miners had left us on the ranch from the mining days. The geologist told us we couldn't do that, we have a solid rock mountain that we're sitting on top of. So we actually formed a mining company and hand blasted the end of the solid rock mountain off and spent the next year hand digging 10,000 square feet of underground wine caverns…And then what we also started looking at was, there was no hospitality center up here, there was no banquet facilities in the county. So it was kind of one of those field of dreams—build it and hope that they come—type of operations. There was nobody in the county or nobody in the Sierras at that time that was doing anything like that. So we saw an opportunity to grow and build something. And, at the time, because there was nothing like this in the county, the county [Board of Supervisors] was very, very amenable and supportive to the fact that we need an anchor on this. When Ironstone was built and opened its doors, there were only 4 wineries and tasting rooms in Murphys. There's 22 tasting rooms in Calaveras County right now. And most of them are located in Murphys because it's kind of become what the town of Napa is to Napa Valley itself. And again the whole perspective of what we were doing was driving agritourism and trying to expand our base to other forms of revenue generation besides the farming itself. Because we knew we could never generate enough revenue off of just cattle, or just the apples, or just the grapes up here.

This quote highlights several important elements of the rationale and beginnings of Ironstone. First, it is striking that in order to get started, the site developers engaged in a comprehensive mining and excavation project. As one respondent put it, Ironstone "cut the end of the mountain off and moved it." The endeavor represents a radical shift to physical topography and landscape. However, the county government was supportive, or at least did not interfere, because, as this respondent put it, "there was nothing like this in the county" and the idea was that a large attraction such as Ironstone would anchor the rest of the area's economy. Moreover, the developers of Ironstone saw themselves as a driving force behind agritourism and building farming viability in the county. And, as one landowner put it, "the survival of all of these Motherlode counties … is in their development of higher gross revenue crops and tourism. Tourism is by far the very best enterprise for these counties, because tourists come, leave money, and go home." In this way, Ironstone was a visionary development for (agri)tourism and a catalyst for a major cultural and economic shift to the town of Murphys and the county more broadly.

3.3.3 Trinitas

The third case features an exurbanite, Mike Nemee, who moved from Stockton, a nearby metropolitan area in California's Central Valley, when he and his family purchased 440 acres of land on the far western edge of Calaveras County. The land had formerly been a sheep ranch, but, from his perspective, was sitting "idle." Although the new landowner's perspective was that the land was "idle," others disagreed. In fact, directly opposite the Nemee property was an active sheep ranch, which had been in operation for a century and which continues its operations today. The difference then is in perception; for a new buyer like Nemee the relatively low return on grazing land would not be sufficient for investment and operation purposes. The family across the street, on the other hand, having owned the land for many years would require a much lower financial return to make continued land ownership and certain management options viable.

Despite differences in perspective regarding various management strategies and their economic viability, the land was clearly agricultural. At purchase, the Trinitas land was enrolled in a statewide agricultural conservation program called the Williamson Act (the common name for the California Land Conservation Act of 1965). The Williamson Act is a California law that provides property tax relief to owners of farmland and open-space land in exchange for an agreement that the land will not be developed or otherwise converted to another use (Department of Conservation 2007a, b). The motivation for the Williamson Act is to promote voluntary land conservation, particularly farmland conservation, while providing a modicum of financial relief to those engaging in conservation (Sokolow 2010; Stumbos 2011).[2] Williamson Act contracts are signed on a rolling 10-year cycle, such that once a property is dis-enrolled from the program, the contract remains in effect for nine more years.[3] In the case of Trinitas, after some time of ownership, the

[2] For a counter argument of the utility of the program see Roberts (2011).

[3] The Williamson Act was defunded in 2009, and, although it remained in effect, counties were no longer reimbursed for lost property tax revenue (Network 2009). As such, the viability of the program came into question. Funding was restored, but then removed again in 2011 before the program was refunded through 2016 (Campbell 2011). The details of the program changed slightly

new landowner wanted the opportunity to shift land uses so he dis-enrolled his land. However, even though he had removed the property from the program, technically its provisions were still in effect for several more years.

The property is located just outside of Valley Springs, in the community of Wallace. The property is nearby a prototypical exurban neighborhood, characterized by 5–10 acre ranchettes and the so-called "hobby" farms, and populated by many in-migrants (Gosnell and Abrams 2011; Taylor 2011). Seeing an opportunity to capitalize on the property's relative proximity to Stockton and its rolling hills, the landowner transformed the site into a unique, sprawling golf course. Indeed, the course was dubbed "Golf's Field of Dreams" by *Golf Today* (Fagan 2007) and was met with much acclaim from the golf world:

> The buzz in the golf industry is all about a golf course in the Sierra foothills that flew so far under the radar no one knew of its existence…Imagine the surprise of golf executives when they encounter someone who has been to Trinitas. This golfer talks of a view of the High Sierra peaks almost from Tahoe to Yosemite, and a golf course that fits so perfectly into the land that it looks like it has always been there…There was no architect. There was no construction company… "You might say God built it," Mike Nemee explained. "The land was that good. The golf course was always there. We just kind of grew some grass." (Salsig 2007)

This description of the course flying under the radar is very fitting because, as the public debate over the course's legality and appropriateness developed, it became clear that Nemee was building it without permission.

Once built, the owner sought to establish an agritourism destination, similar to another well-known operation in the county, Ironstone Vineyards. He built the golfing greens around the historic olive orchard on the property as well as planting additional old-growth olive trees transplanted from a nearby property. Using this century-old, heritage olive orchard, the owner revived olive production and began producing olive oil and "lifestyle products," such as soap and lotion, which are produced off-site. As one family insider put it:

> Our whole concept here was to do the Kautz [Ironstone] formula with a golf course and olive oil rather than a concert hall and grapes. Because our vision would be that a lot of this dry grazing land around us…would be converted to high-density olive plantations, which is just like grapes. They grow just like grapes on trellises…We're trying to create an olive destination, if you would. For olives, olive oil, all the different things you can do with olives. That's why we branched into lifestyle products for the spa, wellness, and all that other stuff. That whole movement.

In addition to the olive production and golfing, the owner also claims to have created wildlife habitat (via the constructed wetlands on the property, which form the water features on the golf course) and a scenic natural area for recreational activities like bird watching and photography (Fig. 3.5).

---

over the past several years as the program was virtually defunded and then restored twice, but the fundamental elements have remained: Tax relief in return for leaving land in agriculture for a specified time with a rolling annual contract (Adler 2011; Campbell 2011; Department of Conservation 2007a, b).

Fig. 3.5 Trinitas golf course, showing a constructed pond and olive trees at the periphery of the greens. A blue heron is just barely visible in front of the olive orchard. *Photo credit*: C. C. Hiner, 2010

While the owner painted a glowing picture of his property and its use and value, pointing out the various sustainability measures he put into place and how the golf course created a veritable "wildlife sanctuary," there were numerous, severe points of disagreement between him and his critics. I focus in depth upon these disagreements because they run the gamut of rural land-use conflicts, ranging from concerns over roads and roadside signage; ecological concerns, such as tree removal and stream (mis)management; varying perceptions of the real and potential economic value of the golf course; and serious concerns over damage to the credibility of the government and the rule of law due to the owners' deliberate or inadvertent negligence. The Trinitas case offers compelling insights into differing perspectives, or environmental imaginaries, of land use in the county, and as the case was the focus of glaring public attention for some time as the legal battle waged on, the ecological implications of differing outcomes are ready for examination.

In the next section, I examine the case of Trinitas more closely. Although all three cases presented offer compelling stories and sites of divergence between belief systems and actions, the sheer volume and variety of contestations emanating from Trinitas makes it an ideal case for developing and applying the concept of *contested ecologies*.

3.4 Trinitas as "Contested Ecology"

3.4.1 Genesis Narratives of the Course: "No Fault" or Deceit

Two distinct narratives emerged from my interviews when I asked about the Trinitas case: one of "no fault" and one of deceit. In the "no fault" telling, the idea is that the golf course "just sort of happened" and/or it was "meant to be." The owner explains that it began as a personal, "friends and family" course (which is allowable on agricultural land), but then he started to see the potential of the course and expanded it to a full 18-hole course. He says he also then began the process of norming his use to local and state land-use regulations. From this perspective, which is not only presented by him but by others in the community as well, he and his family were ruthlessly persecuted by overzealous NIMBY groups, who had a vendetta against him and his dream.

The term "NIMBY," derived from the phrase "not in my backyard," refers to oppositional group activities related to locally unwanted land uses (LULUs) (Dear 1992; Schively 2007; Takahashi and Gaber 1998; Wolsink 2000). "NIMBY" is often used in a derisive sense, as oppositional activities can be perceived to be detrimental to the functioning of cities and municipalities (Dear 1992). While it is true that the so-called NIMBY activity can seriously disturb, delay, or, in some instances, completely thwart public and private efforts at land-use change and development, there are other interpretations of the meaning and value of NIMBY behavior. Some see it as an exercise of the democratic process, as a means for interrupting dominant narratives of use, function, and value (Gibson 2005; Hiner and Galt 2011; Lake 1993). Nevertheless, crying "NIMBY" is a frequent and often effective mechanism for shutting down one's opposition, no matter the merits of either side's case (Lake 1993).

In this context, Trinitas is a "Field of Dreams" story; one man, against the odds, making a dream come true. And many appreciated or bought into this narrative. Golf enthusiasts became members and investors invested. Indeed, the owner even received assurances from the planning director at that time that this was an acceptable use (i.e., agritourism on agricultural land) and that he would ultimately prevail in his zone change and/or permit applications. Multiple accounts note that the landowner was even accompanied by a county official to the bank in Stockton where he received a large loan to move the plan forward.

On the other hand, there is a very different story told by Trinitas opponents. Opponents see the whole incident as a deliberate, manipulative act of deceit. They very much see the project, the process, and the outcome as purposeful. They believe the landowner or "developer"—a term that was generally used by respondents in a pejorative sense—concealed his intentions, plans, and activities deliberately, with the intention that it is "better to ask forgiveness than permission." Nearly 20 % (8 of 45) respondents mentioned this concept in regards to Trinitas' development approach. Here is one representative's take on it:

> I was on the planning commission when Nemee first submitted his proposal. [The land] was in ag[ricultural] reserve [the Williamson Act]. A golf course is an acceptable use if you are talking mowed grass and some holes, but not for commercial uses. A commercial golf course is not acceptable for Williamson Act property…I am not sure, but I speculate that he thought it would be easier to get forgiveness than permission. It was an unwise tactical error.

From this point of view, Nemee is seen as secretly building the course to avoid going through the proper legal processes for such a land-use change. Moreover, this perspective contends that he knew what he was doing—building an illegal golf course—but hid it, hoping to "get away with it." People advancing this perspective suspect he manipulated the process—and the public—in order to get what he wanted despite the evident environmental impacts.

3.4.2 Contested Ecological Viewpoints

In addition to having highly divergent conceptions of how the golf course came about, people on either side of the Trinitas issue also (re)presented contested and varied ecological viewpoints. From the landowner's point of view, the environment of the former sheep ranch located at the far western edge of the county (close to San Joaquin County line and the city of Stockton) benefitted from his management. The site hosted over 1000 old growth olive trees, rolling oak woodland, and a full-fledged golf course. Specifically, Nemee felt he had improved the natural amenities while increasing the economic potential and performance of the property. He transformed an "idle" and marginal landscape into an agritourism destination that utilized the natural amenities—the physical landform, heritage olive orchard, and abundant water supply—to simultaneously promote agriculture and encourage economic development.

On the other side, his detractors presented a starkly different ecology. The naysayer point of view is that viable agricultural land was removed from production, ignoring the land conservation protections in place and disregarding zoning designation. The land was initially under the protection of the Williamson Act, which restricts the land to use directly related to agriculture. However, in operating a commercial golf course, Nemee was, according to opponents of the course, breaking state and county rules that prohibited other commercial activities on land zoned as agricultural.

3.4.3 Negotiating "Acceptable"

Whether accepted by observers or not, the owners of Trinitas claimed to be an agritourism destination, in which case they would have been within the limits of their zoning, but, as suggested, this argument was not universally accepted. In fact, the issue went to court, where a bankruptcy judge ruled that golf is not agritourism. The Nemees appealed to the District Court, but, before a ruling was made, the property was foreclosed and ownership reverted back to the bank, making the agritourism ruling moot (George 2012b; Nichols 2012).

The legality issue aside, from the detractor's point of view, Trinitas represents numerous un-monitored and un-mitigated environmental transgressions, including: the illegal removal of trees; the redirection and cobbling of stream beds without proper regulatory oversight; excessive and irresponsible pumping of groundwater, which led to neighbors' wells running dry prematurely; and concerns about pollution due to the application of pesticides and fertilizers to the golfing greens.

The controversy took on epic proportions and became like a battle of "good versus evil" as both sides claimed the moral high ground on issues that were variously—and sometimes simultaneously—material, environmental, and symbolic (Hiner forthcoming). Many respondents argued that the owner of Trinitas likely would have "gotten away" with asking for forgiveness rather than permission in earlier times. For example, some respondents noted that Ironstone began much in the same way as Trinitas:

> [Ironstone] pretty much started—they did exactly the same game that Mike [of Trinitas] did. Maybe I shouldn't call it a game. You know, they came in and—see, Mike was just coming in and he was building a private golf course. Then, as it turns out, it is commercial. John [of Ironstone] came in and they had 9,000 acres of grapes in the valley, and so he said "I just want to put a little mini winery on the property in Murphys." And that sort of snuck through pretty quick. At the time, they had no grapes planted there. They had a few apple trees and cows. And so, you know, by all standards they don't want you putting in a winery if you don't have grapes. But then they just put in a winery. And then, all of a sudden, [it] was going to become a destination situation. And then they started expanding it. And then that is when they got a little criticism about the roads. But he let it grow so fast that he was skating all over the county.

The above respondent, a long-time observer of county agriculture, describes beginnings of both the Ironstone and Trinitas sites, beginnings which are remarkably similar. He implies that had the circumstances been different, Trinitas may have been able to "sneak" their project through the permitting processes in a similar way by getting in and making something happen so fast that there is no stopping it. However, not only were the merits and circumstances of the project different, the county was different—as was the planning context. This same respondent, a long-time observer and advisor to agricultural policy, continued:

> Probably the other difference is that the county matured a little bit between the two projects. They've had other developers try to [do the same thing]—you need to grow up and start worrying about planning. They haven't yet, but they are.

So, in other words, the county had become more sophisticated in terms of planning and governance between when Ironstone and Trinitas were established.

In fact, several respondents made the Ironstone-Trinitas connection in that Ironstone also radically shifted the use of a piece of agricultural land but has faced little to no backlash. One respondent went so far as to say that Ironstone did not follow the rules completely either and are still in the process of correcting their oversights. In a pointed letter to the editor published in the Calaveras Enterprise, one observer put it this way:

> "Agritourism" is a vague word and the ordinance should be stricken. "Agritourism" has no accepted meaning in the law. It is an invented term—made up—likely coined to support mixed use property such as the Ironstone Vineyards and other vineyard and tasting venues used for concerts and other large social gatherings that have come to the County under the radar. The word itself is a contraction of two words, "tourism" and "agriculture." Golf is tourism. It attracts players from within and without the County. In the western part of the County, it will attract golfers from Stockton and the Bay Area. And, golf is conducted on large acreages of land on which grass is planted, nurtured, and grown. It is Agritourism at its best. Other forms of agriculture that bring in tourists [Agritourism] would be winegrape growing acerages [sic] complete with tasting rooms and concert venues, Christmas tree farms which invite visitors and sell to the public, and pumpkin patches which do the same. The judge is wrong to exclude the golf course. Excluding people because you don't like them is not good government. (Arkin 2011)

This person, evidently, is a Trinitas supporter and he makes his position clear. Indeed, his point that "golf is conducted on large acreages of land on which grass is planted, nurtured, and grown" (Arkin 2011) is accurate. That said, golfing greens do not produce an agricultural product, but rather a landscape on which to play a game. It is striking, though, that he specifically mentions Ironstone in the piece, noting that, in his opinion, Ironstone came to the county "under the radar" (Arkin 2011). In other words, some respondents, like this editorial writer, noticed a similarity between the two sites in terms of their beginnings. Beyond that, however, the similarities end.

3.4.4 Trinitas: The Legal Battle

In their final attempts to save themselves and their "maverick" golf course, the Nemees, via their attorney, Ken Foley, argued in a federal court filing that Trinitas had been the subject of discriminatory treatment by Calaveras County, which had twice denied them the necessary permits to operate legally (Eggleston 2012a). They claim the county government was discriminatory because officials had "look[ed] the other way" when it concerned alleged code violations by Ironstone (Eggleston 2012a). The filing claimed two former planning directors and a former community development director "were told not to take action against the vineyard owner" (Eggleston 2012b). Although two of the three former officials disputed the claims, the one who had been in the position earliest, Ray Waller, did not disagree, saying: "Yes, everybody was told that. The Board of Supervisors didn't want a thing done with this because everybody thought Ironstone was such a great thing for the county. They thought Ironstone's owner would follow up and make things right [i.e., acquire any required permits or permissions], but he never did" (Eggleston 2012b). Mr. Waller refrained from mentioning who had given him this instruction, but the County Supervisor for that District, Tom Tryon, long-standing in his office, denies placing any such restrictions on the staff regarding following up with zoning code compliance. In fact, he said, "I've tried and tried to get them to come into conformance," which, incidentally, implies that they are not currently (or have historically not been) in "comformance" (Eggleston 2012b).

However, John Kautz, the patriarch of the Ironstone family, is quoted saying: "We're not the issue. It's an issue between Trinitas and the supervisors, and we're getting dragged into it" (Eggleston 2012a). The former Community Development Director said: "I think if you look at the testimony, we all agreed Ironstone needed some additional permits, but I had not been told to leave Ironstone alone" (Stephanie Moreno, Community Development Director at the time, as quoted in Eggleston 2012b). Meanwhile, the Kautzes denied the allegations that Ironstone is out of compliance with Calaveras County codes (Eggleston 2012a). The manager, Stephen Kautz (John Kautz's son), when asked if the company was missing any required permits, responded: "Not that we're aware of. We've been doing this same thing as far as the concerts, the vineyard and the winery operations, for 10 or 12 years, and we have all the permits that are necessary" (Eggleston 2012a).

Moreover, "Ironstone Vineyards operates under different zoning than Trinitas and the frequency of its concerts, about eight each summer, doesn't compare to the daily use of a commercial golf course" (Eggleston 2012b). And when John Kautz built the Ironstone winery and tasting room in Murphys in the 1990s, there "was no agritourism ordinance," since the policy was adopted in 2006 (Nichols 2011a). However, county codes requiring building and event permits were in existence. Waller, who was the Planning Director at the time, notes that the Kautzes did follow protocol in terms of building permits, but that event permits were another matter (Nichols 2011a). Indeed, as mentioned by many respondents in my sample, Ironstone events are well known for their associated traffic problems, such as increased car volume on small county roads before events and long lines of cars leaving the area afterward.

The negative impacts aside, Stephen Kautz argues, "Ironstone literally has changed the county and its economic conditions. We've brought people here, filled the motels and restaurants. To tear us down is crazy" (Eggleston 2012a). And, indeed, Ironstone was vindicated when Calaveras County officials denied all of the Nemees' claims "concerning allegedly inconsistent and special treatment of Ironstone Vineyards under agritourism zoning rules" (George 2012a). In fact, the official response to the Nemees' legal brief was that Ironstone was not relevant to their case and their arguments are "creatively misleading at best and intentionally false at worst" (George 2012a), which is an apt summary of how many view the golf course's *modus operandi* since day one.

Nevertheless, as the court documents and arguments proceeded, the facts of the Trinitas case became clear: The owner of the land, whether through deliberate or accidental negligence, built an illegal, aka "non-conforming," golf course on agricultural land. He attempted to come into legal compliance, by seeking permits under the new agritourism ordinance, but was twice denied. Following that outcome, he filed bankruptcy to stay the cease-and-desist orders administered by county officials and took to the courts to (1) seek a ruling in his favor on the agritourism issue and (2) resolve the pending foreclosure with the Community Bank of San Joaquin. Both of these pursued outcomes were feasible at the time due to the circumstances of the lawsuits, as described by a local reporter:

> The agritourism lawsuit is in bankruptcy court, because the Nemees in 2009 filed for bankruptcy. If the lawsuit forces the county to recognize golf as a legal form of agritourism, the Nemees say it will make the business viable and allow them to repay the millions they've borrowed from a bank and private investors, many of them from Stockton. (Nichols 2011a)

The legal ordeal dragged out for 2 years as the "very small one-branch bank" was forced to accept only partial payments on its largest outstanding loan, worth $2.4 million (Nichols 2011b). However, the presiding judge was "not entirely sympathetic and pointed out that the bank played a role in constructing an illegal golf course"; he noted, "Everybody knew what was being built out there was not in compliance with the zoning" (Nichols 2011b). In the end, the stay on a foreclosure auction was lifted as it became clear the Nemees were not likely to prevail. The two parcels making up the golf property were auctioned off in March and April 2012, which then made the agritourism issue moot as the property was no longer owned by the Nemees (Nichols 2012).

Despite the ultimate failure of the Nemees' claim against the County, it is interesting to note that the District Court judge "agrees that the Nemees raise serious questions as to whether the golf course is a permissible use on the property" (George 2012b). So although the agritourism issue seems to have been resolved in terms of its relevance for Trinitas specifically (because the judge was only ruling whether Trinitas fit Calaveras County's particular agritourism guidelines at the time of the dispute), the conceptual issue is not resolved regarding whether golf constitutes agritourism more generally. In short, although the battle is over, the conceptual and ideological conflict over the form and function of rurality—as codified in zoning regulations and their interpretation—continues.

3.4.5 A Case of Divergent Perspectives

Although the property has now been foreclosed, reflecting on the Trinitas supporters and their reasons for support remains a fruitful site of analysis to explore the variety of perspectives observers held about the case. For example, the Calaveras County Chamber of Commerce wrote an opinion piece in 2010 showing support for the owners of Trinitas and their right to due process (Calaveras County Chamber of Commerce 2010). The Chamber's main points were frustration with "being pushed around by special interests"; lodging formally their support for business and investment in the community, as they themselves are business people and investors; registering the opinion that Trinitas "was prematurely tried and convicted by a partisan press, a handful of activist neighbors and a personal agenda by a county employee" in the public hearings on the issue; and that their support stems from the merits of the project, noting: "The project is a good one. It provides a beautiful setting that is a permanent preservation of land. Trinitas is a challenging upscale course that will bring visitors from other counties to spend their money here." Of course, the key to the conflict is a difference in preferences regarding what is worth preserving and how. The Chamber's overarching point was that they support business, the constitution, and the due process of the law:

> When we have a business like Trinitas, and members like the Nemees, that are being deliberately and unfairly targeted for destruction, we must support them. For too long, our business community has stood idol [sic] on important economic issues. This time, the ambivalence has ended. People are fed up with being pushed around by special interests… It's wrong to destroy a business without putting it in front of a fair and impartial entity to make the call…We still believe the best path for our County to take is one of negotiation to try to find a way for Trinitas to work. (Calaveras County Chamber of Commerce 2010)

This sentiment emerged from other sectors of the community as well, who, although upset perhaps by the way the owners of Trinitas had gone about the development of the course, recognized its potential as an asset to the county. Some people argued for negotiation and compromise rather than divisive bickering. Nearly a quarter of my respondents (12 of 51) noted a compromise between the landowners and the county government would be an acceptable or preferable outcome to the controversy. However, ultimately emotions ruled the discourse around the case. Emotional investment is not uncommon in disputes of this kind. As one person noted:

> I think a lot of the cases that are happening in this county in the last, say, 10 years, have developed into more a personal problem, and less so of a real problem…You can take a very small conflict or compatibility issue, and it can turn into a very major thing that can impact state or county policy. But it was more about some personal thing…rather than what the problem was at the outset. But they'll carry that problem saying that's the reason. But the real reason is it'd become personal.

A land-use dispute becoming personal and taking on a life of its own is not unique to Trinitas, but the ways in which such conflicts are resolved can vary (Hiner forthcoming). One suggestion for resolving the Trinitas issue was to come up with a mutually agreeable compromise. Some members of the Ag Coalition worked with Nemee to devise such a compromise:

> "Let's do what we can to make this into a very positive [thing] and minimize those impacts that the people have expressed concern about." And Mike came forward with such a proposal and I don't think it was seriously looked at by the county. Or, I don't know, because there were others involved, it becomes a political thing is what happens. And better reasoning and rationale seem to fly out the window.

Also, fundamentally, the golf course remained a "club good" rather than a public one (Cornes and Sandler 1996; McNutt 2000; Warner 2011), and stakeholders outside of the "club," i.e., the circle of beneficiaries, found it difficult to support.

Some hoped the property would remain a ranch, but without viable compensation in place for ecological services provided by the property and/or a manager with a long-term commitment to ranching as a form of open space (like the manager at Garamendi), ranching as a land preservation strategy is tenuous, particularly since new buyers would likely acquire a sizable debt from the purchase, constraining their options for use. In other words, any buyer must consider what types of uses will generate enough income to sustain the loan payments and their livelihood. Ultimately, in-migrants change the form and function of the landscape in multiple ways, including driving up land values, which, in the end, influences which management strategies are viable. Moreover, even if a buyer was to keep the land in open space, the implication is that, at any time, an investor could "swoop in" with a nice offer and make it "rooftops" (i.e., a subdivision). As such, it is of interest to stop and consider what the alternative to the golf course is: likely more exurban ranchettes instead of productive agriculture balanced with golf. One respondent puts it this way:

> I think that if they did nothing and if they got rid of the golf course and everything, that parcel could be developed into 20-acre lots under current zoning. Would that be better? Going back to the big picture: Cutting that into 20-acre parcels, is that better than having a golf course with 7 home sites clustered in a corner and having the balance of it either habitat or olive orchards? I would say the latter would be preferable to me.

So, while the legal aspects of the case took on a life of their own, the agricultural community (as represented by the Ag Coalition) was generally supportive of having an additional agritourism destination to promote and build the stature of agriculture in the county. And Nemee had in fact worked with the Ag Coalition in support of the "Calaveras Grown" label, an initiative to market local agricultural products (Calaveras Grown 2012). In other words, the site could have been an asset to the community, but instead was embroiled in a bitterly divisive cultural and legal dispute. Nevertheless, a compromise—had it been considered—could have mitigated the concerns of many community members while still serving the purposes desired by the landowner.

Surely some of the environmental claims made by the course opponents had a real (read: observable or objective) ecological or material basis, but, according to inside sources, the landowner pursued sustainability practices to maximize economic potential while minimizing environmental harm. One compromise solution suggested by a prominent agricultural promoter in the county (Respondent 2, January 2010) involved halting all construction, improving habitat conservation measures, increasing the agricultural uses of the property, and, with certain restrictions, allowing for continued use of the course commercially. Such a compromise might have been a "win-win" scenario. Nevertheless, the owner's mitigation efforts, earnest or no, in effect or proposed, were washed away by the anger of his opponents.

> I think, unfortunately, the way that was handled, it's just a mess. … People … they've taken sides on that issue and it's polarized the county. And anytime that happens I think it's bad. If [Trinitas had] been handled right, [and] I think it could have been handled properly both by the county and by the owners, that they might have had a very nice project.

The Trinitas issue was deeply polarizing, and exposed—and perhaps even deepened—the land-use-related ideological perspectives of those involved in and observing the case (Hiner 2015). Some of the opposition might say that keeping the land as grazing ground would be the "best" option, but, as the land sat idle (i.e., not used for any purpose, including grazing) for some years, clearly, it was not a use which was appealing to potential users/suitors/owners. The real problem, it would seem, was the *offense* of the golf course; the fact that the landowner put a golf course on agricultural land without a conditional use permit (basically an exception from his zoning designation) and without adhering to any of the required environmental and regulatory protocols. Not just the offense of neighbors being ignored and cast as irrelevant in the planning process (as demonstrated by the owners' complete disregard for land-use regulations and the public processes that would be necessitated by such a land-use designation change), but also the offensiveness of the change itself, i.e., respondents noting that a golf course does not "fit in" to the rural environment as they envision it (Hiner 2014, 2015, and forthcoming).

In essence, this particular land use sparked opposition based on the environmental ideologies and preferences of local stakeholders. One neighbor, a vehement opponent to the golf course, characterized her desire to live where she does, in a small-parcel subdivision, as a chance to enjoy natural amenities, including:

> the small, the everyday wonders, the everyday miracles that sort of revive the spirits. I mean, gosh, you just walk around and listen to the birds. You watch the bunnies and you feel whole again, whatever else has been going on. Watch the grass come up and the wildflowers. I mean, it's a sense of renewal that you just can't get in a concrete jungle.

This respondent did not perceive the golf course next door as compatible with her version or vision of rural life. She blamed the owner for disrupting nature and destroying natural systems on the property, which had been an unused, open patch of land dominated by annual grasses and forbs. For her, the issue was about someone coming in from the city and ruining the nature—both literal and figurative—of her rural home as he disrupted the ecology of the property and created a draw for non-locals to swoop in and swoop out, causing congestion and increasing traffic and road hazards on roads she must travel every day.

3.5 Conclusion: The Significance of Contested Ecological Perspectives

Trinitas encapsulates the kind of contested ecologies that present themselves in exurbia. There are legal aspects to these contested ecologies and ethical–ecological elements that are not necessarily covered in law, but are the domain of popular struggle. As a Trinitas insider put it, "Logic never plays out in [resource conflicts] that get emotionally and politically charged." In other words, the more emotionally or politically charged an issue becomes, the less likely it is that rational, evidence-based thinking or logic will prevail.

Moreover, the Trinitas case in particular presents numerous environmental claims. I mean "claims" in the sense of ownership or "stake," but also in terms of physical and ecological features that are under debate such as: trees, streams (not just the water in them, but the shape and form of them and whether they have been altered), and plant or vegetation types. In addition, there are other more aesthetic and/or spiritual elements to the controversy. Roads (their adequacy and, presumably, the increased number of people driving on them) and signage have also been cited as concerns. In the case of the roadside signs, the verbiage and claim is about whether they are *legal*, but the real issue is whether they are *acceptable*. Do large, commercial advertisements "fit" in the rural landscape? Who gets to decide? Clearly, aside from the physical material elements of the controversy, there is the larger issue over conception of place (Hiner forthcoming).

This case makes plain that residents in this exurban neighborhood have conflicting views of what the meaning and function of that place is. For example, the local opposition is characterized as activist and "NIMBY," but the actors themselves would not characterize their activities that way. They see their involvement as legitimate and worthwhile community protection, not obstructionism (Lake 1993; Gibson 2005; Hiner and Galt 2011). Moreover, they speak of the solace and calm that they enjoy when seeing "the bunnies hop" and watching the wildflowers grow and are determined to ensure that disruptive activities do not destroy the habitat and environment necessary to enjoy that lifestyle amenity. Theirs is a particular perspective and it leads to a particular conception of the local ecology. This case and the actors involved in it demonstrate that differing perspectives can and do lead to *contested ecologies*.

The cases presented in this chapter demonstrate how different perspectives of the landscape and of the proper role of various social actors—landowners, governments, citizens—can produce very different outcomes. These varying perspectives can lead to actual conflicts in the community, such as fights over appropriate management strategies as demonstrated by the tremendous conflict and contestation over the Trinitas golf course. The sheer volume and variety of contestations emanating from Trinitas is what makes it such a key case in my study. However, it is most fascinating, though, when contrasted with Ironstone, a place that also could have been a site of contestation but has not been. In some cases the landowner prevails and carries through a drastic land-use change (such as in the case of Ironstone), and, in others, the issue becomes so contested that even one's "victory" can taste bittersweet, such as the outcome with Trinitas. In the case of Trinitas, detractors ultimately got what they wanted, but were left with a de-moralized community and cultural landscape scarred by distrust.

In conclusion, land-use change along the rural–urban interface is significant both ecologically and socially as the outcomes shape both the physical landscape and the cultural one. Such land-use change is complicated further by taking place in an exurban context where the relationship between environment and society are in flux and are sometimes in dispute. As such, environmental management regimes in such places, which can be marked by divergent environmental imaginaries and ideologies, pervasive disagreement in terms of values, and differing preferences for environmental management strategies, need to be responsive to both the environmental and social changes taking place.

References

Abrams, J., Gosnell, H., Gill, N., & Klepeis, P. (2012). Re-creating the rural, reconstructing nature: An international literature review of the environmental implications of amenity migration. *Conservation and Society, 10*(3), 270–284. doi:10.4103/0972-4923.101837.

Adler, S. (2011, 26 January). Williamson Act faces a renewed budgetary threat. *The Ag Alert*.
Advocate, F. (2012). *About us*. Retrieved from http://www.freedomadvocates.org/about-us/.
Alcalá, C. (2012, 24 May). El Dorado County folks riled by U.N. agenda for sustainable growth. *The Sacramento Bee*.
Arkin, M. (2011). Letter to the editor: Trinitas through the looking glass. *Calaveras Enterprise*.
Aslan, C. E., Hufford, M. B., Epanchin-Niell, R. S., Port, J. D., Sexton, J. P., & Waring, T. M. (2009). Practical challenges in private stewardship of rangeland ecosystems: Yellow starthistle control in Sierra Nevadan foothills. *Rangeland Ecology & Management, 62*(1), 28–37. doi:10.2111/07-123.
Mintier & Associates, Environmental Science Associates, & Calaveras County Community Development Agency. (2008). *Calaveras county general plan baseline report*, San Andreas, CA.
Ban, H. W., & Ahlqvist, O. (2009). Representing and negotiating uncertain geospatial concepts—Where are the exurban areas? *Computers, Environment and Urban Systems, 33*(4), 233–246. doi:10.1016/j.compenvurbsys.2008.10.001.
Black, R. (1990). 'Regional political ecology' in theory and practice: A case study from Northern Portugal. *Transactions of the Institute of British Geographers, 15*, 35–47.
Blaikie, P. M., & Brookfield, H. C. (1987). *Land degradation and society*. London: Methuen.
Brown, D. G., Robinson, D. T., An, L., Nassauer, J. I., Zellner, M., Rand, W., et al. (2008). Exurbia from the bottom-up: Confronting empirical challenges to characterizing a complex system. *Geoforum, 39*, 805–818. doi:10.1016/j.geoforum.2007.02.010.
Bunce, M. (1994). *The countryside ideal: Anglo-American images of landscape*. New York: Routledge.
Cadieux, K. V. (2008). Political ecology of exurban "lifestyle" landscapes at Christchurch's contested urban fence. *Urban Forestry & Urban Greening, 7*, 183–194. doi:10.1016/j.ufug.2008.05.003.
Cadieux, K. V., & Hurley, P. T. (2011). Amenity migration, exurbia, and emerging rural landscapes: Global natural amenity as place and as process. *Geojournal, 76*(4), 297–302. doi:10.1007/s10708-009-9335-0.
Cadieux, K., & Taylor, L. (2013). *Landscape and the ideology of nature in exurbia: Green sprawl*. New York: Routledge.
Calaveras County Chamber of Commerce. (2010, 8 April). Letter to the editor: Re: Calaveras County Chamber of Commerce support of Trinitas. *The Pine Tree*.
Calaveras Grown. (2012). *Welcome*. Retrieved from http://calaverasgrown.org/
Campbell, K. (2011, 20 July). Governor Brown signs bill to preserve Williamson Act. *The Ag Alert*.
Cornes, R., & Sandler, T. (1996). *The theory of externalities, public goods, and club goods*. Cambridge: Cambridge University Press.
Cromartie, J., & Bucholtz, S. (2008). Defining the 'rural' in rural America. *Amber Waves, 6*(3).
Dear, M. (1992). Understanding and overcoming the NIMBY syndrome. *Journal of the American Planning Association, 58*, 288–300. doi:10.1080/01944369208975808.
DeLyser, D., Herbert, S., Aitken, S., Crang, M., & McDowell, L. (2010). *The SAGE handbook of qualitative geography*. London: Sage.
Department of Conservation, Division of Land Resource Protection. (2007a). *Williamson Act program*. Retrieved from http://www.conservation.ca.gov/dlrp/lca/Pages/Index.aspx.
Department of Conservation, Division of Land Resource Protection. (2007b). *Williamson Act program—Basic contract provisions*. Retrieved from http://www.conservation.ca.gov/dlrp/lca/basic_contract_provisions/Pages/index.aspx.
Eggleston, B. (2012a, 6 January). Ironstone owner claims to have all permits. *Calaveras Enterprise*.
Eggleston, B. (2012b, 3 January). Trinitas appeal claims disputed. *Calaveras Enterprise*.
Epanchin-Niell, R. S., Hufford, M. B., Aslan, C. E., Sexton, J. P., Port, J. D., & Waring, R. M. (2010). Controlling invasive species in complex social landscapes. *Frontiers in Ecology and the Environment, 8*, 210–216. doi:10.1890/090029.
Fagan, B. (2007). Golf's field of dreams. *Golf Today*. Tehachapi, CA: Golf Today.
George, A. (2012a, 20 January). County fires back at Trinitas court appeals. *Calaveras Enterprise*.
George, A. (2012b, 2 March). Trinitas auction goes forward. *Calaveras Enterprise*.

Gibson, T. (2005). NIMBY and the civic good. *City & Community, 4*, 381–401. doi:10.1111/j.1540-6040.2005.00144.x.
Gosnell, H., & Abrams, J. (2011). Amenity migration: Diverse conceptualizations of drivers, socioeconomic dimensions, and emerging challenges. *GeoJournal, 76*(4), 303–322. doi:10.1007/s10708-009-9295-4.
Haaland, C., Fry, G., & Peterson, A. (2011). Designing farmland for multifunctionality. *Landscape Research, 36*, 41–62.
Harrison, S., Massey, D., Richards, K., Magilligan, F. J., Thrift, N., & Bender, B. (2004). Thinking across the divide: Perspectives on the conversations between physical and human geography. *Area, 36*, 435–442. doi:10.1111/j.0004-0894.2004.00243.x.
Harvey, D. (1996). *Justice, nature, and the geography of difference*. Cambridge: Blackwell Publishers.
Hiner, C. C. (2014). "Been-heres vs. come-heres" and other identities and ideologies along the rural–urban interface: A comparative case study in Calaveras County, California. *Land Use Policy, 41*, 70–83. doi:10.1016/j.landusepol.2014.05.001.
Hiner, C. C. (2015) (False) Dichotomies, political ideologies, and preferences for environmental management along the rural-urban interface in Calaveras County, California. *Journal of Applied Geography, 65*, 13–27.
Hiner, C. C. (forthcoming) "Chicken wars", water fights, and other contested ecologies along the rural-urban interface in California's Sierra Nevada foothills. *Journal of Political Ecology*.
Hiner, C. C., & Galt, R. E. (2011). Participation and capacity building in community visioning: NIMBYism and the politics of the rural-urban interface in Elk Grove, California. *Journal of Rural and Community Development, 6*, 104–123.
Huntsinger, L., & Hopkinson, P. (1996). Viewpoint: Sustaining rangeland landscapes: A social and ecological process. *Journal of Range Management, 49*, 167–173.
Huntsinger, L., Johnson, M., Stafford, M., & Fried, J. (2010). Hardwood rangeland landowners in California from 1985 to 2004: Production, ecosystem services, and permanence. *Rangeland Ecology & Management, 63*, 324–334.
Huylenbroeck, G. V., & Durand, G. (2003). *Multifunctional agriculture: A new paradigm for European agriculture and rural development*. London: Ashgate.
Ilbery, B. W. (1998). *The geography of rural change*. Harlow, England: Longman.
Lake, R. W. (1993). Rethinking NIMBY. *Journal of the American Planning Association, 59*, 87–93.
Massey, D. (1999). Space-time, 'science' and the relationship between physical geography and human geography. *Transactions of the Institute of British Geographers, 24*(3), 261–276. doi:10.1111/j.0020-2754.1999.00261.x.
Massey, D. (2001). Talking of space-time. *Transactions of the Institute of British Geographers, 26*(2), 257–261. doi:10.1111/1475-5661.00019.
McNutt, P. (2000). Public goods and club goods. In B. Bouckaert & G. D. Geest (Eds.), *Encyclopedia of law and economics, vol. I: The history and methodology of law and economics* (pp. 927–951). Cheltenham, England: Edward Elgar.
Merriman, P. (2012). Human geography without time-space. *Transactions of the Institute of British Geographers, 37*(1), 13–27. doi:10.1111/j.1475-5661.2011.00455.x.
Momsen, J. H. (1996). Agriculture in the Sierra. In *Sierra Nevada Ecosystem Project: Final report to Congress* (pp. 497–528).
Nelson, A. C. (1992). Characterizing exurbia. *Journal of Planning Literature, 6*(4), 350–368. doi:10.1177/088541229200600402.
Nesbitt, J. T., & Weiner, D. (2001). Conflicting environmental imaginaries and the politics of nature in Central Appalachia. *Geoforum, 32*(3), 333–349. doi:10.1016/S0016-7185(00)00047-6.
Network, C. C. (2009, 10 August). Farmers say elimination of "Williamson Act" is a financial blow. *KPBS*.
Neumann, R. P. (2010). Political ecology II: Theorizing region. *Progress in Human Geography, 34*(3), 368–374. doi:10.1177/0309132509343045.
Nichols, D. (2011a, 15 October). County didn't enforce codes, former officials say: Trinitas owner questions treatment. *The Stockton Record*.

Nichols, D. (2011b, 8 December). Judge holds off Trinitas bank seizure: Golf goes on as foreclosure sale put off until 2012. *The Stockton Record*.

Nichols, D. (2012, 18 April). Bank reaps trinitas property. *The Stockton Record*.

Patton, M. Q. (2002). *Qualitative research and evaluation methods*. Thousand Oaks, CA: Sage.

Paulson, S., & Gezon, L. L. (2005). *Political ecology across spaces, scales, and social groups*. New Brunswick, NJ: Rutgers University Press.

Quast, H., Spangenberg, H., Hannover, B., & Braun, E. (2012). The influence of tuition fees on study aspiration. *Zeitschrift Fur Erziehungswissenschaft, 15*, 305–326.

Reed, M. G. (2007). Uneven environmental management: A Canadian comparative political ecology. *Environment and Planning A, 39*, 320–338. doi:10.1068/a38217.

Robbins, P. (2011). *Political ecology: A critical introduction*. Malden, MA: Wiley-Blackwell.

Roberts, D. (2011, 24 February). Bulldoze Williamson Act subsidies? *Cal Watchdog News*.

Salsig, R. (2007). Silent no more. In *NCGA Golf Magazine*. Pebble Beach, CA: Northern California Golf Association.

Sayer, A. (2010). *Method in social science: A realist approach*. New York: Routledge.

Sayre, N. F. (2006). *Ranching, endangered species, and urbanization in the Southwest: Species of capital*. Tucson, AZ: University of Arizona Press.

Sayre, N. (2011). Commentary: Scale, rent, and symbolic capital: Political economy and emerging rural landscapes. *GeoJournal, 76*(4), 437–439. doi:10.1007/s10708-009-9297-2.

Schively, C. (2007). Understanding the NIMBY and LULU phenomena: Reassessing our knowledge base and informing future research. *Journal of Planning Literature, 21*(3), 255–266. doi:10.1177/0885412206295845.

Sierra Nevada Ecosystem Project. (1996). *Status of the Sierra Nevada: Summary of the SierraNevada ecosystem project report*. Centers for Water and Wildland Resources, University of California.

Sokolow, A. D. (2010). Budget cuts threaten the Williamson Act, California's longstanding farmland protection program. *California Agriculture, 64*(3), 118–120. doi:10.3733/ca.v064n03p11.

Spectorsky, A. C. (1955). *The exurbanites*. Philadelphia: Lippincott.

Stumbos, J. (2011). Loss of Williamson Act would threaten wildlife habitat. In *Western Farm Press*. New York: Penton Media.

Takahashi, L. M., & Gaber, S. L. (1998). Controversial facility siting in the urban environment: Resident and planner perceptions in the United States. *Environment and Behavior*, 30(2):184–215. doi:10.1177/0013916598302004.

Tashakkori, A., & Teddlie, C. (1998). *Mixed methodology: Combining qualitative and quantitative approaches*. Thousand Oaks, CA: Sage.

Taylor, L. (2011). No boundaries: Exurbia and the study of contemporary urban dispersion. *GeoJournal, 76*(4), 323–339. doi:10.1007/s10708-009-9300-y.

Travis, W. R. (2007). *New geographies of the American West: Land use and the changing patterns of place*. Washington, DC: Island Press.

U. S. Census Bureau. (2010). *California*. Retrieved from http://quickfacts.census.gov/qfd/states/06000.html.

Walker, P. A. (2003). Reconsidering 'regional' political ecologies: Toward a political ecology of the rural American West. *Progress in Human Geography, 27*(1), 7–24. doi:10.1191/0309132503ph410oa.

Walker, P., & Fortmann, L. (2003). Whose landscape? A political ecology of the 'exurban' Sierra. *Cultural Geographies, 10*(4), 469–491. doi:10.1191/1474474003eu285oa.

Warner, M. E. (2011). Club goods and local government. *Journal of the American Planning Association, 77*(2), 155–166. doi:10.1080/01944363.2011.567898.

Warrin, D., & Gomes, G. L. (2001). *Land, as far as the eye can see: Portuguese in the old West*. Spokane, Washington: A.H. Clark.

Weltz, M. A., & Dunn, G. (2003). Ecological sustainability of rangelands. *Arid Land Research and Management, 34*(17), 369–388. doi:10.1080/01431161.2013.793867.

Wilensky, S. (2008). *Platform: Agriculture*. Retrieved from http://stevewilensky.com/plat_ag.html.

Wilson, G. A. (2008). From 'weak' to 'strong' multifunctionality: Conceptualising farm-level multifunctional transitional pathways. *Journal of Rural Studies, 24*(3), 367–383. doi:10.1016/j.jrurstud.2007.12.010.

Wilson, G. A. (2010). Multifunctional 'quality' and rural community resilience. *Transactions of the Institute of British Geographers, 35*(3), 364–381. doi:10.1111/j.1475-5661.2010.00391.x.
Wiltshire, K., Delate, K., Flora, J., & Wiedenhoeft, M. (2011). Socio-cultural aspects of cow-calf operation persistence in a peri-urban county in Iowa. *Renewable Agriculture and Food Systems, 26*(1), 60–71. doi:10.1017/S1742170510000505.
Wolsink, M. (2000). Wind power and the NIMBY-myth: Institutional capacity and the limited significance of public support. *Renewable Energy, 21*(1), 49–64. doi:10.1016/S0960-1481(99)00130-5.
Woods, M. (2005). *Rural geography: Processes, responses and experiences in rural restructuring*. London: Sage.
Yung, L., & Belsky, J. M. (2007). Private property rights and community goods: Negotiating landowner cooperation amid changing ownership on the Rocky Mountain Front. *Society & Natural Resources, 20*, 689–703.
Zimmerer, K. S., & Bassett, T. J. (2003). *Political ecology: An integrative approach to geography and environment-development studies*. New York: Guilford Press.

Chapter 4
Rural Residential Development and Its Discontents: A Political Ecology of Sprawl Containment in Wallowa County, Oregon, USA

Jesse Abrams

4.1 Introduction

Wallowa County, Oregon, has remained relatively sprawl-free despite experiencing many of the same pressures that have elsewhere led to widespread patterns of exurban land development. This outcome can be attributed in part to Wallowa County's remoteness and to the fact that it is subject to a statewide system of land-use governance designed to protect farmland by containing development pressure within already-developed areas (Walker and Hurley 2011). Yet these factors are by themselves insufficient to explain the particular land conservation and development trajectory experienced here. A fuller explanation requires attending to the political economy of land use in this remote, rural corner of the Pacific Northwest and to the active engagement of various resident and non-resident populations in the processes of private land governance. In this chapter, I trace the key social and political forces that have shaped the development and conservation landscape in a place that has largely resisted pressures to accelerate sprawl even as it has been transformed in other important ways by processes of rural restructuring.

The Wallowa County of the 1990s and 2000s contained both a growing, if spatially and temporally uneven, amenity economy tied to the consumptive values of a scenic rural landscape and a still-robust productive economy tied to agriculture, livestock production, and forestry and increasingly reliant upon innovative production practices. It also continued to be strongly valued, and to a certain extent occupied and governed, by members of the Indigenous cultural groups that laid claim to this landscape prior to their dispossession by agents of the U.S. government during the period of EuroAmerican settlement. A complex array of local actors organized along the lines of divergent economic and cultural orientations toward the land

J. Abrams (✉)
Institute for a Sustainable Environment, University of Oregon,
130 Hendricks Hall, Eugene, OR 97403-5247, USA
e-mail: jabrams@uoregon.edu

L.E. Taylor, P.T. Hurley (eds.), *A Comparative Political Ecology of Exurbia*,
DOI 10.1007/978-3-319-29462-9_4

attempted to negotiate Wallowa County's social and economic future under the regional-scale influences of statewide land-use policies, middle-class consumption trends in regional urban centers, and the development trajectories of other nearby rural landscapes. Impulses toward sprawl were moderated in part by the ability of a critical mass of heterogeneous interests to mobilize, if only temporarily, around a common set of discourses and actions when confronted with specific proposals for exurban development.

4.2 Setting/Methods

Wallowa County is characterized by, among other things, its remoteness, its scenic beauty, and its historic reliance on a natural resource economy (particularly agriculture, livestock, and forestry). Located in the northeast corner of Oregon, Wallowa County is isolated by distance and topography from regional metropolitan areas (Boise, Idaho; Spokane, Washington; and Portland, Oregon). The county's present-day political boundaries closely match the traditional territory of the Wallowa Band of the Nez Perce (Josephy 1997), dispossessed by white settlers and the U.S. military in the 1870s. Following the Nez Perce war of 1877 and several years in inhumane conditions in Oklahoma, the surviving members of the Nez Perce and allied Plateau Tribes who had taken part in the war were ultimately dispersed to the Nez Perce Reservation in Idaho, the Colville Reservation in northern Washington, and the Umatilla Reservation west of Wallowa County. Nez Perce and other Tribal members from these reservations continued to return to Wallowa County to fish, hunt, and camp for decades following their dispossession, despite being at least initially unwelcome by resident Whites (Marshall 2006).

Today, 58 % of the county's land base is federally owned and managed under the Wallowa-Whitman and Umatilla National Forests and the Hells Canyon National Recreation Area (Abrams 2011). Private land in the county encompasses fertile farmlands in the Wallowa Valley, ponderosa pine and mixed conifer forests along the fringes of federal lands, and arid rangelands dominated by native grasses. The predominant land uses of farming, livestock grazing, and timber production have been present in the county since the beginning of EuroAmerican occupation and settlement in the late nineteenth century. U.S. Census figures from 2013 show a population that is overwhelmingly (93.8 %) non-Hispanic White. The total population in 2013 was an estimated 6814, representing a decline of 5.7 % since the 2000 census. However, the number of housing units within the county increased by 3.9 % over that same time period, reflecting in part the increased demand for rural development among affluent, seasonally resident populations.

Wallowa County was chosen for study because of its unique status as a place that has attempted to proactively engage with the kinds of economic and social challenges typical of the rural West. The people and civil society organizations of Wallowa County have innovated to provide leadership in the face of natural resource crises, initiating numerous formal and informal processes intended to steer the county's development trajectory and ensure that local people benefit from the conservation activities associated with state and federal policy changes (Christoffersen 2011; Waage 2001). Issues related to land ownership, use, conservation, and development were researched here through a mixed-methods case study approach (Yin 2003) that included construction of a county land-use history, key informant interviews, analysis of letters of objection to or support of development proposals, secondary data analysis, direct observation, and a mail-administered landowner survey (see Abrams 2011 for a more complete methodology). In total, 51 interviews were conducted with 70 individuals between 2008 and 2010. Interviews were transcribed and analyzed using iterative coding and analysis techniques inspired by grounded theory (Charmaz 2006; Strauss and Corbin 1998). A 2010 survey of owners of at least 20 acres of Wallowa County land garnered 209 usable responses, representing a 53.3 % response rate. The case study approach was effective for collecting and triangulating diverse forms of social-scientific data; future efforts would be strengthened by more robust incorporation of ecological data as well. This chapter includes limited findings from the survey component and draws more heavily upon the integrated qualitative methods.

4.3 Background

Like many rural, traditionally resource-dependent places in the global North, Wallowa County has experienced substantial restructuring of its economy, demographic profile, and relationships between people and the land in the final decades of the twentieth century and the beginning of the twenty-first (Nelson 2001; Woods 2005). Mirroring patterns seen across much of the American West, restructuring in Wallowa County has been characterized by tensions between multiple regimes of rural capitalism: a "traditional" economy of resource production (logging, grazing, agriculture, etc.), an amenity economy centered on the consumption of scenic and recreational assets by urban and suburban migrants, tourists, and absentee landowners (Sheridan 2007; Travis 2007), and a residential development economy that intersects in important ways with both of the previous two regimes (Walker 2003). For example, many "traditional" resource producer landowners stand to gain from the conversion of agricultural, forest, or range lands to the "higher and better use" of residential real estate, yet the transition to a consumption-oriented landscape threatens to hasten the decline of the remaining production economy (Liffmann et al. 2000).

Amenity migrants, meanwhile, typically value the "unspoiled" (meaning, among other things, lightly developed) character of scenic western environs, yet rely on some level of residential development to allow them access to what were formerly "landscapes of production"—access that necessarily combines lifestyle consumption with productive capital investment (Robbins et al. 2012). Furthermore, in places such as Wallowa County the existence of a productive economy of agriculture, livestock raising, and forestry may itself serve as a desired component of local “authenticity” for amenity migrants, complicating simple narratives of a clash between consumption and production interests (Abrams et al. 2013). In this context, land-use planning emerges as a crucial forum wherein these tensions and contestations play out via the struggles of diverse actors drawing upon various economic, discursive, and political tools (Hurley and Walker 2004; Robbins et al. 2009).

Land-use conflicts in places like Wallowa County occur within a multiscalar institutional context of intersecting county-level governance structures, state-level land-use policies, and national-level constitutional and tribal treaty rights. They are also strongly informed by the local expression of regional-scale rural restructuring processes, in which longstanding production-centered livelihood practices and their associated cultural forms are challenged by the introduction of novel actors, organizations, economies, and ideologies (Brogden and Greenberg 2003; Walker 2003). Oregon’s famously proactive planning framework, centered on concentrating residential development within designated “urban growth boundaries” and restricting subdivision and development outside of them, sets key conditions for the politics of land-use struggles statewide. The framework relies on a power-sharing arrangement between: (1) cities and towns, empowered to demarcate their own urban growth boundaries and make decisions regarding the extension of municipal infrastructure and services; (2) counties, given authority over many rural land conservation and development decisions within their borders; and (3) the state, which typically acts to resolve disputes and ensure that statewide planning mandates are faithfully executed by lower-level governments. This basic structure remains in place despite voter approval in 2004 and 2007 of measures that granted greater rights of development to certain types of landowners and created rights of “compensation” for landowners affected by new land-use regulations (see Walker and Hurley 2011).

In this chapter, I analyze some of the complex and unexpected patterns of land-use activism that emerge from processes of rural restructuring in the U.S. West, taking as points of departure three strands in the political ecology literature. First, Walker and Fortmann (2003), Hurley and Walker (2004), and Hiner (2014), among others, have observed that the politics of land-use governance in the restructuring rural West often do not fall along simple dividing lines of “old-timers” and “newcomers” and their respective economies of resource production and amenity consumption. These authors detail the varied forms of resistance to and promotion of top-down planning frameworks, grassroots or astroturf countermovements, and the building of contingent coalitions among otherwise disparate interests in pursuit of particular social and environmental outcomes. Related to this, a second strand of literature complicates the simple association of smallholders, resource producers, and working-class rural residents with an unreflexive embrace of neoliberal environmental reforms. McCarthy (2002), for example, contends that the moral economies of even the most apparently pro-property rights rural West communities contain within them a communitarian vision that is sharply at odds with the logic of neoliberal environmental governance. Further, Robbins and Luginbuhl (2005), Haggerty and Travis (2006), and Newfont (2012) elucidate the importance of commons-type arrangements, across a variety of land tenure scenarios, to the livelihoods and environmental relations of “traditional” rural residents and interpret diverse forms of rural activism as resistance to commons enclosure. A third strand of literature documents and analyzes the growing engagement and effectiveness of Indigenous populations as actors in land-use governance across ownership categories (Holmes 2002; Larsen 2004; Coombes et al. 2012; Kelly et al. 2013). This literature underscores the capacity for Indigenous groups to draw on a range of resources and tools to influence diverse processes of coalition-building, resistance to the neoliberalization of nature, and defense of important landscapes and tenure arrangements across diverse rural settings.

4.3.1 Rural Restructuring Across the Wallowa County Landscape

Substantial recent changes to Wallowa County's economy, demographics, and land ownership patterns strongly inform the politics of private land development and conservation. Catalysts for rural restructuring in Wallowa County have to some extent been specific to the county's major land ownership categories. Federal lands were most strongly affected by policy changes in the mid-1990s enacted in response to lawsuits brought by environmental advocates seeking greater protection for salmon and other endangered species. These lawsuits came in the context of broader policy reforms that instituted ecosystem management as the dominant emphasis across categories of federal land that had previously been dedicated to timber production. The result was an abrupt end to what had been a reliable, if ultimately unsustainable (Langston 1995), flow of federal timber to local mills, which at the time were some of the county's largest employers. Two sawmills in the town of Joseph and a third in the town of Wallowa had all shuttered by 2007 after years of struggling to secure a steady supply of material (Christoffersen 2005). These policy changes were and continue to be widely viewed by many long-time local residents as an unjustified reassignment of access to benefits associated with public lands, and as prime contributors to contemporary local economic malaise. Very little timber management activity occurs on the local national forest unit, despite a regional shift in public lands policy and discourse toward restoration and fire hazard reduction. Local forests under private corporate ownership currently account for the bulk of the county's timber harvest; however, the turnover of roughly 150,000 acres of forestland from traditional industry ownership to portfolio investor ownership in the early years of the twenty-first century raised concerns regarding the future trajectory of the county's private forest base (cf. Bliss et al. 2010).

A complex suite of influences is responsible for driving restructuring processes on private ranchland and agricultural parcels. Developments in the macrostructure of U.S. agriculture have served to marginalize many independent producers through agribusiness consolidation and the globalization of agricultural production (Buttel 2003; McMichael 2003; Robbins and Luginbuhl 2005). Wallowa County beef cattle producers have also struggled with reduced access to forage on federal lands, which often make up substantial proportions of functional ranches in the West (Gentner and Tanaka 2002). The recent appearance of Canadian gray wolf populations that migrated from reintroduced packs in Idaho added further strain and prompted Wallowa County ranchers to organize in defense of their interests at local to state levels. Intergenerational transfer of agricultural land—never an easy process even under the best of circumstances—became even more difficult as the market value of rural properties skyrocketed in the early 2000s (Abrams 2011).

Restructuring on private agricultural estates has manifested via a variety of changes to longstanding patterns of land access and use. Many younger multigenerational producers have shifted to operations conducted largely or entirely on properties rented from other landowners—often amenity-oriented and/or absentee landowners. Some agricultural and livestock producers here have supplemented their practices with—or converted altogether to—niche and specialty production in response to both local challenges and wider crises within their respective markets. Examples of innovative production and marketing practices include grass-finished or specialty beef production, certification of farm products by third-party auditors such as Food Alliance, no-till wheat production, and production of non-GMO certified crops. Some of the more entrepreneurial producers have actively cultivated relationships with buyers, both individuals and high-end restaurants, in regional urban areas such as Portland, Oregon, and Seattle and Spokane, Washington, creating unique short-commodity-chain connections between urban and rural spaces. In doing so, they have capitalized upon rapidly growing demand for products deemed "local" and "natural" among regional middle-class populations. In other cases, landowners have sought ways to supplement on-farm income by developing scenic parcels for residential real estate or have sold out entirely to amenity buyers lacking agricultural backgrounds (at least some of whom continue to lease their properties to local producers). As the case studies in this chapter illustrate, however, the overall persistence of a viable production economy has contributed in important ways to land conservation.

Patterns of restructuring in Wallowa County also reflect the important gains in political efficacy made by regional Indigenous populations (Winchell 2013). The Nez Perce Tribe, Confederated Tribes of the Umatilla Indian Reservation, and Confederated Tribes of the Colville Reservation have all increased their role and visibility in land management and governance in Wallowa County in recent decades. The Nez Perce Tribe leveraged reserved rights guaranteed under treaty language that protected tribal members' access to "usual and accustomed" places off-reservation to establish a robust native fish recovery program in Wallowa County, facilitated by the opening of a tribal fisheries office in the town of Joseph. The Tribe also took ownership of a 15,000 acre parcel of land along Wallowa County's Joseph Creek as a result of a settlement with the Bonneville Power Administration over violations of these same rights to "usual and accustomed" places. In 1995 the Wallowa Band Nez Perce Homeland Project opened an interpretive center in the town of Wallowa and subsequently purchased a 320-acre site where the Tamkaliks cultural celebration has been held every year since 1998. There are no official reservation lands within Wallowa County, and as of 2015 only a handful of tribal members live year-round within the county, yet the Nez Perce and other Tribes maintain a number of institutional tools for defending rights of land access and use across a variety of public and private ownership categories, as explained in greater detail below.

4.3.2 The Influence of Amenity Migrants and Landowners

Urban-to-rural migration and the purchase of farm, forest, and/or rangeland by amenity-oriented populations (whether seasonally or permanently resident) are common components of rural restructuring processes (Shumway and Otterstrom 2001). Both amenity migration and amenity land ownership carry the potential to influence contestations over sprawl and land conservation (Walker and Fortmann 2003; Hurley and Walker 2004; Gosnell et al. 2006; Hurley 2013; Hiner 2014). In Wallowa County, amenity-based migration and amenity land ownership have been occurring since at least the 1940s, but remained at relatively low absolute numbers for most of the twentieth century as compared to many places with greater access to urban areas or better-developed recreational infrastructure. This all changed by the 1990s and 2000s, as reflected in a 2006 *New York Times* Real Estate profile that described the recent influx of amenity migrants from urban areas buying up scenic ranch estates at "bargain" prices (Preusch 2006). Between 2000 and 2008, over 130,000 acres of Wallowa County land was purchased or otherwise transferred to owners who maintained a permanent address outside the county (Abrams et al. 2013). A total of 37.7 % of the landowners surveyed in 2010, representing 25.6 % of the county's nonindustrial private land, reported that recreational and residential amenities were more important than agricultural or forestry production as reasons for owning land. Well over half of these "amenity" landowners held a permanent address outside Wallowa County.

The capital that flooded into Wallowa County in the 1990s and 2000s in pursuit of landed assets had its roots in larger macroeconomic forces. Most important was a national-scale inflation of land and real estate markets as abundant global speculative capital sought a class of investments that were believed to be both incapable of long-term decline and, to some extent, protected and propped up by favorable governmental policies. This generalized trend intersected with a set of more nuanced regional-scale dynamics in the Northwest. Chief among these were: (1) a cultural predilection toward the consumption of particular kinds of scenic rural landscapes among particular, mostly urban, social classes; (2) "push" factors in regional urban and suburban areas, which had seen sustained population increases for several decades; (3) substantial equity gains among homeowners in regional urban and suburban areas as a result of both population increases and inflated real estate markets; and (4) "push" factors in some of the region's existing rural amenity destination communities, as prior waves of amenity migration and rural gentrification began to degrade many of the qualities that had previously encouraged affluent migrants to relocate from urban areas or to purchase properties for seasonal residence. In short, capital in search of new arenas of investment intersected with populations in search of new arenas of consumption. Or as one amenity landowner quipped in describing his reason for purchasing a several hundred-acre estate in rural Wallowa County, "investing in this is … not that much worse than the stock market and a hell of a lot more fun."

Migration and absentee land ownership patterns in Wallowa County both demonstrate strong regional influences (Abrams 2011). Survey data show that nearly all absentee landowners held their permanent residence in Oregon (58.7 % of the total) or in neighboring Washington (32.0 %) or Idaho (5.3 %). A total of 16.5 % of seasonal landowners reported owning a permanent residence in the Portland, Oregon metropolitan area (approximately 5 h distant by car). Of landowners who had permanently relocated to Wallowa County, 42.0 % previously lived elsewhere in Oregon, 18.8 % lived in California, 13.0 % lived in Washington, and 7.2 % relocated from Idaho. A substantial proportion of Wallowa County migrants and seasonal landowners hailed from other rural communities in the Northwest—communities such as Sandpoint, Idaho, Walla Walla, Washington, and Bend, Oregon.[1] In general, Wallowa County lacks the developed recreational infrastructure of some of these more populous rural communities; however, it does offer migrants and seasonal property owners tranquility, outstanding scenery, backcountry (undeveloped) recreation in canyon and mountain environments, excellent hunting and fishing, and relatively low land prices. For some migrants, it is precisely Wallowa County's status as an "unspoiled" and "authentic" landscape, in contrast to many already-gentrified rural environments, that serves as a primary attractant (Abrams et al. 2013).

4.4 Planning For and Against Sprawl

Because Oregon's statewide land-use policy was structured in large part to respond to concerns over the loss of productive farmland, it is typically quite difficult for landowners to gain planning approval to build new residences outside of established urban growth boundaries unless those structures are specifically related to farming or ranching activities. As Walker and Hurley (2011) observe, this approach to private land conservation may rankle those whose chief interest is in land speculation, subdivision, or development, but it offers numerous benefits to those who make their living from farming, ranching, or forestry, providing a potential base of political allies for continued sprawl containment. Yet the productive economy and the amenity economy intersect in complex ways: as the profitability of productive activities declines, producer landowners increasingly look for other ways to gain income from their holdings. The political economy of the amenity landscape, meanwhile, relies on both land conservation (to provide valued scenic amenities) *and* development (to provide valued residential spaces) for its realization (Robbins et al. 2012). The political landscape of development planning is further complicated by the

[1] Bend is not, strictly speaking, rural, but rather a high-amenity, recreation-oriented community that has become more urban in recent decades as its population has boomed.

growing influence of actors who fail to fit neatly into either the traditional "production" or the amenity "consumption" economies, such as members of the Indigenous groups described above. Planning frameworks, operating at nested county and state scales (and informed by quintessentially American ideals of landowner autonomy (Freyfogle 2003)), play crucial roles in mediating the transfer of these competing impulses to privately held terrain.

Two examples illustrate the complexity of the politics of private land conservation in Wallowa County—politics informed by the intersection of diverse forms of rural capitalism (and resistance thereto), restructuring rural economies, and the tools provided by a multiscalar institutional framework. The first example took place on the urban fringe of the town of Joseph, Wallowa County's tourism and amenity hub, over several years starting in the 1990s. The second resulted from the county's 2001 decision to allow farmland owners to subdivide and develop parcels deemed not to be contributing to farm production.

4.4.1 Marr Ranch/Iwetemlaykin

The most divisive Wallowa County land use dispute in recent memory surrounded the fate of a roughly 62 acre parcel known locally as "Marr Ranch," named for the EuroAmerican family that owned it for several decades prior to its sale in 1991. The land in question is located on the outskirts of the town of Joseph (2000 population: 1054), adjacent to residential neighborhoods and to undeveloped lands used primarily for livestock grazing. Even more importantly, the parcel sits close to Wallowa Lake, which serves simultaneously as the scenic centerpiece of Wallowa County's amenity economy, the source of much of the water used in the county's irrigated agriculture, and a place of immense cultural and historical importance to the Nez Perce and other Plateau Tribes. Tribal oral history recognizes the Marr Ranch site as an important location for catching and preparing sockeye salmon on the oceangoing species' return migration to the lake (Oregon State Parks 2009). The property lies within Joseph's urban growth boundary, the spatial zone designated for future residential development, but long remained in an undeveloped state and was never annexed by the town. The parcel is lightly forested, covered largely in native grasses and forbs, and lacks permanent habitations. Two important irrigation canals traverse the property, but the land itself is not suitable for intensive agriculture due to its rocky surface geology; it is situated on the terminal moraine of the glacier that carved Wallowa Lake during the last ice age. It also sits immediately adjacent to the final resting place of the famous Nez Perce Chief known as Old Chief Joseph (tıwi·teq̓ıs). The 5-acre gravesite is managed by the National Park Service and serves as the starting point of the Nez Perce National Historic Trail. The body of Old Chief Joseph was moved to this location next to Wallowa Lake in 1926 after it was repeatedly desecrated at its original burial site further west, and following the landowner's complaints that the gravesite made agricultural operations difficult (Butterfield 1945).

The Marr Ranch property was purchased in 1991 as a roughly 70-acre parcel by the same southern California family that also purchased and reopened one of Wallowa County's shuttered sawmills following the federal timber shutdown of the early 1990s (this mill would close in 2001 after years of financial struggle). The previous owners had traditionally maintained Marr Ranch as open space accessible to the local community for access to fishing, hunting, and other light recreational pursuits. Like much of the lake moraine area and other local private lands, it was treated as a kind of commons, accessible for low-impact uses by local residents despite its private tenure status. The new owners, however, embarked upon efforts to capitalize on the development potential of the site soon after taking ownership. In 1994 the owners proposed and received preliminary approval from the county to rezone and subdivide the property into 69 separate 1-acre residential lots. Fearing possible impacts to the Chief Joseph gravesite and surrounding undeveloped lands, the Nez Perce Tribe protested the proposed development and called upon the federal government to purchase the property, using condemnation[2] if necessary. In turn, the property owners enlisted the support of Oregon Senator Mark Hatfield and Congressman Bob Smith, both of whom promised to block any federal condemnation of the property. An apparently satisfactory resolution was reached in 1997 when the federal government purchased an 8-acre buffer between the gravesite and the remaining developable portion for $250,000. A second similar development proposal was forwarded 3 years later for the remaining 62 acres, which would have relied on annexation of the property by the city of Joseph and extension of city water and sewer infrastructure. City councilors voted narrowly against annexation after the plan stirred local controversy.

The landowners' next attempt to develop the site came in 2003 when they submitted a new proposal to subdivide the property into 11 parcels for upscale home development (Fig. 4.1); the county granted development permission in February 2004. This plan was quickly appealed to the state by the city of Joseph, a coalition of three Joseph residents, and three Plateau Tribes: the Nez Perce, Confederated Tribes of the Colville Reservation, and Confederated Tribes of the Umatilla Indian Reservation. The Joseph citizen group opposing the proposal consisted of a mix of long-term local residents and newer migrants; all three held strong sympathies with tribal concerns regarding the site. The appeal filed by the city of Joseph focused on the lack of appropriate infrastructure to support residential expansion; city councilors objected to the fact that the approval did not take into consideration a set of recommendations made by the city. Specifically, the councilors were concerned that the terms of the county's approval precluded future parcelization to allow for denser development and that there would be no sidewalks, no bike lanes, and a lack of public access to subdivision streets. The city's appeal also argued that the county had violated an intergovernmental agreement by approving the development proposal without first seeking the city's consent.

[2]The condemnation process draws upon the state's "eminent domain" authority over property by forcibly transferring ownership rights away from the titleholder at a government-approved market rate.

Fig. 4.1 A 2004 photograph of Marr Ranch shows the anticipated parcelization of the property into exurban homesites. *Photo credit*: Leon Werdinger Photography

The Tribes' involvement in a land dispute at this jurisdictional level was very unusual, as they had traditionally preferred to resolve land and resource conflicts directly with the federal government via their treaty-recognized government-to-government relationship. The Tribes' appeal rested on concerns regarding the archeological and cultural importance of the land in question. The Tribes were particularly alarmed by the possibility that earlier generations of their people may have used the site for burials, and that construction of houses and associated infrastructure would disturb any existing remains. In response, the landowners pointed to previous archeological surveys that turned up only lithic scatters, with no compelling evidence of more intensive occupation and use. Ruling on the challenge, the Land Use Board of Appeals (LUBA), Oregon's dedicated land-use adjudication body, overturned the county's approval due to noncompliance with archeological protection requirements and other violations of state planning requirements. The Oregon Court of Appeals later upheld LUBA's decision.

As tensions regarding the fate of Marr Ranch continued to escalate, third-party organizations began to search for compensated solutions in which the landowners would voluntarily sell the land at a price that reflected, at least to some extent, its speculative value for development. The Trust for Public Land offered to purchase the property in late 2003 for $1.2 million and again in 2004 for $1.4 million but the landowners rejected both offers. A federal purchase option was again floated, this time with both of Oregon's senators joining Idaho's delegation[3] in support of a fed-

[3] Idaho's interest related to that of the Nez Perce Tribe, which has a reservation and most of its population in the state.

eral buyout. The landowners went on record stating that they would refuse to sell to the federal government or to any entity using federal funds. In light of the landowners' continued attempts to develop the property, the Confederated Tribes of the Umatilla Indian Reservation petitioned the Oregon State Historic Preservation Office (SHPO) to designate the entire site as archeologically significant. Were this designation granted, it would have sharply restricted any development options that threatened to degrade the property's archeological values.

The dispute escalated to the point that it began to attract statewide attention, including a 2005 editorial in *The Oregonian*, the state's largest newspaper, calling upon the Governor to step in to resolve the crisis. During this time, the Wallowa Land Trust was created by a small group of local residents, including those with long-term standing in the county as well as newer arrivals, and began to work quietly in the background to broker a deal that would protect the site while providing a fair purchase price to the landowners. A particularly important role for the Wallowa Land Trust was its work in establishing and representing local sentiment regarding the dispute; this was necessary due to concern in the Governor's office that any state action to purchase the land could be met with local backlash. In addition to advocating for—and working to settle on the terms of—a possible state buyout, the Wallowa Land Trust helped to provide legitimacy for the buyout process as a whole. The young organization was able to bring together financial support from amenity owners and non-residents with a significant degree of moral support from local producers and long-time residents as well as the trust of Tribal interests. Although the local long-term population was traditionally very defensive of private property rights, overall it did not make common cause with the Marr Ranch landowners, for two principal reasons. First, many saw the owners' actions as an attempt to enclose and profit off of what had long been seen and used as a local commons. Second, the owners did not ingratiate themselves well with the local population, who continued to see them as wealthy and abrasive "outsiders" lacking legitimate membership in the local community. In short, the rural residents' moral economy accorded greater value to the community interests associated with this particular piece of land than to the title owners' property rights in the abstract.

The Marr Ranch conflict escalated even further in 2005 in response to three separate events. Pointing to the potential for an influx of affluent recreation-seekers, the landowners announced new plans to convert the property into an upscale recreational vehicle resort on 150 separate lots. In late August, the landowners bulldozed a road across the property in what was widely viewed as the first step in converting the site to residential or commercial space. Two days prior to this action, the landowners had been awarded a claim under Oregon's recently voter-approved Measure 37. This law required state and local governments to monetarily compensate landowners for any regulation enacted since the time of property purchase that reduced property values, or else waive the enforcement of those regulations (in practice, this meant waivers of regulation enforcement in the case of every Measure 37 claim in Oregon except one; see Walker and Hurley, 2011). For the owners of Marr Ranch, a successful Measure 37 claim meant that only those constraints and regulations already in existence when they purchased the property could be enforced; any restriction on development enacted after that time was effectively null and void. Public outcry boiled over in the form of dueling public rallies on and near the property, with the landowners and their supporters defending development plans via a discourse centered on the sanctity of private property rights. Members of the Nez Perce Tribe and both tribal and non-tribal allies gathered near Chief Joseph's grave to protest the desecration of the archeological and cultural values associated with the parcel in question.

Although subsequent archeological surveys turned up only limited evidence of past human habitation, the Tribes continued to press for formal recognition of the land's cultural significance. Meanwhile, the landowners continued to plan for development options, including various proposals to divide and develop the property. The Oregon Governor's office appeared sympathetic to tribal concerns, but SHPO reported that they did not have sufficient evidence to justify designating the entire parcel as a site of archeological significance. Following a spate of threatened lawsuits in both directions, the standoff was resolved in 2007 when the landowners agreed to sell the undeveloped property to the state of Oregon for $4.1 million. Of the purchase price, $3.2 million came from lottery-funded state park and recreation funds and the balance was provided in equal parts by the Nez Perce Tribe, Confederated Tribes of the Umatilla Indian Reservation, and Confederated Tribes of the Colville Reservation. The Tribes' contribution was particularly significant in that it required a departure from Tribal policy by helping to fund the purchase of land that would not be owned by the Tribes themselves (Oregon State Parks 2009). The site was dedicated on October 10, 2009 as the Iwetemlaykin State Heritage Site in a ceremony featuring Tribal leaders, Oregon Governor Kulongoski, and local elected officials. The name was suggested by the Tribes in recognition of their own linguistic appellation for the site; it roughly translates to "at the edge of the lake" (Oregon State Parks 2009).

The ultimate outcome of this land-use dispute reflected in large part the agency of local and regional actors, principally the Nez Perce, Confederated Umatilla, and Confederated Colville Tribes and the nascent Wallowa Land Trust, strategically utilizing a variety of institutional resources and strategies to crystallize support both within and outside the local community. Despite the somewhat divergent interests animating their conservation activism, the Tribes and land conservation advocates were able to draw upon complementary tools and discourses, including tribal claims to the site as an area of archeological concern, legal appeal to the state Land Use Board of Appeals, and local-level challenges to county decisions. Local (city and county) governance of development proposals within this traditionally pro-property rights social environment was complicated by the activism of small numbers of local non-Tribal residents, some with multigenerational ties to the county and some falling into the "amenity migrant" category. While these acts of protest and administrative appeals were not by themselves sufficient to halt the development process, they were successful in slowing development progress and raising the political profile of the dispute to the point that state-level actors were compelled to devote financial and political capital to achieving a solution.

A burgeoning amenity economy created the demand for scenic rural residential development to which the property owners' proposals responded, yet existing amenity populations also contributed to efforts to preserve the site in its undeveloped state. By attempting to convert the social and cultural benefits that had long been shared at a community level into an extractive commodity, the new property owners ran afoul of the local moral economy and thereby lost potential allies. At the same time, the property's high public visibility and cultural importance allowed for the development of an important, if underappreciated, dimension of land-as-power: the property owners were able to command fantastic prices simply through the threat of converting open space into exurban development. While the ultimate conversion of this contested piece of property into "Iwetemlaykin" suggests a reterritorialization in favor of Tribal and conservation interests, it was a reterritorialization paid for at extortion-level prices. The ultimate outcome in this case demonstrates that the fundamental expectation of the Marr Ranch landowners—that they deserved to profit from rights of development—was ultimately ratified.

4.4.2 Oregon House Bill 3326

A second vignette of the political ecology of sprawl in Wallowa County comes from the county's experience with a little-known but nonetheless important state land-use policy, House Bill 3326. This policy, passed in 2001, gave eastern Oregon county governments new institutional tools to attempt to resolve the contradictions between the productive economy, the amenity economy, and real estate development interests. Eastern Oregon counties that opted-in to the policy could offer farm owners the ability to identify up to two parcels on their estates as "generally unsuitable for production" and thereby available for the kind of subdivision and residential development otherwise prohibited by Oregon land-use law. Under HB3326, such developments can only be approved if they are not expected to increase the cost of farm or forestry uses on adjacent lands and if approval would not "materially alter the stability of the overall land use pattern of the area." An important dimension of the law is its deference to county governments to define exactly what it means for land to be "generally unsuitable for production."

The Wallowa County commission's decision to opt-in to the development possibilities offered by HB3326 spurred landowners and other parties into action in defense of their own interests within a restructuring social-economic environment. At least four distinct, but semi-overlapping, political orientations with respect to HB3326 emerged over time: (1) farmland owners who hoped to benefit from the ability to subdivide and develop farmland parcels; (2) neighboring agricultural and livestock producers opposed to the proliferation of nonagricultural households across the "working" landscape; (3) amenity landowners interested in defending their rural idylls by restricting residential development within their own viewsheds; and (4) large absentee landowners who hoped to leverage HB3326 into a tool to create expanded opportunities to profit from real estate development. What is significant about these categories is that they do not divide easily along old-timer/newcomer lines (cf. Hiner 2014). The first two categories both represent landowners whose personal fortunes are primarily tied to the traditional productive economy, yet who held divergent positions relative to development prospects under HB3326. The latter two categories were also often opposed to one another, although in some cases incoming amenity owners purchased rural parcels created under HB3326, and both groups tended to be affluent, absentee, and economically independent of the production economy. The conflict over HB3326 in Wallowa County was rooted in local experiences with restructuring, particularly the intersection of declining on-farm profitability with expanded opportunities to realize financial gains from the emerging amenity economy.

What started as a relatively conflict-free process under HB3326 soon evolved to become highly contested. In the 7-year period inclusive of the years 2002–2008, a total of 23 development proposals, representing 43 nonfarm parcels, were proposed in Wallowa County under HB3326 (Abrams and Gosnell 2012). During the first 4 years of the policy's implementation, nearly every proposal for subdivision and development gained approval from the county planning commission, despite the commission's receipt of several letters of opposition concerning particular applications. Starting in 2006, however, proposals began to garner greater scrutiny and controversy and became more likely to be rejected by the planning commission. The commission denied five of the nine applications submitted between 2006 and 2008 on the grounds that lands proposed for development did not meet the still-evolving criteria of land "generally unsuitable for production;" two of these denials were later overturned on appeal. The county's pivot away from near-blanket approval of HB3326 applications came in response to three intertwined factors: countywide concern regarding the loss of the "working landscape" of agricultural production, exacerbated by the existence of other factors such as the recent approval of Measure 37 (see Sect. 4.4.1 and Walker and Hurley 2011); growing resistance from both amenity and producer landowners to specific development proposals; and difficulty differentiating lands "generally unsuitable for production" given the prevalent ecological and economic conditions of productive agriculture in a largely arid, rocky, and high-elevation county.

Landowners petitioning for rights of development under the policy were not a homogenous group. Some were producers who planned to keep the bulk of their land in production but who saw the policy as a means to amass additional income or to provide housing for family. A handful of proposals came from agricultural or livestock producers (or their heirs) who hoped to use the policy to maximize the exchange value of the land before selling out entirely. Still other petitioners were absentee landowners seeking otherwise elusive exurban development opportunities. In contrast to the Marr Ranch case, these development proposals were not discursively defended through recourse to abstract notions of private property rights. Rather, they were more often framed by the petitioners as vital to helping producers weather the misfortunes of high production costs and low commodity prices. Echoing the language of the policy itself, nearly all petitioners emphasized the unproductive nature of the parcels in question, framing the issue as primarily one of land suitability. Some producers claimed that development would allow farmers to keep their operations going by monetizing "worthless" terrain. Agricultural producers petitioning for rights of development often recognized the changing production economy as a factor pushing them to find new sources of income. According to one, "Farming has changed dramatically with the ever increasing costs of doing business and low commodity prices … the farmer is forced to work in town in order to pay the bills … I can't afford to keep all the land. It is too much work and I should not be forced to farm at a loss."

It was almost exclusively other landowners, most often neighbors, who went on record expressing objections to particular HB3326 proposals. These objections came from long-time agricultural producers, amenity-oriented landowners, and some small-scale producers whose livelihoods and practices represented alternatives to both traditional commodity production and amenity landownership. Objectors typically framed the developments in question as representing privatized gain (for the petitioner) along with externalized costs to neighboring producers and the county as a whole (Abrams and Gosnell 2012). Typical concerns raised by producers in the vicinity of proposed HB3326 subdivisions included the incompatibility of residential development with the land-use patterns of a productivist landscape, threats of land prices increasing well beyond the productive value of the land, and the potential for trespassing on private roads and properties. Reflecting an understanding of the working landscape as a kind of multi-landowner "commons" (cf. Haggerty and Travis 2006), some neighboring producers made strong distinctions between development intended for family members and development intended for open market sale. For example, one letter of objection read, "… as it is written, [the neighbor's HB3326 petition] will allow the sale and use of the land, to non-family. Thus it could, and would be used as a precedent for opening up the county's historical grazing land to subdivision. We feel the county needs tighter restrictions to prevent this happening, and this petition changed to family use only." Objections by producers were often tied to specific concerns: road access, water drainage, and the introduction of populations unfamiliar with and potentially hostile to agricultural practices.

Amenity owners' objections, meanwhile, were typically more generalized in their opposition to exurban development and often highlighted specific legislative language in their complaints. At the same time, these objections—like those filed by producers—often decried the possible loss of productive or "working" landscapes through the introduction of incompatible populations and land uses. Both producers and amenity owners expressed fears about the potential for runaway development changing the character of the countryside and potentially encouraging an influx of wealthy outsiders. Beyond this, their objections reflected different understandings of the landscape, with amenity owners generally opposed to further development in the abstract and producers more often distinguishing between proposals that were compatible and those that were incompatible with a productive landscape benefiting local families. Yet, as in the Marr Ranch case, strange bedfellows made up a "contingent coalition" largely aligned against exurban development.

Questions of legitimacy plagued Wallowa County's implementation of HB3326. Given that the policy was seen by many, including those tasked with implementing the law as written, as intended to benefit struggling producers, the proliferation of subdivision and development petitions by non-producers (e.g., absentee landowners) or by apparently successful producers was problematic. Local opposition to these petitions proved difficult for the county planning commission to resolve, given that eligibility for development under HB3326 was based on characteristics of the land itself, not of the landowner. Objectors typically refuted petitioners' claims of agricultural marginality by observing that the county's dominant agricultural practices (especially extensive livestock grazing) take place across large expanses of relatively unproductive terrain. Several individuals interviewed for this study insisted that nearly any piece of land can potentially be useful as part of an overall agricultural operation. For example, rocky, infertile ground may be a valuable place for livestock to congregate when the surrounding pastures are saturated, as often occurs in springtime (Abrams and Gosnell 2012).

The local controversy over HB3326 was situated within a larger debate regarding residential development, economic development trajectories, and property rights. Measure 37, passed as a state ballot measure by Oregon voters 3 years after the legislature passed HB3326, loomed as an even greater dilemma, promising economic opportunities for some while appearing to others as a threat to the integrity of the rural landscape. The construction of ostentatious homes on farm properties (marketed to wealthy exurbanites) and the opening of high-end shops and restaurants within the county appeared to signal a turn away from the production economy and toward a gentrified amenity-based economy. The controversy and heightened scrutiny brought about by local opposition to HB3326 forced the county government to revisit the concept of agricultural marginality. After the approval of 32 nonfarm parcels during the first 5 years of the policy's implementation in Wallowa County, the county substantially tightened its definition of lands "generally unsuitable" for production following the advice of an ad-hoc group consisting of local natural resource and planning experts chartered to resolve this difficult definitional issue. HB3326 petitions that would have been approved under previous guidelines were now being denied, signaling to other landowners that their development options had diminished. As the real estate boom of the early 2000s gave way to the Great Recession, HB3326 petitions for subdivision and development came to an abrupt halt.

Paralleling the Marr Ranch case, public opposition to specific HB3326 development proposals did not by itself hold sprawl at bay. In this case, what prevented an acceleration of exurban development more than anything else was the onset of the Great Recession shortly after Wallowa County was "discovered" as an arena of consumptive real estate investment. However, objections to specific HB3326 petitions served several important purposes. They highlighted contradictions in the (implicit or explicit) policy narrative that held that farm properties could be converted to residential properties without triggering impacts on surrounding landowners, users, and community members. They communicated the breadth of public discomfort with the liberalization of land development policy (demonstrating that both newer amenity migrants as well as longstanding residents had vested interests in restricting residential development in rural areas). And they indicated that the county's implementation of HB3326 would be closely scrutinized, including the precise way in which the county determined what lands would be considered "generally unsuitable for production." Disputes over rights of development under HB3326 exposed the complex fault lines running across the changing social landscape of Wallowa County, with some amenity migrants and some producers working in parallel to oppose exurban development while other landowners and residents attempted to valorize their properties supported the development proposals of friends and neighbors, or sat quietly on the sidelines. The narrowing over time of the county's definition of lands "generally unsuitable for production" was an attempt on the part of county planning and government officials to reconcile the tensions between HB3326 as written, the expectations of individual landowners, and broad community interests in maintaining a "working landscape" in the face of strong market demand for scenic arenas of investment and consumption.

4.5 Conclusions

Planning processes encompass diverse influences from multiple scales; in the case of Oregon, they can entail local actors engaged in contestation over development proposals governed by county-level officials, all with strong oversight by state-level entities empowered by a comprehensive land-use planning framework. These local-to-state-level deliberations are themselves situated within a set of national-level legal and cultural institutions in which landowner rights and freedoms are considered paramount, often to the exclusion of wider societal interests (Freyfogle 2003). Yet these planning processes also act as opportunities for social organization and action in furtherance of material and cultural interests. In the Marr Ranch case, neoliberal notions of landowner sovereignty to develop were counteracted to varying degrees by the use of institutional tools supporting cultural landscape meaning on the part of Indigenous tribal groups, as well as by non-Tribal partners and sympathizers. Tribal interests in this case successfully defended their own claims to the important site, first using legal channels and then using political, discursive, and economic tools and working closely with non-Tribal conservationists. In both the Marr Ranch and HB3326 cases, neighboring landowners and other affected parties drew upon institutional resources provided by county- and state-level land-use governance policies to defend particular land uses and landscape visions in the context of a restructuring rural economy. In neither case were these forms of activism sufficient by themselves to contain the pressures toward exurban development, but they were used effectively to heighten public scrutiny, enlist actors at non-local scales (e.g., the Oregon Governor's office, the Land Use Board of Appeals), and slow the pace of development until more permanent solutions could be reached.

These contestations regarding development were informed by the structural positions of relevant social groups. The outcome of the Marr Ranch dispute was strongly influenced by efforts on the part of the Nez Perce Tribe, Confederated Tribes of the Umatilla Indian Reservation, and Confederated Tribes of the Colville Reservation to rebuild political and economic power over the decades since their dispossession and subjugation at the hands of the American state. The Tribes' effectiveness in this case reflects their changing position within a regional economy and regional to national political context. Amenity landowners and migrants operated from their positions as affluent and politically engaged actors and in furtherance of their own particular imaginaries regarding their adopted landscape (Taylor 2011). They played key roles in both the Marr Ranch and HB3326 disputes, working alternately in defense of Tribal interests in sacred site conservation, agricultural interests in "working landscape" conservation, and their own interests in scenery, privacy, and a sense of rural authenticity.

The continued presence of a viable production economy (including a growing value-added sector) both created a source of local allies for "working landscape" conservation and served to bolster discourses that held exurban development to be a threat to the local economic foundation. While the production economy as a whole in Wallowa County had been strained by its peripheral position within a national to global commodity supply chain, it remained buoyant enough to reinforce a local moral economy regarding land use. As McCarthy (2002) observes, the moral economy of rural Western producers is complex and resists simple categorization as an unreflexively "conservative" political orientation. The property rights defended by long-time local populations did not necessarily include the abstract rights of non-locals, particularly those intent on restricting community access to valued places or introducing uses or users at odds with continued productive uses of the land.

These observations underscore the fact that the containment of exurban sprawl in Wallowa County is far from a finished project, and that prospects for continued efficacy will change as rural restructuring proceeds. While Marr Ranch was ultimately conserved as Iwetemlaykin without a single home or RV pad being constructed, the protection of this 62-acre parcel required the expenditure of a tremendous amount of financial and political capital. Equivalent amounts of capital are unlikely to be available for the conservation of other parcels within this two million acre county. The HB3326 saga ultimately quieted down when the real estate boom went bust, but this is likely only a temporary resolution. The future of the rural landscape in Wallowa County will ultimately depend not only on the political and economic potency of regional Tribal interests and amenity landowners, but also on the prospects for existing and novel forms of agricultural and livestock production as these articulate with a political and economic context that extends far beyond county boundaries.

References

Abrams, J. (2011). *Restructuring of land and community in the remote rural west: The case of Wallowa County, Oregon*. Ph.D. dissertation, Oregon State University, Corvallis, OR.

Abrams, J., Bliss, J. C., & Gosnell, H. (2013). Reflexive gentrification of working lands in the American West: Contesting the 'middle landscape'. *Journal of Rural and Community Development, 8*(3), 144–158.

Abrams, J., & Gosnell, H. (2012). The politics of marginality in Wallowa County, Oregon: Contesting the production of landscapes of consumption. *Journal of Rural Studies, 28*(1), 30–37. doi:10.1016/j.jrurstud.2011.09.004.

Bliss, J. C., Kelly, E. C., Abrams, J., Bailey, C., & Dyer, J. (2010). Disintegration of the US industrial forest estate: Dynamics, trajectories, and questions. *Small-Scale Forestry, 9*(1), 53–66. doi:10.1007/s11842-009-9101-7.

Brogden, M. J., & Greenberg, J. B. (2003). The fight for the West: A political ecology of land use conflicts. *Human Organization, 62*(3), 289–298.

Buttel, F. H. (2003). The impact of global economic practices on American farming. In D. L. Brown & L. E. Swanson (Eds.), *Challenges for rural America in the twenty-first century* (pp. 77–89). University Park, PA: The Pennsylvania State University Press.

Butterfield, G. (1945). Old Chief Joseph's Grave. *Oregon Historical Quarterly, 46*(1), 70–73.

Charmaz, K. (2006). *Constructing grounded theory: A practical guide through qualitative analysis*. Thousand Oaks, CA: Sage.

Christoffersen, N. D. (2005). Wallowa resources: Gaining access and adding value to natural resources on public lands. In B. Child & M. W. Lyman (Eds.), *Natural resources as community assets: Lessons from two continents* (pp. 147–180). Madison, WI: The Sand County Foundation and The Aspen Institute.

Christoffersen, N. D. (2011). Collaboration: A catalyst for restoration. In D. Egan, E. Hjerpe, & J. Abrams (Eds.), *Human dimensions of ecological restoration: Integrating science, nature, and culture* (pp. 93–105). Washington, DC: Island Press.

Coombes, B., Johnson, T. J., & Howitt, R. (2012). Indigenous geographies I: Mere resource conflicts? The complexities in Indigenous land and environmental claims. *Progress in Human Geography, 36*(6), 810–821. doi:10.1177/0309132511431410.

Freyfogle, E. T. (2003). *The land we share: Private property and the common good*. Washington, DC: Island Press.

Gentner, B. J., & Tanaka, J. A. (2002). Classifying federal public land grazing permittees. *Journal of Range Management, 55*(1), 2.

Gosnell, H., Haggerty, J. H., & Travis, W. R. (2006). Ranchland ownership change in the Greater Yellowstone Ecosystem, 1990–2001: Implications for conservation. *Society and Natural Resources, 19*(8), 743–758. doi:10.1080/08941920600801181.

Haggerty, J. H., & Travis, W. R. (2006). Out of administrative control: Absentee owners, resident elk and the shifting nature of wildlife management in Southwestern Montana. *Geoforum, 37*(5), 816–830. doi:10.1016/j.geoforum.2005.12.004.

Hiner, C. C. (2014). 'Been-heres vs. come-heres' and other identities and ideologies along the rural–urban interface: A comparative case study in Calaveras County, California. *Land Use Policy, 41*, 70–83. doi:10.1016/j.landusepol.2014.05.001.

Holmes, J. (2002). Diversity and change in Australia's rangelands: A post–productivist transition with a difference? *Transactions of the Institute of British Geographers, 27*(3), 362–384. doi:10.1111/1475-5661.00059.

Hurley, P. T. (2013). Whose sense of place? A political ecology of amenity development. In W. P. Stewart, D. R. Williams, & L. Kruger (Eds.), *Place-based conservation: Perspectives from the social sciences* (pp. 165–180). Dordrecht, The Netherlands: Springer.

Hurley, P. T., & Walker, P. A. (2004). Whose vision? Conspiracy theory and land-use planning in Nevada County, California. *Environment and Planning A, 36*(9), 1529–1547. doi:10.1068/a36186.

Josephy, A. M. (1997). *The Nez Perce Indians and the opening of the Northwest*. Boston, MA: Houghton Mifflin.

Kelly, E. C., Bliss, J. C., & Gosnell, H. (2013). The Mazama returns: The politics and possibilities of tribal land reacquisition. *Journal of Political Ecology, 20*, 429–443.

Langston, N. (1995). *Forest dreams, forest nightmares: The paradox of old growth in the Inland West*. Seattle, WA: University of Washington Press.

Larsen, S. C. (2004). Place identity in a resource-dependent area of northern British Columbia. *Annals of the Association of American Geographers, 94*(4), 944–960. doi:10.1111/j.1467-8306.2004.00442.x.

Liffmann, R. H., Huntsinger, L., & Forero, L. C. (2000). To ranch or not to ranch: Home on the urban range? *Journal of Range Management, 53*, 362–370.

Marshall, A. G. (2006). *An ethnohistoric assessment of Nez Perce relationships with the Marr Ranch property, Wallowa Lake, and the Wallowa Country*. Interim Report to The Nez Perce Tribe Office of Legal Counsel, Lewiston, ID.

McCarthy, J. (2002). First world political ecology: Lessons from the wise use movement. *Environment and Planning A, 34*(7), 1281–1302. doi:10.1068/a3526.

McMichael, P. (2003). The impact of global economic practices on American farming. In D. L. Brown & L. E. Swanson (Eds.), *Challenges for rural America in the twenty-first century* (pp. 375–384). University Park, PA: The Pennsylvania State University Press.

Nelson, P. B. (2001). Rural restructuring in the American West: Land use, family and class discourses. *Journal of Rural Studies, 17*(4), 395–407. doi:10.1016/S0743-0167(01)00002-X.

Newfont, K. (2012). *Blue ridge commons: Environmental activism and forest history in Western North Carolina*. Environmental History and the American South. Athens, GA: University of Georgia Press.

Oregon State Parks. (2009). *Iwetemlaykin state heritage site master plan*. Salem, OR: Oregon State Parks.

Preusch, M. (2006, May 26). Rugged, Western and still a bargain. *The New York Times*. Retrieved from http://www.nytimes.com/2006/05/26/realestate/26havens.html.

Robbins, P., & Luginbuhl, A. (2005). The last enclosure: Resisting privatization of wildlife in the Western United States. *Capitalism Nature Socialism, 16*(1), 45–61. doi:10.1080/10455750520 00335366.

Robbins, P., Martin, S., & Gilbertz, S. (2012). Developing the commons: The contradictions of growth in exurban Montana. *The Professional Geographer, 64*(3), 317–331. doi:10.1080/0033 0124.2011.601193.

Robbins, P., Meehan, K., Gosnell, G., & Gilbertz, S. J. (2009). Writing the new West: A critical review. *Rural Sociology, 74*(3), 356–382. doi:10.1526/003601109789037240.

Sheridan, T. E. (2007). Embattled ranchers, endangered species, and urban sprawl: The political ecology of the new American West. *Annual Review of Anthropology, 36*, 121–138. doi:10.1146/ annurev.anthro.36.081406.094413.

Shumway, M. J., & Otterstrom, S. M. (2001). Spatial patterns of migration and income change in the mountain West: The dominance of service-based, amenity-rich counties. *The Professional Geographer, 53*(4), 492–502. doi:10.1111/0033-0124.00299.

Strauss, A. L., & Corbin, J. M. (1998). *Basics of qualitative research: Techniques and procedures for developing grounded theory*. Thousand Oaks, CA: Sage.

Taylor, L. (2011). No boundaries: Exurbia and the study of contemporary urban dispersion. *GeoJournal, 76*(4), 323–339. doi:10.1007/s10708-009-9300-y.

Travis, W. R. (2007). *New geographies of the American West: Land use and the changing patterns of place*. Washington, DC: Island Press.

Waage, S. A. (2001). (Re)claiming space and place through collaborative planning in rural Oregon. *Political Geography, 20*(7), 839–857. doi:10.1016/S0962-6298(01)00026-9.

Walker, P. A. (2003). Reconsidering 'regional' political ecologies: Toward a political ecology of the rural American West. *Progress in Human Geography, 27*(1), 7–24. doi:10.1191/0309132503ph410oa.

Walker, P. A., & Fortmann, L. (2003). Whose landscape? A political ecology of the 'exurban' Sierra. *Cultural Geographies, 10*(4), 469–491. doi:10.1191/1474474003eu285oa.

Walker, P. A., & Hurley, P. T. (2011). *Planning paradise: Politics and visioning of land use in Oregon* (Society, Environment, and Place). Tuscon, AZ: University of Arizona Press.

Winchell, D. G. (2013). Indigenous rural renewal in the inland Pacific Northwest: The emergence of American Indian tribes in regional economic development. *Journal of Rural and Community Development, 8*(3), 174–197.

Woods, M. (2005). *Rural geography: Processes, responses and experiences in rural restructuring*. London: Sage.

Yin, R. (2003). *Case study research: Design and methods* (3rd ed.). Thousand Oaks, CA: Sage.

Chapter 5
Competing or Compatible Capitalisms? Exurban Sprawl and High-Value Agriculture in Southwestern Oregon

Innisfree McKinnon

5.1 Introduction

> In 1973 the first land-use laws were enacted and my dad was furious. He was so unhappy about it. It was just that someone was going to be telling him what to do with his property. This property has been in the family since 1902 and they were the ones who were making decisions about it. He did have a brother who built a small subdivision on his property, so I don't know if my dad had visions of that. Because he liked doing that a lot better than farming and in his diary he talks about how he divided his farm up amongst his children and he went to Medford and started building commercial buildings. That was more lucrative and more what he liked to do. That was the main thing. Just taking away some rights they felt they had.—*Small fruit grower, discussing her father's attitudes towards the passage of Oregon's statewide land-use planning system. Interview, 2012*

The passage of Oregon's Senate Bill 100, which enacted statewide land-use planning in 1973, provides an intriguing context for the study of land-use change. This regulatory system strongly limits real estate development outside urban growth boundaries. Although one of the primary goals of the planning system was to limit the conversion of agricultural land to urban and residential uses, Oregon's farmers were divided in their attitudes toward it, right from the start. While some farmers in the Willamette Valley were strong proponents of planning to limit sprawl onto farmland, many others were concerned that increased government regulation would limit their ability to make decisions about their land. Farmers in more rural parts of the state, in particular, have expressed ongoing resistance and resentment toward the imposition of regulations viewed as enacted by urban outsiders. This case study reveals some of the roots of this discontent in southern Oregon (see Fig. 5.1).

I. McKinnon (✉)
Department of Social Sciences, University of Wisconsin, Stout, Menominee, WI 54751, USA
e-mail: mckinnoni@uwstout.edu

L.E. Taylor, P.T. Hurley (eds.), *A Comparative Political Ecology of Exurbia*,
DOI 10.1007/978-3-319-29462-9_5

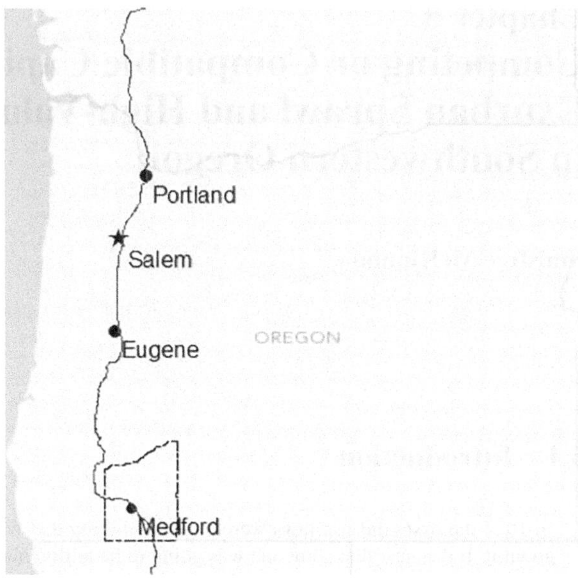

Fig. 5.1 Jackson County denoted by the dashed black outline. Produced by I. McKinnon data via Oregon Geospatial Gateway and ESRI

In the context of broader research agendas that examine socioeconomic, political, and ecological upheavals in communities that have traditionally been dependent on natural resources, this chapter addresses Robbins et al.'s (2009) question, "what is the relationship between extractive development and amenity economies and how smoothly can they be combined?" My study reveals a significantly different political and social response to the rise of the amenity economy than has been seen in other parts of the West (Walker and Fortmann 2003). This can be attributed to Oregon's strong regulatory regime, which directs conflict around land management and environmental values toward the state-mandated planning system and away from conflicts between locals. Additionally, the particularities of the historical development of local land-use patterns and the specific forms of industry involved have created a different set of relationships between extractive and amenity economies than seen in locales dominated by ranching or mining. It is these historical and socio-ecological particularities that I focus on in this chapter, providing a nuanced picture of the complex intertwining of farming and real estate development in Jackson County and uncovering how early growth of the agricultural sector in the region was fueled by real estate speculation, tourism, and the arrival of urban to rural migrants with access to external capital. In this case, rather than conflicting with extractive economies, amenity-based economies provided capital and labor for their growth. As I outline in this chapter the political economy and aesthetic characteristics of fruit growing, the dominant form of agriculture in the region proved to be compatible with rather than competing with the amenity values of the landscape.

Because of the early development of amenity-based industries in the region, in the form of tourism and scattered residential development in rural areas, the passage of statewide land-use regulation represented a major shift in the environmental management regime of the region and a disruption of the ability of rural landowners to engage in both extractive and amenity economies. In the context of contemporary political conflicts over land-use planning in Oregon, this case provides insight into the socioeconomic patterns of a part of the state that has been largely dismissed in policy circles as backward. Opposition to land-use regulation from rural parts of the state has largely been dismissed as ideological devotion to private property rights and greed by individual landowners. Yet the impact of Oregon's statewide land-use regulations on the lives and livelihoods of rural landowners has largely been ignored and the struggles of some farmers to make a living from the land dismissed. When family farms remain intact over the generations, they are celebrated in the popular press and through the awarding of official centennial or century farm designation, yet when a farmer sells their land, there is little acknowledgement of the economic pressures and regulatory limitations that factored into their decision.

Oregon's land-use regulations were designed to ensure the sustainable use of Oregon's resources for agricultural production and to protect specific environmental and socioeconomic values. But as this system of regulations has developed, rural voices often have been marginalized and dismissed (Walker and Hurley 2011). My intention in this chapter is to take seriously the local historical context within which complaints about contemporary planning developed. In doing so, I illustrate how agriculture and amenity development grew in tandem and how this in turn shaped local views of state land-use regulations. Political ecology maintains that it is impossible to fully understand ecological systems without also examining their interactions with social, political, and economic systems. Thus environmental sustainability can only truly develop hand in hand with social justice. The first goal of Oregon's statewide planning system is citizen involvement, yet as recent work by Walker and Hurley (2011) has shown, rural communities often feel that state regulators do not take their concerns seriously.

5.2 Literature Review

I situate this chapter within the broad interdisciplinary literature on the economic, political, and cultural transformations faced by the American West during the latter part of the twentieth century. While I acknowledge that there have been significant socio-ecological shifts in the American West in recent decades, I argue that if we look carefully into the history of the American West, it is possible to see, in some places, the existence of characteristics more typically associated with the New West and long before its widespread emergence as an increasingly dominant force on the Western landscape. Historical evidence from Jackson County aligns with the work of historians such as Limerick (1987) who have long pointed to the continuities of the West in contrast to characterizations that emphasize recent changes (see also

Taylor 2004; Hyde 1998). Robbins et al. (2009) also point out that the idea of a distinctly New West may not hold up under careful scrutiny.

In Jackson County amenity economies emerged slowly, alongside extractive industries, beginning in the first decades of the twentieth century. Exurban style real estate development and tourism were already present during this time, alongside the timber and agricultural land use that dominated economic growth in the region. Thus, growth in the fruit growing industry was linked to rapid growth in rural population and the residential housing industry. The passage of comprehensive land-use planning in the 1970s, which curtailed rural real estate development in the region, represented a distinct shift in the existing land-use management regime that had been relatively stable for a number of decades. Oregon's land-use planning system limited residential development outside of urban growth boundaries and directed population growth into urban centers. This represented a major shift for rural landowners in agricultural areas of the county who had historically mixed development of small farms and homesteads with commercial agriculture. The picturesque valley landscape, filled with flowering fruit trees, and the promise of a rural lifestyle had proved a powerful draw for urban residents throughout the twentieth century. Lack of a major urban center in the region meant that concern over loss of farmland to development really did not develop in the region until almost 20 years after the passage of the statewide land-use planning system (Rogue Valley Council of Governments 2000).

Much of the prior research on exurbanization and amenity migration focuses on the migrants and their characteristics as a group distinct from longtime rural residents, particularly in terms of differing cultural and environmental values (Hines 2010a, b; Smith and Krannich 2000). When I began my work in Jackson County in the winter of 2009, one of the first surprises for me was that contemporary conflicts over land-use planning in the region did not seem to fit well with this narrative of newcomers and old-timers. The commonly accepted narrative is that urban migrants bring with them new values resulting in acrimonious conflicts over land-use and environmental regulation (Haggerty and Travis 2006; Jones et al. 2003; Travis 2007). Whereas longtime residents view the landscape as a resource—a working environment—new migrants value the aesthetics of the landscape, the picturesque rivers, mountains, forests, and bucolic Old West towns.

Scholars embracing the newcomer thesis assume this cultural split has significant implications for natural resource management in the region, arguing that the management practices of individual landowners and new political alliances result in conflicts over regulation of the landscape. Recent research on land ownership has shown a significant shift in land ownership in many parts of the rural West in recent decades. Gosnell et al. (2006) show a major shift in the ownership of ranchlands in Montana between 1990 and 2001, with a large portion of new owners more interested in the amenity values of sensitive habitats over the utility of the lands for livestock grazing. This shift has been shown that the management choices of individual landowners have significant implications for the management of a variety of resources including elk (Haggerty and Travis 2006) and fisheries (Gosnell et al. 2007). Walker and Hurley (2004) examine a similar shift in ownership in the Sierra Nevada foothills that triggered new forest management plans and conflict over control of local government institutions. Writing on northeastern Oregon, Abrams et al. (2013) state that the wave of amenity-oriented migrants that began there in the 1990s has a more complex and reflexive relationship to long-term residents and productivist practices; Abrams et al. (2013) show that amenity migrants in northeastern Oregon attempt to balance their own desire to visually consume the landscape with concern for the preservation of "working lands."

Assuming that newcomers and longtime locals represent different cultures and socioeconomic positions is problematic, though. Cases, because of a focus on new arrivals, often assume that longtime locals are culturally and socioeconomically homogenous. My chapter, which outlines the historical development of Jackson County, aligns well with research such as Smith and Krannich's (2000) who question the idea of a "culture clash" in their work in rural communities in the Rocky Mountain West. Nelson (2002) suggests negative attitudes toward change have less to do with long-term resident versus newcomer status than with economic status. Low-income residents report higher levels of anxiety about changes, perhaps because of lack of economic resources and support services. And while it is easy to imagine that cultural differences lead to disagreements over land management, a cause and effect relationship cannot be assumed. It might equally be argued that economic conflicts and outcomes result in the adoption of conflicting cultural identities (Robbins et al. 2009). Indeed, the ideals of the Old West—beauty, freedom, wide-open spaces, and caring communities—that have attracted so many "amenity" migrants in recent decades may represent a fiction that resonates with longtime residents as well, making them nostalgic for a past that never existed (Hyde 1998; Limerick 1987).

To avoid the conceptual confusion created by relying on the assumed cultural and socioeconomic contrasts between newcomers and locals, this chapter draws on the competing rural capitalisms framework developed by Walker and Fortmann (2003). Rather than focusing on questions of clashing cultures or ideologies, this work examines the economics of the Old and New Wests and the ways economic ties to the landscape and efforts to extract value from these landscapes shape particular political engagements. Walker rightly pointed to the relationship between these two economies, rather than focusing his analysis on a clash of cultures: "The literature of the 'New West' that frames the resulting conflicts as clashes of cultures or ideologies misses the point that these conflicts reflect underlying tensions between competing capitalisms that commodify nature in incompatible ways" (Walker 2003). Yet a careful review of the various industries associated with the New West and the Old West reveals varying levels of incompatibility and complementarity. Robbins et al. (2009) take an open approach to understanding the relationships between various forms of rural capitalism asking, "what is the relationship between extractive development and amenity economies and how smoothly can they be combined?" Indeed, a broader examination of the relationships between resource-based industries and amenity economies in different times and places might reveal a complex dynamic in which compatibility is contingent on a number of human and environmental factors. While there is no doubt that the rural West has experienced significant economic and demographic change in the last 50 years, as rural restructuring has led to a decline of Old West industries in many locations, the emphasis on amenity economies as a new phenomenon in the region and resource economies as "traditional" obscures the complex and shifting relationships between these two forms of rural capitalism.

5.3 Methods/Study Area

5.3.1 Research Methodology

This chapter, which focuses on the history of the region, emerged within the context of my examination of contemporary land-use management during which I interviewed 52 key informants, reviewed thousands of pages of planning documents, and observed more than a dozen public hearings. Interviews with landowners and farmers, in particular, provided insights into the shifting political economies of agriculture in the region. I collected the historical data for this chapter largely from local historical archives where I accessed primary documents from the period such as local newspapers, promotional materials printed by local business leaders, early *Sunset* magazines, promotional and cadastral maps, and oral histories.

Because I am relying on archival evidence in order to understand development in Jackson County during the early part of the twentieth century, it is not possible to neatly divide new arrivals into those who migrated primarily to enjoy natural amenities and those who primarily valued "working landscapes" (Abrams and Bliss 2013). Rather, there is ample evidence that new arrivals to the region during this period shared many of the general characteristics that today are associated with amenity migration. Abrams et al. (2012, p. 270) define amenity migration as "the movement of largely affluent urban or suburban populations to rural areas for specific lifestyle amenities, such as natural scenery, proximity to outdoor recreation, cultural richness, or a sense of rurality." According to this definition then, there are several defining characteristics of this migrant group: they are, at least in general, affluent; they formerly resided in urban or suburban areas and they enjoy the visual aspects of the landscape along with recreational experiences it can provide. Because I cannot measure the attitudes and values of these historical migrants directly, my focus in this paper is on the role taken by boosters and real estate developers who were equally fevered in their attempts to sell land as capitalist investment and to promote an idealized rural lifestyle.

There is good reason to assume that migrants came to the valley both to make money by growing fruit and to enjoy the many amenities promoted to them by developers. In discussing the horticultural boom in California during this period, Vaught argues that orchardists in particular had a different attitude toward the relationship between the city and the countryside than other rural industrialists:

> Horticulture was a way of life and a business. . . . A specialty crop community, they firmly believed, was a virtuous place somewhere between the isolated and self-sufficient Jeffersonian rural order and the market-dominated, impersonal industrial city. It was a place where educated, land-owning families live on small, orderly, and prosperous orchards

> or vineyards in close proximity to one another. It thus fostered neighborliness, strong local social, cultural, and political institutions, and economic progress, all in an environment that was aesthetically pleasing as well. (Vaught 1999, p. 53)

Mechling (1999, p. 136) describes a similar early promotion strategy in Florida and Southern California, where developers and promoters marketed orange growing specifically to wealthy urban and suburban businessmen who would be interested in country clubs, tennis courts, and golf courses. In this study, a similar pattern emerges in Jackson County; as the growing of orchard fruits becomes the dominant industry in the region, urbanites who value particular environmental and cultural amenities come to dominate the valley both culturally and economically.

Gosnell and Abrams (2011) note in their review of the literature on amenity migration that while British scholars have studied urban to rural migration in the UK since the early nineteenth century, US scholars only began studying the phenomenon in the 1970s and focused on occurrences of this pattern only since the 1950s. In contrast, my study examines urban to rural migration and the role of "rural idyll" in the development of industrial agriculture in the early twentieth century. By looking back at the history of amenity land uses in the American West, we can better understand the complexity of the relationship between what have been conceptualized as two separate and competing economies—rural resource-based industries and service-based industries focused around urban to rural migration and tourism.

The term "exurbanite" is sometimes used interchangeably with amenity migrant. However, recent literature often centers on exurbia as a place or exurbanization as a process (Taylor 2011; Cadieux and Hurley 2011). Exurbia is also associated with rural gentrification, as in Spectorsky's (1955) original characterizations of exurbanites as wealthy urbanites who move to the country but retain their cultural, economic, and political connections to their urban roots. But exactly who is an exurbanite and where is exurbia located remain somewhat unclear. Exurban settlement is generally placed outside the outer suburban zone and is characterized by low-density settlement, sometimes defined as one household every 2–20 acres (Theobald 2001). This pattern has also sometimes been referred to as rural sprawl, in contrast to urban sprawl, because it is characterized by low-density housing often intermixed with "rural" land uses such as farming, ranching, and logging (Theobald 2003). However, the term "exurban" has been used to describe a wide range of conditions in which people with few economic ties to rural economies settle outside of cities. In some demographic schemas, counties are classified as exurban if they are within metropolitan areas but most of their population lives at rural densities (Berube et al. 2006). Yet efforts to use particular landscape metrics to define exurbia are only of limited utility since none of these metrics are able to capture the diversity of exurbanization processes. While amenity migrants are often associated with exurbanization, amenity migrants may live in areas officially categorized as urban, rural, or anywhere in-between.[1]

[1] Around Medford, the largest city in Jackson County, amenity migrants commonly live within the city limits, in small towns, or are intermixed into rural areas.

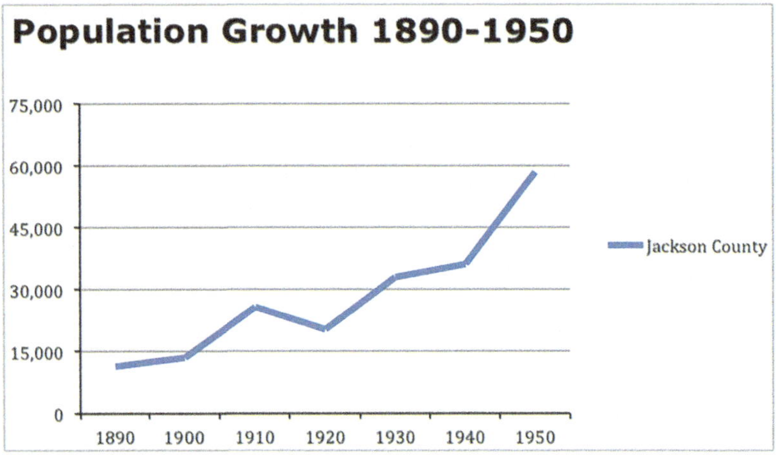

Fig. 5.2 Population growth in Jackson County. *Source*: US Census

5.3.2 Study Area and Contemporary Context

In many ways Jackson County, Oregon, resembles other so-called Old West communities. The county (see Fig. 5.2) is characterized by a predominately mountainous, forested landscape. Agriculture in the region is concentrated in several valleys of the tributaries of the Rogue River, the largest of which is the Rogue Valley. The topography of the county is the result of the complex interactions of mountain building and weathering by glaciers and rivers, with the Rogue Valley itself measuring about 10–15 miles across east to west and about 25 miles north to south. When the first Europeans settled in the valley, agriculture and forestry largely served the booming gold mining economy, but over time, these industries replaced mining as the primary economic drivers of the region. This rugged topography and the distinctive ecosystems it contain have long attracted the interest of those seeking to manage key resources, both for extraction and for tourism. Today, land in Jackson County is about 80 % forested: to the south lies the Klamath National Forest and Oregon Caves National Monument, to the east the Cascades and Crater Lake National Park, and to the west and north the Siskiyou Mountains and the Rogue River Siskiyou National Forest.

Unlike the vast Central Valley of California, or even the smaller Willamette Valley farther north in Oregon, only a small portion of the county consists of rich loam soils. Agriculture in the county is also limited by water availability, microclimate, and soil type. Much of the Rogue Valley consists of gentle slopes and mixes of heavy clay, beds of rock, and gravel. The complex topography of the region allows unpredictable spring frosts and hailstorms to damage crops in one field while those a mile or two down the road remain untouched. The dry Mediterranean climate means that access to irrigation water is a key-limiting factor in the development of agriculture.

Today the vast majority of the population of the county, over 200,000, lives within the agriculturally productive valley. The Interstate-5 corridor, the primary north-south transportation corridor between Oregon and California, also runs through this valley. Medford, the largest city, serves as the hub of the service economy of a vast rural region stretching over much of Northern California and Southern Oregon, primarily focused around retail and health care. These regional services, along with tourism, have largely eclipsed forestry and agriculture as economic drivers. A proportionally large percentage of the population consists of retirees, many of whom have moved to the region upon retiring. The Ashland Shakespeare Festival and the Britt Music Festival, along with a variety of outdoor recreational activities including hiking, skiing, rafting the Rogue River, and fly fishing, draw tens of thousands of tourists to the region each year.

Like many resource-dependent communities in the West, up until World War II, the population of Jackson County followed a boom-bust pattern tied to the exploitation of various resources, with rapid growth in some decades and little to no growth in others (see Fig. 5.3). Growth from 1900 to 1940 consisted of an even mixture of people in rural areas and town centers. During the war, an army-training base, Camp White, trained more than 40,000 soldiers. The infrastructure of Camp White supported a postwar boom in the timber industry and rapid population growth in the region. In the post-1945 era, urban population growth in the region began outstripping rural growth. In 1973, Oregon passed Senate Bill 100, which set up a system of statewide regulations designed to limit development on "high-value farm and forest land." It took 10 years, however, for Jackson County to create and put into place a comprehensive plan that met statewide goals, and the 1970s marked a peak in population growth rate in the valley. Starting in the 1980s urban growth boundaries and regulations restricting residential development in rural areas zoned for farm or forest uses limited population growth outside of established cities and towns in Jackson County.

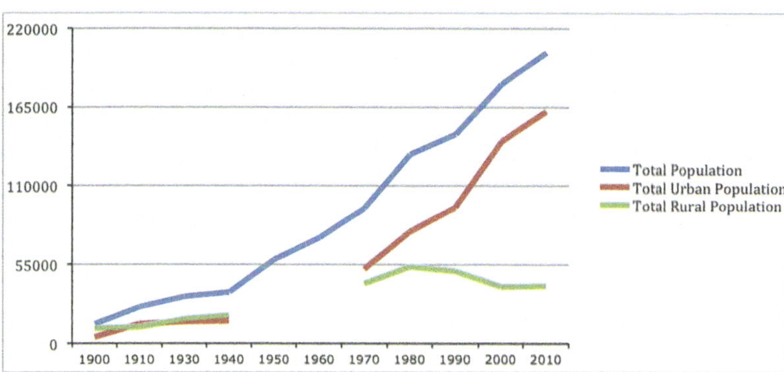

Fig. 5.3 Jackson County population growth since 1900. Note the decline in rural population after 1980 when the enforcement of comprehensive planning began due to limits on the creation of new housing in rural areas and annexation of rural zones near cities into urban growth boundaries. *Source*: US Census

Evidence that comprehensive planning disrupted an established environmental management regime can be seen in Fig. 5.3. Until the 1980s when the first comprehensive plan was implemented, rural and urban populations roughly remained in equal proportions as they grew. Since the 1980s rural population growth in the valley has been limited by state regulation, while towns and cities have continued their rapid growth. Today, the valley floor remains a mix of commercial agriculture and large areas of residential development outside of cities at both suburban and exurban densities, frustrating contemporary planning efforts that seek to enact a strong separation between rural and urban land uses.[2] These existing rural residences have remained as artifacts of earlier growth in the valley despite limitations on rural development over the past 35 years.

5.4 A Historical Political Ecology Case Study of Agriculture and Amenity Development

5.4.1 The Arrival of the Railroad and Creation of an Orchard Industry

A boom and bust economy has dominated Jackson County since Europeans first settled it in the 1850s. A gold rush began in the region only a few years after gold was discovered in California. What marked these early years was a willingness by settlers to make use of this new territory in whatever ways would net them a hefty income. Many arrived with the intention of making their fortunes mining, but quickly realized that a profit could also be made in supplying goods and services to the growing camps and towns. The first white farmers in the region claimed the valley bottomlands where soils were rich and there was access to stream water. Markets for agricultural products in the isolated valley were limited to supplying fresh fruits and vegetables to nearby mining camps and growing wheat, which could be shipped over long distances. Despite farming's limitations in the region—labor shortages, lack of irrigation, and the isolated location of the valley (far from urban markets)—the promise of fruit growing began to be realized beginning in the 1890s. The idyllic valley, with its picturesque landscape, would be well suited for attracting settlers, and the favorable climate and alluvial soils grew apples and pears that rivaled any on the continent. Yet without access to labor and a reliable way to get that fruit to market, most farmers chose wheat growing, which had the advantage of being relatively mechanized and transportable by wagon. The 1880 census reflects the relatively large average farm size (332 acres) required for one family to make a living growing wheat.

Thirty years after the first land claims were made in Jackson County, the arrival of the railroad in the 1880s set in motion a major social and ecological transformation

[2] Suburban densities are commonly single family homes on 1/4 to 1/2 acre, while exurban settlement is often one home per 1 acre to 20 acres.

in the county. A rail line connecting the Rogue Valley to Portland facilitated transport of both commodities and migrants into and out of the region. The recent invention of the refrigerated rail car allowed the transportation of perishable goods to major urban markets across the country, opening the region to new forms of investment and settlement.

The 1890s saw the completion of the Siskiyou rail line connecting Jackson County to Sacramento; the Siskiyou opened the valley to the Transcontinental Rail Line and markets in the eastern USA. The arrival of the rail line in Jackson County signaled the real opening of the region for development in two significant ways. First, the rail line brought thousands of visitors, many of whom eventually became new residents. Second, the train allowed farmers to ship perishable produce to large urban markets across the country. The first enterprising growers had small shipments of apples and pears ready to ship to California in early 1888, just a few weeks after the line between Oregon and California was complete (Results of Fruit Culture in Southern Oregon 1888). The success of these first fruit shipments coincided with a widespread depression in global wheat prices and led to rapid growth in orchards in the region. Growth in the fruit industry prompted massive land speculation and rapid parcelization of agricultural lands in the valley in the first decades of the twentieth century. Fruit growing prompted reduced farm sizes for several reasons. First, fresh fruits require relatively large amounts of skilled labor to cultivate, harvest, and ship. Second, because they are perishable and require these large amounts of labor, the value of the crop per acre farmed is high. Since each acre generated higher levels of revenue, farmers were able to earn enough from a small parcel to support a family. By 1940, average farm size was 112 acres, somewhat smaller than the most recent averages according to the 2007 Agriculture Census. This transformation of the landscape and the regional economy could not have taken place without a massive influx of outside capital and migrants into the region.

5.4.2 Capital and Labor for Fruit

The fact that other researchers (Vaught 1999; Mechling 1999) have found a pattern of lifestyle migration from urban centers in other fruit growing regions is no accident. The cultivation requirements of tree fruits are both capital and labor intensive. Orchard trees take 8–10 years to come into full harvest and cannot be grown from seed. In order to grow well, fruiting tops must be grafted onto vigorous rootstocks. Then the trees must be nurtured and pruned during the intervening years. For fruit to be marketed in major urban centers thousands of miles away, the delicate fruit must be carefully packaged and shipped. Fruit growing favored those who had sufficient capital to pay not only for land but also for seedling trees, irrigation equipment, and a packinghouse. The capital also had to be patient; fruit farmers had to have some other source of income that would support them during the decade during which they were waiting for their trees to produce.

Fruit farming also required much larger numbers of labor hours per acre than wheat production. Because the need for labor in farming is highly seasonal, in order to be successful a farmer needed labor to be available at the right times. As the fruit industry grew, so too did its need for labor, leading to an urgent need to attract additional workers to the region.

In order to develop a successful fruit growing district, boosters had to attract both capital and labor. In general, it was only urban migrants who had the capital to invest. The early decades of the twentieth century in the Rogue Valley are described by local historians as the "Orchard Boom." The period was characterized by rapid growth in both population and in the fruit industry. Cheap land and the prospect of a stable, long-term investment drew many easterners to invest in the Rogue Valley. The demographic characteristics of these easterners were in many ways similar to those of urban to rural migrants today. In some cases growers were wealthy investors from Portland, San Francisco, Seattle, or Chicago, looking for a summer home where they could also play at being a gentlemen farmer. Other new arrivals included older professionals looking for a rural lifestyle and young college graduates supported by income from family members in urban centers in the east. Wealthy investors provided capital to develop the industry, and the many new exurbanite families provided local labor pools sufficient to maintain the orchard-filled landscape.

5.4.3 Early Twentieth-Century Real Estate Speculation and Tourism

Land speculation was rampant in the early twentieth century. The relatively high returns per acre of fruit growing led to rapidly rising land prices, fueling speculative land grabs. The Southern Pacific railroad combined forces with local boosters and real estate developers to promote the region for both tourism and settlement. Prior to the 1930s, the industry relied almost exclusively on local labor; being located so far from large population centers, attracting new settlers was a key element in building the industry. As noted, the arrival of the railroad in the Rogue Valley created a market for fruit from the Rogue Valley, but it also greatly increased the number of visitors to the area, opened up the region to its first wave of tourism, and encouraged population growth in the region. The Southern Pacific railroad conducted widespread promotion of rail vacations and settlement in the West during this period. Thousands of people passed through the region by train every year while traveling between Portland and Sacramento. Many local leaders and businessmen were eager to promote the region for tourism (Oregon's Greatest 1911). The regions' many natural attractions, including nearby Crater Lake, abundant fishing and hunting opportunities, and volcanic hot springs, provided impetus for the early efforts to attract tourists to the region.

The establishment of Crater Lake National Park in 1902 first put the region on the tourist map. During the earliest years of the park, the road up to the lake was slow and rough, limiting the number of visitors.[3] The trip from Medford was 83 miles and took about 10 h by car in 1910 (Oregon's Greatest 1911). In 1912, the Park Service estimated 5770 tourists visited the park (Special Population Report 1913). But over the decade, as travel and facilities improved, visitation rose to more than 16,000 by 1919 (Unrau and Mark 1987). A few years after the establishment of Crater Lake National Park, the picturesque evergreen forests and mountain streams surrounding the valley were also made into national parks; Klamath National Forest to the south, the Rogue-Siskiyou National Forest to the West, and Umpqua National Forest to the North were all established between 1907 and 1908. Early magazine and newspaper articles frequently featured locals and visitors enjoying the fresh air, hiking, skiing, hunting, and fishing (see Fig. 5.4).

Many boosters also had their eyes on promoting the health-giving properties of the region's many mineral springs. Ashland, at the southern end of the valley, promoted a European-style health spa based around the health-giving properties of the sulfurous waters of Lithia Springs (Holt 1915). In 1903, a promotional piece in a Portland paper reported that Ashland hosted an "extra" 1000 visitors over the summer looking to "take the waters" (Ashland, Oregon 1903),[4] a significant number for a town whose population was only 2634 in 1900. The city government passed a bond for the development of the springs into a full resort. Although the plan was never realized because of political wrangling and mismanagement, the city did build a large park around the springs and began hosting summer theater productions.[5]

It wasn't until the 1960s that the early city leaders' vision of a tourism-based economy began to reach fruition. However, the process was one of slow growth in the tourist sector over many years rather than a sudden transformation to a tourist-centered economy in the post-WWII era. While revenue from tourism was certainly a small portion of the overall economy, visitors and promotional materials had a large impact on the growing farm sector. Many who happened to stop in Medford on their train trip later decided to buy land in the region. Others were directly attracted by reports of opportunities for profitable farming and a pastoral lifestyle in national newspapers and *Sunset* magazine. New arrivals included settlers with horticultural experience and expertise, while many others were urban businessmen, professionals, and recent college graduates with no agricultural background.

---

[3] Superintendent Arant estimated the number of visitors to Crater Lake in 1905 to be between 1200–1400 (Report of the Superintendent 1913).

[4] "Taking the waters" was a common phrase referring to the purported healing properties of bathing in mineral hot springs.

[5] The summer Chautauqua productions eventually led to the development of the now famous Oregon Shakespeare Festival that runs practically year round in Ashland and attracts approximately 90,000 visitors a year (Oregon Shakespeare 2012).

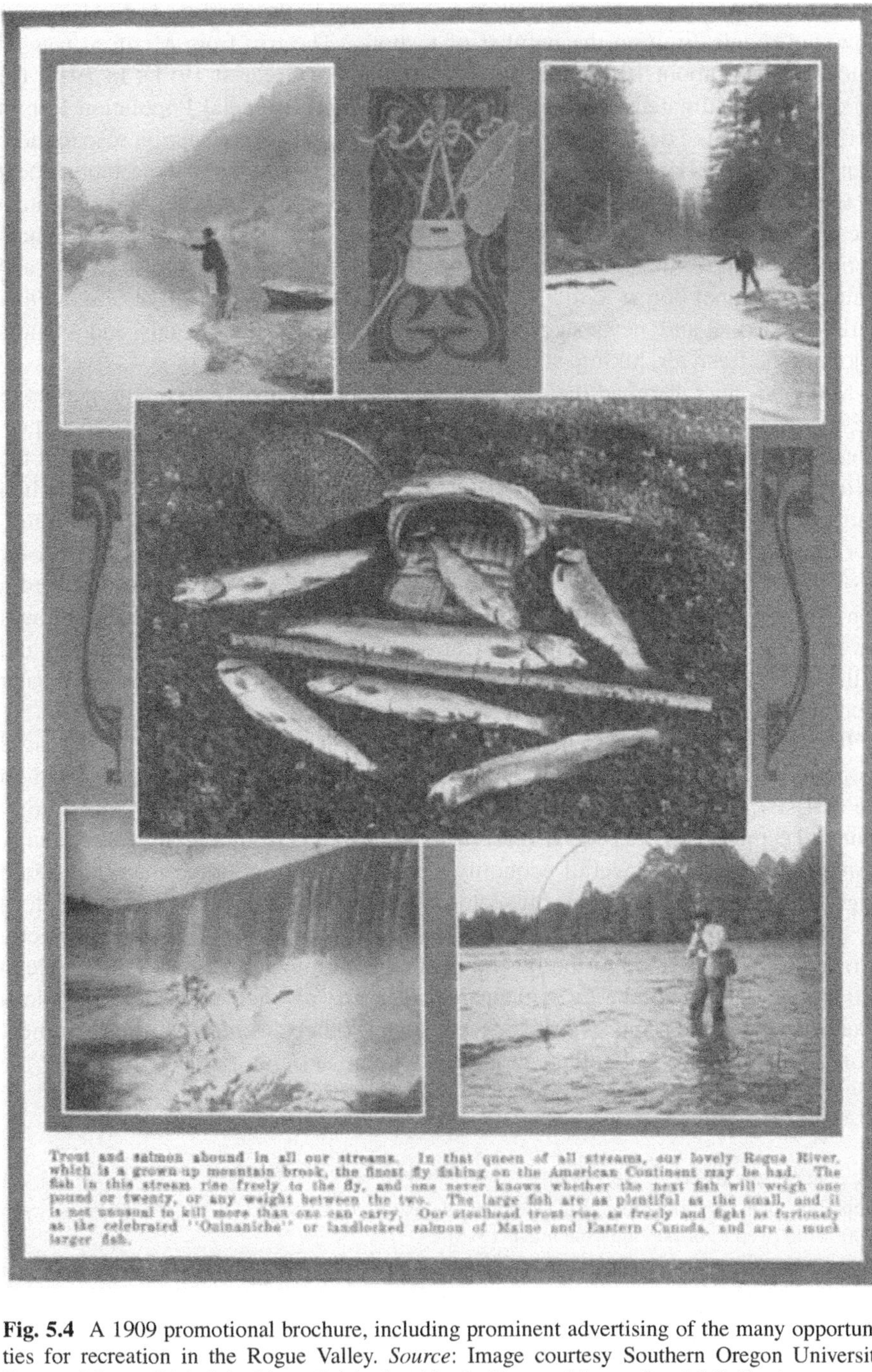

Fig. 5.4 A 1909 promotional brochure, including prominent advertising of the many opportunities for recreation in the Rogue Valley. *Source*: Image courtesy Southern Oregon University Hannon Library Digital Archives. Original resides in Hannon Library Special Collections

5.4.4 Selling the Dream of the Gentleman Farmer

Promotional materials and activities in the Rogue Valley were not limited to promoting outdoor recreation; abundant natural amenities were just one part of the gentile rural lifestyle offered to potential settlers by boosters and developers. The Medford Commercial Club, a local booster organization, built an exhibition at the Medford rail station showing the bounty of the valley, both promoting fruit from the Rogue Valley to tourists and promoting the region as a pastoral paradise for homebuyers (see Fig. 5.5). *Sunset* magazine, owned by Southern Pacific Railroad Company, also promoted the valley through multipage inserts. In these publications, the beauty of the valley, its pleasant climate, and many natural and cultural amenities were prominently featured along with the supposed ease of successful fruit growing. The promotion of the Rogue Valley produced a population boom as urban migrants flocked to the area. During the first decade of the twentieth century, the population of Medford grew from 1791 to 8840 (US Census). The pattern of rural subdivision seen today in the valley began with the rapid growth and real estate speculation stimulated by the Orchard Boom. By 1893, real estate speculation and the subdivision of large farms had begun in the Rogue Valley. The 214-acre Nickell farm was one of the first recorded examples of a farm subdivided into small parcels; it was subdivided into 1-acre tracts with 30 acres set aside for streets. These were sold to railroad employees and tourists enchanted by their travels through the valley. Such parcels were promoted to potential buyers as homes, as secure real estate investments, and as potential sources of agricultural income.

Land planted in mature orchards fetched some of the highest prices in all of Oregon at the height of the boom. In 1909, mature orchards were selling for $2300 an acre, while unplanted land sold for $150–250 an acre (Cordy 1977). The high prices being obtained in eastern markets for fruit from the region (Nuggets of Gold 1910) fueled a speculative bubble for land planted in orchards, but the high prices obtained for orchard lands also reflected both the capital investment required to bring the land to bearing and their potential for future earnings. Orchard development required not only an investment in planting but also labor to nurture trees to a bearing age. Once an orchard had been grown to maturity, it could be expected to bear fruit for 40 or more years.

As the boom continued, real estate speculation and subdivision became a much simpler way to make money than waiting for a fruit harvest. This real estate boom relied on the idea that a family could make a good living from a small acreage of fruit trees, and this vision of the small independent grower was widely promoted by real estate developers. It was the image of life as a gentleman farmer that proved key to growth, much more than the reality. Rural life as a productive small farmer was a key aspect of the entire lifestyle “package” being sold.

Once lands suitable for orchards became scarce, speculators had no compunctions about buying up rocky lands with lean soils in the northeast corner of the valley to sell as orchard homes though they were clearly not suitable and impossible to irrigate. In some places real estate companies used dynamite to create holes large

Fig. 5.5 A promotional brochure showing the many excellent orchard homes in the valley. *Source*: Image courtesy Southern Oregon University Hannon Library Digital Archives. Original resides in Hannon Library Special Collections

enough to plant trees in the cement-like soil. One of the most well-known developers was John Westerlund, a Chicago real estate dealer who in 1903 founded the Western Oregon Orchard Company and began speculation on Rogue Valley orchard lands. While the company name implied a connection to orchards, Westerlund was selling the idea of the independent gentleman farmer. Newspapers from the period were filled with advertising from companies such as Westerlund's offering small orchard parcels for sale, often sight unseen. In some cases buyers were encouraged to send money to the company and in exchange their orchard would be cared for until it matured (Here's Further 1910) at which point the buyer could presumably move out to Oregon, build their house on the land, and live comfortably off the income the orchard provided. In some cases, it was not until the buyers started showing up in the valley that it became clear that large portions of such orchard subdivisions had not been planted and were unsuitable for fruit production.

5.4.5 Gentlemen Farmers: Small and Large

In 1893 Frederick Jackson Turner gave his famous speech on the closing of the American frontier. In some ways, that speech marked the line between living in the "Old West" and romanticizing it (Limerick 1995). By 1910, when the boom in orchard subdivisions was reaching its height, an increasingly urbanized America was already full of nostalgia for farm living. The orchard boom attracted many of both wealth and modest means with little to no experience with agriculture, but full of the dream of the gentleman farmer.

One of the most prominent and influential new arrivals in the Rogue Valley was the Palmer family, the millionaire owners of the famed Palmer House hotel in Chicago (Potter Palmers' 1913). Mrs. Palmer and her sons promoted the area to other wealthy Chicagoans, and in 1911 the Chicago Record Herald reported on a "millionaire colony" in Medford. Other millionaire investors came from Seattle. Reginald Parsons, prominent financier and philanthropist, bought the HillCrest Orchard in 1908 as a summer home and investment after hearing about the profitability of fruit growing in the Rogue Valley from business associates in Seattle. Wealthy eastern families sent their sons to the valley to make their fortunes as horticulturalists. Some of these "remittance men" were Ivy League graduates who received monthly income from their wealthy families on the East Coast (Many College 1910).

5.4.6 Fruit Growing and Parcelization

> The census of 1920 ought to show a population of 150,000 for Jackson County. Its area and resources will amply support many times this number of people. The entire Rogue River valley should be an immense, continuous orchard, with a family upon every ten acres. Thousands of men should be employed manufacturing lumber, quarrying granite and marble, manufacturing lime and cement, and thousands more in mining.—*Editorial article advocating growth, Medford Mail Tribune, December 7, 1910, p. 4*

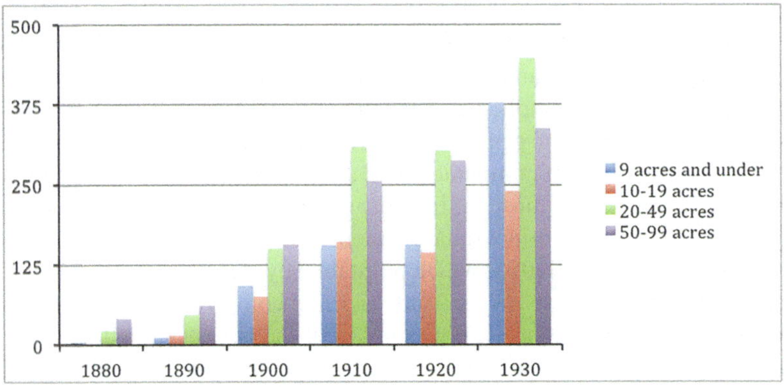

Fig 5.6 Jackson County farm sizes over time. *Source*: US Census

The full impact of subdivision on this region in this early period is difficult to estimate. However, examining the agricultural census can make some sense of its impact. According to the US Census between 1880 and 1930, the population of Jackson County grew from 8154 to 32,918. During this period, there was massive growth in the number of small- and medium-sized farms in the region, with the most rapid growth happening in farms less than 50 acres in size. In 1880, there were 30 farms of less than 50 acres. By 1930, there were 1066 farms of less than 50 acres, including 618 farms under 20 acres (see Fig. 5.6). Obviously, farm size is not a direct indication of an individual landowner's orientation toward land management, but the growth in small and very small farms in the region is significant for two reasons. First, industry experts estimated that even during the early twentieth century, a farmer would have needed at least 40 acres to earn enough to support one household[6]; regardless of intention, then, it is likely that most of these small farms relied on additional sources of income to supplement farm earnings. Second, parcelization and the mixing of farming and "hobby farm" populations are major concerns in relation to exurban development today, and both were clearly present in the valley by 1930. Recall that a parcel size between 1 and 20 acres per residence is one common measurement of an exurban settlement pattern (Theobald 2003).

5.4.7 *Benefits of "Hobby Farmers" for the Industry*

The arrival of these aspiring farmers had several positive impacts for the fruit growing industry. First, they provided an influx of capital investment that poured into planting and nurturing orchards to maturity. Most growers, both wealthy and those of modest means, over time discovered that making a profit growing fruit was

[6] While farm size is often used to estimate income and indicate farm type, earnings per acre vary widely depending on the crop grown.

more complicated than the boosters and promotional materials made it out to be. Successful growers needed horticultural skills, knowledge of marketing, business savvy, and enough capital or outside income to get them through years when drought or late frosts ruined crops. While the most well-managed orchards on the best land could make a significant profit, particularly in good years, many orchardists struggled. The bankruptcy of less-savvy growers provided opportunities for successful growers to buy land with mature trees from the bank or creditors for much less than it would have cost to develop a new parcel.[7]

Second, the many small growers in the region also provided a ready source of labor to large orchards for harvest and processing. Some estimated that a grower would have needed an orchard of at least 40 acres in order to make a living solely as a farmer. Since a large number had parcels less than that, they and their families instead patched together a living from a mix of industries, providing labor on large farms, in fruit-packinghouses, in mining, and in lumber. Women in particular were thought to make the best packers, while young people were commonly enlisted to help with the harvest, often involving the entire family in the fruit growing industry.

Finally, while small growers provided labor for the industry, wealthy large growers with summer homes on their orchards provided the capital for the development of the infrastructure needed by the growing industry. These industrialists had the capital to expend on packinghouses and rail cars. Their most important contribution, however, was investment in the building of the many irrigation canals that eventually covered the valley. Irrigation proved key to the continued success of fruit growing in the region. Because the Rogue Valley only received on average 20 in. of rain per year, irrigation in drought years was essential. Large investments were required to build these canals, which then returned profits through subdivision of newly irrigated lands and the selling of water and land to growers (Rogue River 1910).[8]

5.5 Discussion

5.5.1 Remittance Men, Hobby Farmers, and Other Urban Outsiders

> The district will not be the home of the workman who spends his days in smoke-begrimed and dirty factory buildings, living in unhealthy tenements with hundreds of his fellows, and barely earning just enough to live, but the home of prosperous and happy men and women, who work in the sunshine, live in modern bungalows, making their living from the orchards which they own.—*Rogue River Valley Canal Co. Advertisement, Medford Tribune January 4, 1911*

[7] The Democratic Times of Jacksonville, July 14, 1898, reported one of these sheriff sales, in which the 214 acres of orchard land belonging to the Orchard Home Association was sold to the Portland Trust Co. for $5500.

[8] The first irrigation project in Jackson County, the Fish Lake Water Company, was privately financed. In 1910, Pat Welch, a wealthy contractor from Spokane, Washington, bought the company and began promoting and selling irrigated small parcels to aspiring fruit growers.

During the years between 1870 and 1920, the number of people in the USA who lived in cities grew from 10 million to 54 million. The Progressive Movement that was actively lobbying to improve the health and safety of America's cities and suburbanization was well underway (Hayden 2004). The dream of a healthier, more wholesome life available in the West had already become firmly established in the minds of Americans (Limerick et al. 2009). Many new arrivals to the Rogue Valley were enchanted by the marketing promises on the pages of *Sunset Magazine* and reports of an easy country life among the apple and pear trees.

Yet, during this period, there is no evidence that the preponderance of what might now be called hobby farmers or exurbanites had a significant negative impact on the growth of the fruit growing industry in the valley. While there were undoubtedly tensions among growers, these seem to have been based largely on class differences and the stresses of an industry in which a mix of cooperation and competition, or "co-opetition" (Larsen and Hutton 2011), was required. Fruit growers were in competition with each other but were often forced to cooperate in order to successfully process and market their products. Additionally, poor orchard care by one farmer may result in pests and diseases to spread to neighboring orchards, so growers made education a key priority in the industry. Real estate development and speculation represented a competitive pressure on land while at the same time providing labor and capital for the expanding industry.

5.5.2 Farmers as Land Owners

The speculative buying and selling of land has been an integral part of the economy of southern Oregon since Euro-American settlement began in the 1850s. In this context, the passage of statewide land-use regulation in 1973 represented a disruption of the existing environmental management regime of the region under which farming, mining, and forestry had coexisted with tourism and real estate development for many years.

Oregon's statewide regulations strongly restrict parcelization and the building of new residences in farming zones. This has caused vocal resistance in the region, particularly among fruit growers,[9] the very group the laws were, in principle, designed to protect. This paradox has often been dismissed as greed by supporters of statewide regulation or an illogical belief in private property rights. However, this attitude ignores the dual role that farmers hold as landowners and a misunderstanding of the economic realities of farming. For this farmer (and many like them), the land itself represents an investment with the potential for appreciation:

> The appreciation of the land was part of the reason for being in agriculture—because if agriculture didn't make any money, at least the land values would be there. So you could sell and retire. Now you can sell and go starve to death, live on Social Security. It's terrible now. We've been able—because of our marketing and culture—we've been able to survive. It certainly hasn't been easy. (Interview with independent fruit grower 2011)

[9]Fruit growers are by no means united in opposition to statewide planning. Many have concerns about specific elements of the regulations that they believe limit their ability to be economically successful, but express strong support for farmland conservation. Some believe current planning regulations are not protective enough.

By limiting development on agricultural lands, statewide planning holds down agricultural land values, which potentially makes it easier for new farmers to buy land; however, this does not, in itself, solve the problem of farm succession. Even farmers like the one quoted above, who has a son who wants to continue farming, expressed the difficulty of financing retirement or increasing farm income to accommodate the transition between generations without either selling part of the farm for development or building additional residences on the farm, both of which are limited by Oregon's land-use regulations. Growers relied on the ability to sell less productive land for development in order to finance capital improvements, replanting of aging orchards, the creation of a comfortable retirement fund, or to facilitate the transfer of the business to a new generation.

5.5.3 How New Is the New West?

The story of Rogue Valley over the last 50 years could be understood as similar to so many other rural communities in the American West: a story of transition from an economy based on extractive industries to one focused around service, retail, and health-care industries. This story of the transition between the Old West economy and the New West has been told numerous times (Hines 2010a, b; Ghose 2004; Travis 1997, 2007). The small town of Jacksonville, now thoroughly tourist in its orientation, started out as a gold rush town, and its economy languished for much of the twentieth century. Now, dilapidated nineteenth-century architecture has been revisioned for the twenty-first-century amenity economy. The nearby town of Ashland is also thoroughly reliant on a New West consumptive economy, largely built around a theater festival that runs through most of the year, supplemented by outdoor recreation including a small ski resort. But Ashland's attempts to build an economy around tourism and recreation began not in the 1960s or 1970s but in the early twentieth century with dreams of a health spa, the establishment of Lithia Park, and summer Chautauqua theater productions. Other towns have begun taking advantage of recent growth in tourism and exurban migration while contrarily continuing to hold up the mythology of their "Old West" roots.

In the Rogue Valley, both new arrivals and longtime residents are invested in nostalgia for the Old West. The literature on the New West has been focused on a specific type of place with a distinct history, in which the development of amenity-based economies has been relatively recent and sudden, because such places have the propensity to produce conflict. In contrast, the history of Jackson County provides a picture of a different historical trajectory in which rural real estate and tourism developed slowly alongside a form of agriculture more compatible with their development than ranching or mining. This points to the potential variety of relationships between New West and Old West economies, based on the specific characteristics of both. Hines (2010a, b) describes the creation of a "New West Archipelago," islands of postindustrial space in an industrial sea. However, it would be instructive to examine the diversity of relationships between extractive industries and consumptive ones rather than focusing only on the distinctive characteristics of such islands. Otherwise we risk assuming that these islands are surrounded by an undifferentiated sea of static, unchanging industrial landscape.

5.6 Conclusion

In this chapter I have emphasized the way that the Rogue valley and surrounding region were marketed for their many amenities, in addition to their potential to produce financial gain. To understand the drivers of exurban sprawl and social and environmental impacts that have increased amenity-based real estate development worldwide, the researcher must move beyond an emphasis only on "amenity migrants" as individual actors, to understand real estate development and associated activities as an industry with connections to and conflicts with other rural industries.

Discussions with individual landowners drew my attention to the flexibility of livelihood strategies and the complexity of motivations for settling in the Jackson County. Continued parcelization and rural real estate development are ultimately not a sustainable way for farmers to raise the capital needed to maintain or adapt their production since eventually each farm family would be left with too small a parcel to farm. At the same time, sustainability and social justice must go hand in hand. Farmers in Jackson County often express that the current regulatory regime limits their ability to adapt their farming operations to changing market conditions by curtailing subdivision, strictly regulating the construction of any new homes or buildings, and restricting the development on nonfarm businesses. If Oregon's land-use planning system is to be sustainable, it must consider both environmental and social justice concerns. In Jackson County and other rural parts of the state, that means taking the economic concerns of farmers and other rural residents seriously and finding new ways to help farmers adapt to changing ecological and market conditions.

This historical case also opens up a number of questions, which have yet to be thoroughly examined in the literature on the political ecology of exurbia. What has been the relationship between amenity economies and resource-based production in different times and places? Under what conditions do these industries produce conflict? Thus far, little work has focused on a sustained examination of the role of tourism and the real estate industry in the development of resource-based industries prior to the most recent waves of exurban development. For example, little has been written so far about the history of Dude Ranching (Borne 1983). How large a role did this traditional amenity-based industry play in rural areas and how was it related to production-oriented ranching?

Studies of contemporary political ecologies need to take into account the role of natural amenities in the broader historical context of settler colonialism. By looking into scholarship on landscape painting (Ferber 2009) and literature (Marx 2000), we can see the importance of the pastoral ideal to the settlement of the West. Euro-American settlement of the Western USA was driven by capitalist industries selling a dream of a better life to migrants from the east. The beauty of the Western landscape and the opportunity for health, recreation, and enjoyment of that landscape were key elements in that process (Krueckeberg and Ward 1998). As Limerick (1987) points out, taking land, dividing it up, and selling it for a profit were one of the first and most profitable capitalist activities that Euro-Americans undertook when they arrived. I am not suggesting that there have been no significant changes between the nineteenth and the twenty-first century in terms of aesthetic valuation and real estate development of land in the West. In the nineteenth and early twentieth centuries, only a relatively few had the luxury of enjoying the Western landscape purely for its aesthetic qualities. Technological and economic developments during the twentieth century allowed many more Americans to enjoy and value the recreational and aesthetic values of the West than ever before (Rome 2001; Sutter 2004), without needing to extract their incomes directly from the land.

What is striking about this Rogue Valley case study is the specific combination of soil and climate conditions and geography that facilitated the joint growth of fruit cultivation and real estate development. Successful commercial orchard growing is an industry reliant on particular microclimate conditions to be successful, and orchards are certainly not a widespread land use in the Western USA. Jackson County was situated on a major transportation corridor for many years, which meant that while it was located far from any major population centers, it received an ongoing stream of visitors, facilitating the growth of tourism and real estate development. An interesting comparison could be made to other Western regions dominated by orchard landscapes, for example, Orange County and Santa Clara County in California, Hood River County in Oregon, or Yakima County in Washington in order to better understand the interactions between orchards, tourism, and parcelization.

Shifting the focus from analyzing exurbanites as consumers to the production of amenity landscapes through the real estate industry puts exurban development in a different light. Yet few studies to date have focused on the ways that exurban development is driven by particular capitalist industries[10]: tourism, real estate, health care, and other service industries. Viewing amenity-based industries as potentially one of the "traditional" industries associated with the American West since Euro-American settlement also moves us from discourses of newcomers versus locals or Old West versus New West economies to an acknowledgement of the complex and shifting economic relationships that continue to drive land-use change in the Western USA.

References

Abrams, J., & Bliss, J. C. (2013). Amenity landownership, land use change, and the re-creation of 'working landscapes'. *Society & Natural Resources, 26*(7), 845–859. doi:10.1080/08941920.2012.719587.

Abrams, J., Bliss, J., & Gosnell, H. (2013). Reflexive gentrification of working lands in the American west: contesting the 'middle landscape'. *Journal of Rural and Community Development, 8*(3), 144–158.

[10] One exception is Robbins et al. (2012).

Abrams, J., Gill, N., Gosnell, H., & Klepeis, P. (2012). Re-creating the rural, reconstructing nature: An international literature review of the environmental implications of amenity migration. *Conservation and Society, 10*(3), 270. doi:10.4103/0972-4923.101837.

Ashland, Oregon has very many attractions as a summer resort. (1903, April 4). *The Morning Oregonian*. Retrieved from http://oregonnews.uoregon.edu/.

Berube, A., Katz, B., & Lang, R. E. (2006). *Redefining urban and suburban America: Evidence from census 2000, volume three (redefining urban and suburban America)*. Washington, DC: Brookings Institution Press.

Borne, L. (1983). *Dude ranching: A complete history*. Albuquerque, NM: University of New Mexico Press.

Cadieux, K. V., & Hurley, P. (2011). Amenity migration, exurbia, and emerging rural landscapes: Global natural amenity as place and as process. *GeoJournal, 76*(4), 297–302. doi:10.1007/s10708-009-9335-0.

Cordy, C. B. (1977). *History of the Rogue Valley fruit industry*. Central Point, OR: n.p.

Theobald, D. (2003). *Defining and mapping rural sprawl: Examples from the northwest U.S.* Fort Collins, CO: Growth Management Leadership Alliance white paper.

Ferber, L. S. (2009). *The Hudson River school: Nature and the American vision*. New York: Skira Rizzoli.

Ghose, R. (2004). Big sky or big sprawl? Rural gentrification and the changing cultural landscape of Missoula, Montana. *Urban Geography, 25*(6), 528–549. doi:10.2747/0272-3638.25.6.528.

Gosnell, H., & Abrams, J. (2011). Amenity migration: Diverse conceptualization of drivers, socioeconomic dimensions, and emerging challenges. *GeoJournal, 76*(4), 303–322. doi:10.1007/s10708-009-9295-4.

Gosnell, H., Haggerty, J. H., & Byorth, P. (2007). Ranch ownership change and new approaches to water resource management in Southwestern Montana: Implications for fisheries. *Journal of the American Water Resources Association, 43*(4), 990–1003. doi:10.1111/j1752-1688.2007.00081.x.

Gosnell, H., Haggerty, J. H., & Travis, W. R. (2006). Ranchland ownership change in the Greater Yellowstone Ecosystem, 1990–2001: Implications for conservation. *Society and Natural Resources, 19*(8), 743–758. doi:10.1080/08941920600801181.

Haggerty, J. H., & Travis, W. R. (2006). Out of administrative control: Absentee owners, resident elk and the shifting nature of wildlife management in southwestern Montana. *Geoforum, 37*(5), 816–830. doi:10.1016/j.geoforum.2005.12.004.

Hayden, D. (2004). *Building suburbia: Green fields and urban growth, 1820–2000*. New York: Vintage.

Here's Further Evidence for You. (1910, March 6). *Medford Mail Tribune*. Retrieved from http://oregonnews.uoregon.edu/.

Hines, J. D. (2010a). In pursuit of experience: The postindustrial gentrification of the rural American West. *Ethnography, 11*(2), 285–308. doi:10.1177/1466138110361846.

Hines, J. D. (2010b). Rural gentrification as permanent tourism: The creation of the New West Archipelago as postindustrial cultural space. *Environment and Planning D: Society and Space, 28*, 509–525. doi:10.1068/d3309.

Holt, E. (1915, February). Building a city on health. *Sunset Magazine, 34*(2), 356–358.

Hyde, A. F. (1998). Nothing New Under the Sun: Continuities in the West. *Pacific Historical Review, 67*(3), 393–399.

Jones, R. E., Fly, J. M., Talley, J., & Cordell, H. K. (2003). Green Migration into Rural America: The New Frontier of Environmentalism? *Society and Natural Resources: An International Journal, 16*(3), 221. doi:10.1080/08941920309159.

Krueckeberg, J., & Ward, S. (1998). *Selling places: The marketing and promotion of towns and cities 1850–2000*. London: Routledge.

Larsen, S., & Hutton, C. (2011). Community discourse and the emerging amenity landscapes of the rural American West. *GeoJournal, 77*(5), 651–665. doi:10.1007/s10708-011-9410-1.

Limerick, P. (1987). *The legacy of conquest : The unbroken past of the American West*. New York: Norton.

Limerick, P. (1995). Turnerians all: The dream of a helpful history in an intelligible world. *The American Historical Review, 100*(3), 697–716. doi:10.2307/2168601.

Limerick, P. N., Cowell, A., & Collinge, S. K. (2009). *Remedies for a new West: Healing landscapes, histories, and cultures*. Tucson, AZ: University of Arizona Press.
Many College Men Buying Homes in Rogue River Valley. (1910, September 13). *Medford Mail Tribune*. Retrieved from http://oregonnews.uoregon.edu/.
Marx, L. (2000). *The machine in the garden: Technology and the pastoral ideal in America* (35th ed.). Oxford: Oxford University Press.
Mechling, J. (1999). Oranges. In D. S. Wilson & A. K. Gillespie (Eds.), *Rooted in America: Foodlore of popular fruits and vegetables* (pp. 120–141). Knoxville, TN: University of Tennessee Press.
Nelson, P. B. (2002). Perceptions of restructuring in the rural West: Insights from the "cultural turn". *Society and Natural Resources, 15*, 903–921. doi:10.1080/08941920290107648.
Nuggets of Gold on Fruit Trees in Rogue River Valley. (1910, January 2). *Medford Mail Tribune*.
Oregon Shakespeare Festival. (2012). *Oregon Shakespeare Festival: State and local economic impact*. Retrieved from http://oregonnews.uoregon.edu/.
Oregon's Greatest Natural Wonder, Drawing Card for Tourists from Over the World. (1911, August 6). *Medford Tribune*. Retrieved from http://oregonnews.uoregon.edu/.
Potter Palmers' Planting Largest Pear Orchards in World. (1913, January 1). *Medford Mail Tribune*. Retrieved from http://oregonnews.uoregon.edu/.
Crater National Forest Special Population Report. (1913). Rogue River National Forest Historical Records Collection: A-6.
Report of the Superintendent of the Crater Lake National Park to the Secretary of the Interior, United States. (1913). *Superintendent of Crater Lake National Park*. Washington, DC: G.P.O.
Results of Fruit Culture in Southern Oregon. (1888, January 5). *The Oregon Sentinel*. Retrieved from http://oregonnews.uoregon.edu/.
Robbins, P., Martin, S., & Gilbertz, S. (2012). Developing the commons: The contradictions of growth in exurban Montana. *The Professional Geographer, 64*(3), 317–331. doi:10.1080/00330124.2011.601193.
Robbins, P., Meehan, K., Gosnell, H., & Gilbertz, S. J. (2009). Writing the new West: A critical review. *Rural Sociology, 74*, 356–382. doi:10.1526/003601109789037240.
Rogue Valley Council of Governments. (2000). *Our region*. Medford, Oregon.
Rome, A. W. (2001). *The bulldozer in the countryside: Suburban sprawl and the rise of American environmentalism. Studies in environment and history*. Cambridge, England: Cambridge University Press.
Smith, M. D., & Krannich, R. S. (2000). 'Culture clash' revisited: Newcomer and longer-term residents' attitudes toward land use, development, and environmental issues in rural communities in the Rocky Mountain West. *Rural Sociology, 65*(3), 396–421. doi:10.1111/j.1549-0831.2000.tb00036.x.
Spectorsky, A. C. (1955). *The exurbanites*. Berkeley, CA: Berkeley Publishing.
Sutter, P. S. (2004). *Driven wild: How the fight against automobiles launched the modern wilderness movement*. Seattle, WA: University of Washington Press.
Taylor, J. E. (2004). The many lives of the new West. *The Western Historical Quarterly, 35*(2), 141–165. doi:10.2307/25442968.
Taylor, L. (2011). No boundaries: Exurbia and the study of contemporary urban dispersion. *GeoJournal, 76*(4), 323–339. doi:10.1007/s10708-009-9300-y.
Theobald, D. M. (2001). Land-use dynamics beyond the American urban fringe. *Geographical Review, 91*(3), 544–564. doi:10.2307/3594740.
Travis, W. R. (1997). *Atlas of the new West: Portrait of a changing region*. New York: W.W. Norton.
Travis, W. R. (2007). *New geographies of the American West: Land use and the changing patterns of place*. Washington, DC: Island Press.
Unrau, H. D., & Mark, S. (1987). *Crater Lake National Park: Administrative history*. Denver, CO: Crater Lake Institute.
Vaught, D. (1999). *Cultivating California: Growers, specialty crops, and labor, 1875–1920*. Baltimore, MD: Johns Hopkins University Press.
Walker, P. (2003). Reconsidering "regional" political ecologies: Toward a political ecology of the rural American West. *Progress in Human Geography, 27*(1), 7–24. doi:10.1191/0309132503ph410oa.

Walker, P. A., & Fortmann, L. (2003). Whose landscape? A political ecology of the "exurban" Sierra. *Cultural Geographies, 10*(4), 469–491. doi:10.1191/1474474003eu285oa.

Walker, P. A., & Hurley, P. T. (2004). Collaboration derailed: The politics of "community-based" resource management in Nevada County. *Society and Natural Resources*.

Walker, P. A., & Hurley, P. T. (2011). *Planning paradise: politics and visioning of land use in Oregon*. Tuscon: University of Arizona Press.

Chapter 6
"In the Real Estate Business Whether We Admit It or Not": Timber and Exurban Development in Central Oregon

Brent Olson

6.1 Introduction

Exurban landscapes emerge from transition. Often this transition is most visible as a shift from productive to leisurely uses, but the transition is also built on a strategic intersection of cultural change, government engagement, and industrial reinvention. The 1967 USDA Yearbook of Agriculture slightly shifted its focus away from sustainable timber yields and farmland economics toward the growing industry of outdoor recreation. The cover of the handbook prominently featured the title "Outdoors USA" and a Kodachrome image of a group of men in cowboy hats (one with a guitar) gathered around a campfire with a tent in the background. The handbook highlighted the relationship between outdoor recreation and existing agricultural resources and offered tools to help farmers and ranchers capitalize on the recreational opportunities of the land they owned. Throughout the report, the USDA spoke to "farmers and rural developers interested in profit making recreational enterprises" (United States Department of Agriculture 1967, p. 3). Orville Freeman, the secretary of agriculture, linked amenity development and its potential for profit to a new kind of rural America, one that would be "synonymous with good living." A key part of that good life was the ability to "see the mountaintop from your backyard patio" (USDA 1967, p. iv). The view was only one of the amenities at the heart of amenity development. Developers looked for sites to build communities with easy access to green space, golf courses, tennis courts, and a wide array of recreational opportunities. At the same time, the USDA was developing recreational opportunities on Forest Service lands; they began encouraging and supporting the efforts of private landowners to capitalize on the recreation boom and the natural landscapes beyond the suburbs.

B. Olson (✉)
Environmental Studies, Westminster College,
1840 South 1300 East, Salt Lake City, UT 84105, USA
e-mail: bolson@westminstercollege.edu

L.E. Taylor, P.T. Hurley (eds.), *A Comparative Political Ecology of Exurbia*,
DOI 10.1007/978-3-319-29462-9_6

The private production of those amenity landscapes would occur at a profoundly local level and depend upon complicated existing relationships between landowners, local and state governments, and particular landscape histories. Today, almost 50 years later, Deschutes County is described by Oregon's Chamber of Commerce as "The Recreation Capital of Oregon." But in 1967, Deschutes County still relied heavily on timber production. Views of the mountains from backyard patios were a secondary perk to those residents living in the area, many of whom worked in natural resource-related industries. The history and political ecology of amenity development in Deschutes County is the history of the production of a new kind of space; one still grounded in the natural resources of the region but built on emerging meanings of nature and landscape and the potential profits those meanings might contain.

In this sense, Deschutes County's transition from an economy based on timber production to one centered on amenity development and outdoor recreation is not unique but part of a larger process of transformation that has often been discussed under the banner of a transition to the "New West" (Travis and Robb 1997; Travis 2007; Hines 2010). Recent debates, however, have questioned just how far those differences extend (Taylor 2004; Robbins et al. 2009; Bryson and Wyckoff 2010). If there is a distinction between the New West and any number of "Old Wests," it may lie largely in the degree to which the aesthetic conditions of the landscape have been remade and capitalized upon as natural resources through processes of abstraction, quantification, and commodification (Olson 2010; Bridge 2013). Paradoxically, then, the population and economic booms of the 1990s associated with the New West are tightly bound to the modernist, scientific, and capital intensive developments undergirding the booms, busts, exploitation, and environmental degradation often associated with resource use in the "Old West." In Jay Taylor's terms, the New West has begun to look "a helluva lot like the Old West" (Taylor 2004, p. 145). Further, the roots of the clear demographic and economic changes in the 1990s existed decades before as an integral part of the longer resource history of the region. While the results of the economic shift in the west away from timber and mineral extraction may resemble the Old West, the new migrants to the region have brought with them specific sensibilities in regard to both urban landscapes and those landscapes beyond cities' boundaries. Considering this cultural shift, historians Liza Nicholas, Elaine Bapis, and Thomas Harvey argue that "The 'New West' is less about the loss of feed stores and the proliferation of espresso shops, as the *Atlas of the New West* (Travis and Robb 1997) would have it, and more about the circulation of certain kind of knowledges, an emergent taken-for-granted-ness of 'the way things ought to be' particularly with regard to western space" (Nicholas et al. 2003). Those knowledges and ingrained values percolate through debates over the appropriate uses of public lands, the common goods and private property, resource management, and suburban and exurban development.

The political ecology of amenity development in the New West revolves around the production of this space, the contestation of its meanings, and the institutions and processes that support its production (Hurley and Walker 2004; Post 2013; Abrams 2016, this volume). These transformations are the result of decidedly cultural processes built on broad historical trends and with implications for relationship between the country and the city and the production of space that contains them both (see also Walker and Hurley 2011). The economic development of the region continues to depend upon its resource hinterland but with a vastly different economy valuing nature for its amenity rather than extractive value—in this case recreational landscapes produced as spaces of cultural consumption and identity. "Consequently, the political struggle over the construction of the landscape shapes the emerging economic and cultural values for private property, assets and resource access relative to an amenity rich public space" (Larsen and Hutton 2012, p. 652). Put another way, the aesthetic encounter with nature in the New West revolves around complex interactions between public and private space and the conflicts they contain. In many cases, emerging values of the landscape merely camouflage the resource histories and similarities between the New West and the Old West, between golf courses and timberlands, between ATV trails and logging roads, between campgrounds and timber camps, and between ski resorts and mining communities. These resource histories and the political ecology of amenity development are deeply implicated in the histories of the firms that managed them and strengthened the links between the city and amenity-rich countryside.

Resource and amenity rural capitalisms are more similar than portrayed in the literature describing the violent social transition from Old to New West. In each case, discrete pieces of nature were abstracted from their particular contexts and brought into a larger market to serve the particular needs of a changing culture. Michael Redclift (2006) writes that tourist spaces are "full of ambiguities. First, there is the ambiguity of abandoned places that open the door, as it were, to new discovery, settlement, and occupation" (p. 188). These discoveries erase some histories just as they illuminate others. The timber history of Deschutes County, however, is only partially erased by the new resort developments, breweries, and golf courses. It was written over by the same agencies, firms, and processes that first produced it. I argue in this chapter that recreational development of the New West in general, and Central Oregon in particular, is built upon this strategic reproduction of resource space on the urban fringe. In the process of transforming existing resource landscapes and natural amenities into housing developments and icons of particular lifestyles, the narratives surrounding those landscapes and the cultural values underpinning them were remade to support new exurban development. Whether in terms of brownfield redevelopment in Spokane or the transformation of timberlands to suburbs outside of Seattle, the conversion of industrial landscapes to residential ones requires a reimagination of the place and a reconfigured understanding of the value of it (Bryson 2010; Klingle 2007). In reimagining these landscapes, corporations reforge connections between the city and the countryside in continued development of the region's amenities. The current reorganization of space in Central Oregon and the knowledge and capital that produce it are following the well-worn trails of the timber production that dominated Deschutes County's development during the first half of the twentieth century.

In this chapter, I consider the contrasting fates of two companies in their transition from Old West to New West capitalisms. Brooks Scanlon, Oregon's largest timber company, is an example of successfully negotiating the reterritorialization process, moving from timber to recreational use: they led the way in remaking local politics resulting in profitable revaluation of a considerable portion of their properties. Boise Cascade, on the other hand, failed to effectively navigate the process of reterritorialization, seeing their property value largely eliminated through environmental designation and their profits lost due to impatient capital. These companies are actors in the story of the broader shift in rural capitalisms being felt across the West and elsewhere in the rural USA but also shapers of the landscapes of value being produced as evidence of a supposedly "New West."

6.2 "Outdoor Recreation Is a Salable Product"

The 1960s saw a considerable downturn in the extraction and forestry industries that had bolstered Oregon's economy for much of the twentieth century (Robbins 2004). In order to help homeowners discover the recreational potential of the rural lands, the Oregon Extension Service began in 1967 to distribute pamphlets to landowners throughout the state. The series of publications, generally titled "PROFIT" (Planning Recreation Opportunities for Income and Tourism), aimed to provide those people and firms with the tools to capitalize on the economic potentials of outdoor recreation and rural tourism on lands that were no longer providing an income through agricultural or resource production. In a report for PROFIT, Wilder (1970a) writes:

With the continuing high demand for outdoor recreation opportunities by an increasingly urbanized and affluent population along with the changing pattern of land-use and public policy, we find many landowners throughout the state of Oregon considering carefully the development of private recreation enterprises. It is the special hope of the author that this publication will encourage potential operators to explore the various aspects of private enterprise operation in detail (Wilder 1970a, p. 2).

PROFIT provided tools and assistance for landowners to inventory their property, the potential market, and ways their property might be improved to increase its profitability as a tourist resource. One survey, meant to be an "idea generator" (Wilder 1970b, p. 2), asked landowners to consider the hunting, fishing, and resort potential of their lands. "Recreation is a salable product," the report concluded. "Thousands of people are making a good income from operating or working in recreation enterprises" (Wilder 1970b). Those that started small could imagine getting bigger. Finally outdoor recreation and rural tourism might be their primary source of income. Other reports provided a bibliography for interested landowners and a lease template for those interested in leasing their property. Throughout Oregon and, indeed, the American West, open rural lands presented a new opportunity for capitalist development as new technologies, and a shift away from primary production allowed greater mobility among those who could afford second homes. The viewsheds and recreational opportunities of the rural countryside offered new meanings for both visitors and landowners. New meanings meant a changed relationship to the landscape. For many, that changed relationship meant potential profits.

PROFIT was primarily geared toward small agricultural landowners who might be able to use a portion of their property to make some money from urbanites seeking to explore the great outdoors. Large landowners in Oregon, however, also began to wonder if they could take advantage of the shift in thinking about their landscapes and the vacationers will to spend money on Western recreational opportunities. By 1968, Michael Hollern and the other executives at Brooks Scanlon, Central Oregon's largest timber company, began to consider the recreational potential of the 201,335 acres they owned. While they were the biggest timber operator in Central Oregon, Brooks Scanlon never ventured far beyond their operational base in Bend, resulting in very close ties between the city and the corporation. They had owned the same 201,335 acres for 30 years, neither buying nor selling any acreage. The amount of saleable timber on these lands had declined significantly as the company had harvested far more than sustainable harvesting practices would dictate. Hollern decided that the company needed to begin buying, selling, and leasing property. "It was pretty clear that we were in the real estate business whether we admitted it or not" (Hollern 2008). Suddenly Brooks Scanlon's greatest asset wasn't the trees that grew on their lands but the land itself. Brooks Scanlon's real estate activities would move beyond simply buying and selling property because they assumed the role of developer of the land they owned. In order to understand just which lands might be the most valuable, the timber company contracted a group of consultants to complete their own recreational resource inventory. Brooks Scanlon, with their vast holdings of ponderosa forest both near Bend and throughout Central Oregon, hired a consulting firm, Fanning and Fenton, from Seattle to survey the land and provide an inventory of recreational development possibilities.

The report found considerable lands that would be amenable to recreational development, particularly those "very substantial" Brooks Scanlon acreages that met five basic criteria: (1) near a primary highway, (2) near a river or other recreational facility, (3) improved roads to the highway, (4) fairly level topography, and (5) some availability of water. The consulting planners who authored the report noted that these conditions didn't "indicate a very sophisticated or highly selective market" (Fanning and Fenton 1969, p. 23), though that would change as population pressures increased. Fortunately for Brooks Scanlon in this endeavor, the characteristics of ponderosa forests and the practice of ponderosa logging allowed them to consider previously logged lands as potential recreational developments. Unlike timber operators on the west side of the Cascades, who could clear-cut high-value Douglas fir, ponderosa forests were harvested through a selective logging approach, because younger trees needed the shade of more mature trees in order to grow. While the resulting stands no longer contained the majestic older trees, the multiple aged stands maintained an aesthetic value that was lost through logging operations in other parts of the state. The patchwork landscape was categorized into different areas to be targeted for different kinds of development. They predicted an increase in population in the area, and that outdoor recreation and tourism would contribute substantially to Deschutes County's economy in the coming years. The Black Butte

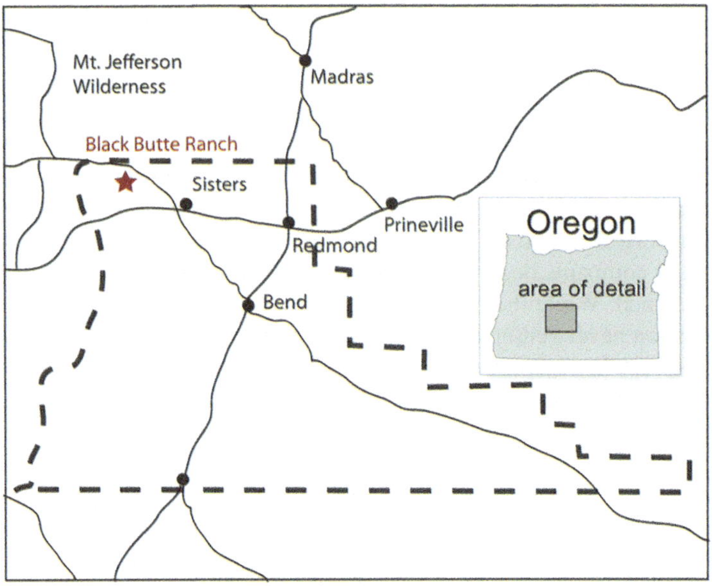

Fig. 6.1 Black Butte Ranch and Deschutes County, Oregon, Produced by B. Olson

Ranch land was the parcel that most clearly met these criteria, and the report urged prompt development of the land (Fanning and Fenton 1969, see Fig. 2 therein). The site with the most development potential was an old ranch in the northwest corner of the county, under the shadow the iconic Black Butte. The upscale development of Black Butte Ranch would be the first project the company undertook (see Figs. 6.1 and 6.2).

Following the report of Fanning and Fenton, Hollern hired Bill Smith and Jeff Carl, two business school interns from Stanford, to ascertain the best way to go about developing Black Butte Ranch. It quickly became apparent that planning and managing a recreational development, and the opening of more recreational developments of Brooks Scanlon lands, would require a different corporate infrastructure than what was required to run a timber company (Carl and Smith 1969; Smith 1969). In Hollern's understated words, "the real estate business is different" (Hollern 2008). In order to more carefully address concerns specific to the development business, Brooks Scanlon spun off "Brooks Resources" and granted the new company control of the nonproductive timberlands identified as having the most development potential. Brooks Resources would be responsible for developing and managing the company's new recreational and real estate holdings. Free from the demands of timber management and lumber production, the new company was charged to find ways to capitalize on the recreational opportunities of the region and to develop a plan for recreational development (Black Butte Ranch Corporation 1970). The company sought to build new resort developments with close relationships to the towns

Fig. 6.2 Black Butte and Mt. Jefferson. Before the golf courses and homes, Black Butte Ranch was a meadow and timber lands with views of its namesake butte. *Source*: Bakowski (1910). Reproduced with permission from the Gerald W. Williams, Oregon State University Libraries Special Collections & Archives Research Center

of Central Oregon and, perhaps most importantly, the dramatic landscapes and recreational opportunities on the National Forests and BLM lands in the region (Harrison 1969a).

Brooks Resources was hardly a trailblazer in transforming timberlands to resorts. Boise Cascade, another large timber company, had attempted to reinvent itself in a similar manner, investing in its timber lands to produce resort communities in Southern Oregon and Northern California. On a tour of Boise Cascade's resorts in the early 1970s, the director of Boise Cascade's development program carefully counseled Hollern to "Never consider real estate with prime beauty, the Sierra Club will kill it" (Hollern 1971). The Sierra Club had emerged by this time as the dominant environmental voice on the West Coast, and their individual chapters, supported by the national office, discovered that the threat of litigation could effectively delay or stop development on undeveloped natural landscapes. "Environmentalists argued that they [the public] had environmental rights as well as the right to property ownership and freedom from physical harm that could be protected by legal action" (Hays 1987, p. 484). The Sierra Club viewed Boise Cascade's development plans in California as an infringement upon environmental rights. The planned sites constituted ecological and aesthetic commons, viewsheds, and natural landscapes to be set aside for the common good. The Sierra Club launched a series of lawsuits against Boise Cascade arguing that their developments failed to comply with the *National Environmental Policy Act*.

The Sierra Club and other organizations had discovered that litigation was a powerful tool for preserving "pristine" landscapes, slowing the process of development and making the whole process considerably more expensive. Often, these lawsuits revolved around the necessity for, or inadequacy of, the Environmental Impact Statements (EIS) mandated by the *National Environmental Policy Act.* Sierra Club attorney James Moorman commented in retrospect: "had Congress understood what the EISs would be and what the lawsuits over EISs would do, I doubt if they would ever have passed NEPA" (Moorman 1994, p. 57).

The Sierra Club's proactive use of NEPA to protect lands from development, logging, or other degradations demonstrates their capacity to produce the reconfigured resource spaces of the New West. NEPA, while focused on federal projects, did have an effect on private projects as well. Specifically, if projects had potential impacts on federally regulated air or water, if they involved federal loan guarantees, or if it requires passage across federal lands. These clauses within NEPA framework afforded advocates like the Sierra Club opportunities to sue developers of private land even as they slowed federal projects. NEPA became a tool through which environmental advocates enacted a set of values that sought to protect a nature free from obvious human interference, a perspective on conservation that broadened the appropriate uses of natural resources to include aesthetic and ecological concerns. For timber, or real estate companies, the resource space of the timberlands continued to provide opportunities for investment and profit, but the ways in which profit could be made was expanding into recreational home development. This was evidenced by second home development in many parts of the West. At the same time, the field of dispute opened new opportunities for environmental organizations to limit the behaviors of firms even on their private lands. Even before the spotted owl controversies in Oregon's woods (Proctor and Pincetl 1996), the federal government had passed significant environmental and land-use legislation that would have a major effect on the workings of Brooks Scanlon. At the national scale, NEPA provided environmental organizations a tool to limit the behavior of firms as they sought to capitalize on their undeveloped lands (Hays 1987; Rothman 2000; Rothman and Nash 1998).

From a corporate perspective, the New Western political economy nearly ruined Boise Cascade, which had been so successful under the old regime. Where Brooks Scanlon successfully transitioned through an internal restructuring to manage investment and profit taking, mismanagement haunted Boise Cascade throughout the late 1960s, but "it was recreational real estate that put Boise on the ropes" ("Cinderella" 1972, p. 73). After the Boise Cascade fiasco, Hollern was justifiably careful about Brooks Scanlon's own recreational endeavor, yet the two scenarios are less synonymous than they would initially seem (Hollern 1971). First, while environmentalists held up some of Boise's projects, they were hit hardest by their own accounting practices. Boise Cascade had already entered into large amounts of debt to finance their recreational project and the delay made recovering that debt nearly impossible. A series of court decisions regarding the accounting practices and the hard sell pitches of their real estate agents would further limit their recreational investments ("Cinderella" 1972, p. 74).

Unlike Boise Cascade, Brooks Scanlon and Brooks Resources faced little opposition from environmental groups in their plan to develop Black Butte Ranch. Finally and perhaps most importantly, Brooks Scanlon was a relatively small, extremely local company. Brooks Scanlon could trace their corporate lineage directly to the timber companies that had first established the town in 1905, and they had established a long history in the area as a civic benefactor. Their reputation in the town—the social capital they had built with local civic and political leaders—helped them to make the case to environmentalists and others that their resort developments would have fewer environmental impacts than their logging activities and would provide a sustained economic base for the region in the midst of a timber downturn (Hollern 2008).

Further, Brooks' development approach would seek to learn from the successes of other developments in the West. The original plan for Black Butte Ranch consisted of a set of condos grouped around a central club and restaurant. The plans included a championship golf course and would feature an open meadow that had been a distinctive part of the land when it had been a working ranch, as well as dramatic views of the Three Sisters and the iconic butte from which the community took its name. While Hollern was responsible for examining Boise Cascades' recreational developments in California, Bob Harrison, the director of the newly formed Brooks Resources, took his own tour of recreational developments in Colorado. He returned with a commitment to ensuring that the condos and homes of Black Butte Ranch would be placed close to local recreational amenities, including the golf course and horse riding areas. He was particularly impressed with the "highly sophisticated job" the developers of Snowmass in Aspen had done in planning the central area of the community (Harrison 1969b). William Janss, the developer of Snowmass, sought to integrate all aspects of the resort business. Rothman (1998) puts it this way:

The Janss developments created a primacy for the one developer resort, the ordered, structured location that functioned as a result of predetermined precepts and that followed the patterns of planned communities. This highly structured and controlled environment, company owned or tightly zoned elsewhere, served as a precursor of later housing developments across the West.(Rothman 1998, p. 237)

Brooks Resources intended for Black Butte Ranch to fit within that successful mold.

Brooks Resources' marketing materials promoted the meadow, the golf course, the views, and the chance to be in close contact with nature both on the ranch itself and throughout the region. The hiking, camping, and fishing opportunities of the Deschutes National Forest and, more commercially, the continued expansion and growth of Mount Bachelor, the region's largest ski area (Hollern 1968a), provided the prime recreational amenities. Hollern himself had long recognized that the future of Black Butte Ranch would be linked to the recreational and cultural resources throughout the county. The chief among these cultural resources was the town centers of Sisters and Bend.

The small town of Sisters, just east of Black Butte Ranch, was in the midst of significant economic hardship due to the timber decline of the late 1960s. As part of an effort to construct an identity for the region as a whole, the company offered up to $5000 and architectural help to merchants to redesign their storefronts to create an Old West image for the town. Most businesses in town took the company's offer. And so, just outside the newly built, stylish, and modern resort community of Black Butte Ranch, the smaller timber and ranching community of Sisters refashioned itself as an Old West town with clapboard buildings, wide streets, and fake hitching posts (Deschutes County Community Development Department 2009; Nave 2002, p. 1). The play on an imagined cowboy past was built into the landscape in support of the wealthy enclave just up the road, even as it provided a distinct new identity for the town of Sisters.

By 1975, units were selling at Black Butte Ranch so quickly that salesmen had to be cautioned against overemphasizing the popularity of the place. Reporting back from a homeowners meeting, Brooks Resources' employee Bill Smith noted: "I get the feeling that those in attendance at the meeting would be quite happy if we cease selling and building immediately. Can we avoid the subject? Rather than brag about it?" (Smith 1975). They never stopped building or selling but continued to expand Black Butte Ranch and to open other resort developments throughout the county. Their former logging lands and the development opportunities these lands presented offered a chance for the company to recapitalize on the landscape. But critically the value of that land also depended upon access to the National Forest and the city of Bend, dramatic vistas of the Central Cascades, and recreational opportunities. Together, these amenities indeed proved to be a salable resource.

Hollern's plan for the future of Black Butte Ranch and Brooks Resources involved a dramatic reworking of the landscape of the entire county. That transformation would involve more than simply the construction of condos, golf courses, and ranchettes. Unlike the other big resort communities of the West, including Snowmass and Sun Valley, Black Butte Ranch lacked a ski resort in its immediate back yard. Mt. Bachelor, the nearest ski resort, was fresh off a wave of modest expansion and was a full 40 min away. Hollern approached Bill Healy, the founder and owner of Mt. Bachelor, in 1972 about buying the resort. Hollern was disappointed with the ski area's modest pace of expansion and saddened by Healy's decision to eliminate the on-site lodging at the base of the mountain. Healy, however, spurned Hollern's offer to buy a controlling interest in Mt. Bachelor. Healy remained committed to controlling Mt. Bachelor himself with a comparatively slower approach to implementing changes in order to maintain its local feel (Hollern 1968b; see Fig. 6.3). Hollern proceeded to look elsewhere.

Inspired by Harrison's reconnaissance trip to Colorado, the managers of Brooks Resources looked for ways to recreate the relationship places like Vail or Aspen had to their recreational hinterlands. In their plan, the city of Bend would act as a hub for the ski resort, golf and condo communities, and hiking, camping, and fishing opportunities of the Deschutes National Forest (Smith 1972). Hollern and later Bill Smith focused their energy on the transformation of Bend itself from a provincial timber town to a small cosmopolitan hub, which would attract an endless stream of tourists and second home owners from across the country. Bend, as Deschutes

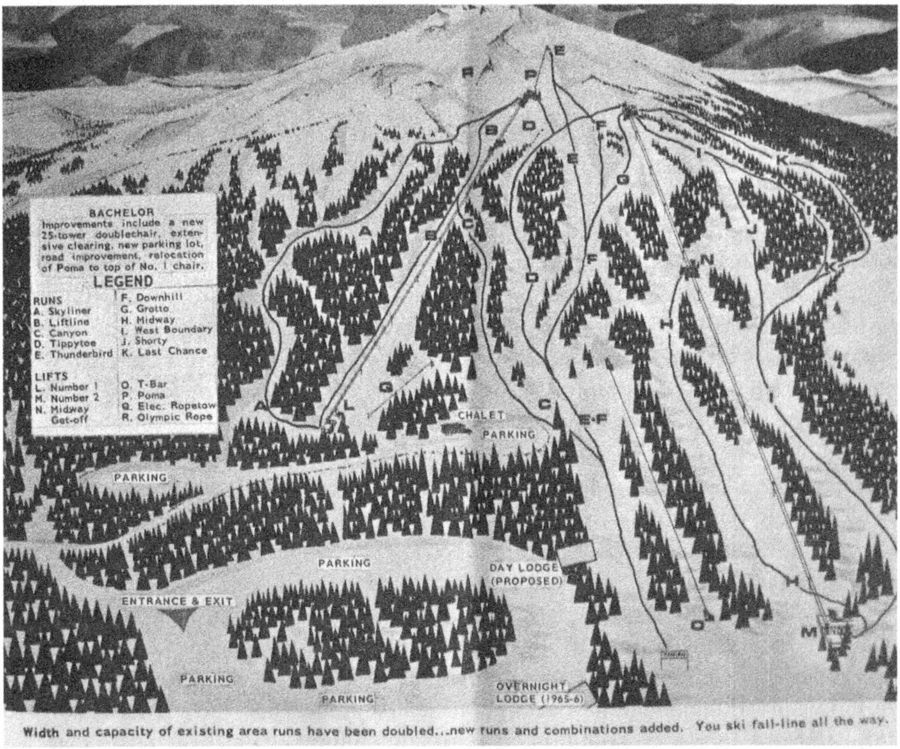

Fig. 6.3 Map of Mt. Bachelor Expansion, 1965. Healy remained committed to slowly and responsibly expanding recreational opportunities on the mountain in ways that would limit its isolation from the town. The overnight lodging on the base remained open from 1966 to 1971. *Source*: Central Oregon Vacationer (1965). Reproduced with permission from Mt. Bachelor, LLC

County's largest city with a population of 13,700 at the time (Bend's current population is over 80,000), captured Black Butte Ranch within its orbit. While Sisters offered charm and character under the guise of the Old West, Bend would provide the hub for the upscale shopping, dining, and entertainment that national visitors and exurban residents desired. Bend residents, though proud of the small town feel of the city and the intimacy they felt with fellow residents, recognized the economic imperative to develop the cultural institutions that would bring people from Portland, Seattle, and San Francisco to invest in Brooks Resources' exurbs (Kleinsasser 1969, p. 44). In a fit of "self interested altruism" (Hollern 2008), Hollern funded the "Bend Foundation" to promote the arts within Bend, and he encouraged the board members of both Brooks Scanlon and Brooks Resources to involve themselves in the downtown redevelopment project the city embarked on in the 1970s (Hollern 1973). Both companies and individuals from the corporate boards played significant roles in the project, which reimagined the downtown core along the lines that Harrison found exciting about Colorado's ski communities.

6.3 Conclusion: Natural Resource Production and the "New West"

The early success of Black Butte Ranch, the changes in Sisters, and the redevelopment of the central business district in Bend together demonstrate the links between the country and the city in the region's transformation from a timber shed to "the Recreation Capital of Oregon." The early days of Deschutes County's progression toward an economy centered on amenities and outdoor recreation may be read simply as a story of boosterism at a time of economic hardship and of big capital demonstrating flexibility in continuing accumulation by promoting the region's natural characteristics and investing in resources that would capitalize on them. Yet, the story can also tell us much more than that, in particular as we consider the political ecology of amenity development in the locally specific roles that individual and corporate actors play as they respond to broad cultural changes. Brooks Resources' success at Black Butte Ranch was contingent on the kinds of landscapes they controlled, their particular historical (social, political, natural) relationships to the region, the natural viewsheds, and the relatively remote and unused recreational amenities they afforded.

Many discussions about the growth of the exurban West following World War II focus on incursions of the city into the country, including unchecked sprawl and accompanying ecological and social devastation. Important discussions about loss of farmlands, fragmentation of ecosystems and communities, the cost of expansive infrastructure, increased pollution, and difficult cultural changes are represented in these arguments (Daniels 1999, p. xii; see also Ghose 2004; Robbins et al. 2009; Walker and Hurley 2011; Robbins 2004). In this chapter, the history of resort development, and the relationship of that development to Bend, points to a different analysis. This analysis does not necessarily look past those environmental impacts but examines the impacts of the city and the country alongside corporate and governmental motivations and entanglements. Hollern himself describes his altruism as "self-interested" in that it helped produce an image of Deschutes County that would draw people to it. The institutions he supported through the charity work of Brooks Resources would provide the cultural landscape to go with the natural landscapes his resorts depended upon. The recreational capital developments are wedded to the changing meanings and value of rural landscapes in the West. Those changing meanings provided new opportunities for existing firms to *once again* transform rural amenities into natural resources. As Brooks Resources placed greater emphasis on building homes and golf courses than on harvesting timber, they nonetheless continued their work of capitalizing on the region's resources, refashioning space and the sense of place in the region through both a naturalization of the capitalist landscape and the production of a new kind of resource space.

That resource space not only represents a reorganization and development of private property but also a reevaluation of the commons represented by the National Forests, the viewsheds, and many of the other natural amenities associated with recreational resorts. Karen Bakker (2004) has described water as an "uncooperative commodity" that makes governing its use, conservation, and capitalization difficult:

Water is a flow resource over which it is difficult to establish private property rights; is characterized by a high degree of public health and environmental externalities—the costs of which are difficult to calculate and reflect in water prices; and is a partially non-substitutable resource essential for life with important aesthetic, symbolic, spiritual, and ecological functions which render some form of collective, public oversight inevitable.(Bakker 2007, p. 441)

Recreation resources, like water, pose challenges for private development. They have important aesthetic, symbolic, and ecological functions. Indeed, much of their value is in those cultural implications. Unlike water, however, they are not flow resources. They are fixed geographically in specific landscapes though they are impacted by the movement of nature (e.g., deer or wildfire) through the resource landscapes. Capitalizing upon them requires bringing people to them, developing means of access, and creating value within that access. For Brooks Scanlon and later Brooks Resources, this meant converting former timberlands into a new resort development. The economic success of Black Butte Ranch depended not only on the construction of homes or new golf courses but on the character of the countryside beyond the development's borders and the built landscape of the town.

Amenity development in Deschutes County is built upon a resource that is at once symbolic and material. If "recreation is a salable product" (Wilder 1970b), then that product is derived from the natural resources upon which that recreation depends. Those resources are geographically grounded, produced from meadows transformed into golf courses, and made up of trees, lakes, rivers, snow, sun, and views of the mountains. They are, in short, simultaneously cultural and natural landscapes, profoundly symbolic, material, and historical. The story of Black Butte Ranch is one in which a timber company opened new regional markets through the transformation of the meaning and value of one of its key assets, land. But it is also the story of how the transformation of that asset extends far beyond the boundaries of their property geographically, socially, politically, and ecologically.

References

Abrams, J. (2016). Rural residential development and its discontents: A political ecology of sprawl containment in Wallowa County, Oregon, USA. In L. E. Taylor & P. T. Hurley (Eds.), *A comparative political ecology of exurbia: Planning, environmental management, and landscape change*. New York: Springer.

Bakker, K. (2004). *An uncooperative commodity: Privatizing water in England and Wales*. New York: Oxford University Press.

Bakowski, B. (1910) Black Butte and Mt. Jefferson. Oregon State University Libraries, the Gerard Williams Collection. Eugene, OR. Reproduced with permission from the Gerald W. Williams, Oregon State University Libraries Special Collections & Archives Research Center.

Black Butte Ranch Corporation. (1970). *Special meeting of the board of directors of Black Butte Ranch Corporation*. Brooks Scanlon Papers. Oregon Historical Society, Portland, OR. On loan to the Des Chutes County Historical Society, Bend, OR.

Bridge, G. (2013). Resource geographies II The resource-state nexus. *Progress in Human Geography, 38*(1), 118–130. doi:10.1177/0309132513493379.

Bryson, J. (2010). *The nature of gentrification: Urban environments and redevelopment in the Inland Northwest.* Ph.D. Dissertation. Syracuse University, Syracuse. Supervisor: R. Wilson.

Bryson, J., & Wyckoff, W. (2010). Rural gentrification and nature in the old and new Wests. *Journal of Cultural Geography, 27*(1), 53–75. doi:10.1080/08873631003593232.

Carl, J., & Smith, B. (1969, July 23). [Report RE: Recreation homes presented to the Management Group: July 17, 1969]. Brooks Scanlon Papers, Oregon Historical Society, On Loan to the Des Chutes County Historical Society, Bend, OR.

Central Oregon Vacationer (1965). Des Chutes Historical Society, Bend, OR.

Cinderella. (1972, November 15). *Forbes Magazine,* 72–74.

Daniels, T. L. (1999). *When city and country collide: Managing growth in the metropolitan fringe.* Washington, DC: Island Press.

Deschutes County Community Development Department. (2009). *Deschutes County Comprehensive Plan: Section 4.7, Resort Communities.* Deschutes County Planning Records. Bend, OR.

Fanning, W., & Fenton, H. (1969). *A study for Brooks Scanlon, Inc: Land use analysis of company holdings and economic study of Central Oregon.* Seattle, WA: Fenton, Conger, & Ballaine. The Brooks Scanlon Papers. Oregon Historical Society. On Loan to the Des Chutes County Historical Society, Bend, OR.

Ghose, R. (2004). Big sky or big sprawl?: Rural gentrification and the changing cultural landscape of Missoula Montana. *Urban Geography, 25*(6), 528–549. doi:10.2747/0272-3638.25.6.528.

Harrison, R. (1969a, November 10). [Memo. R. L. Harrison to Michael Hollern, RE: Property status reports on Brooks Resources Corp., LaPine, Black Butte Ranch. November 10, 1969]. Brooks Scanlon Papers, Oregon Historical Society, On Loan to the Des Chutes County Historical Society, Bend, OR.

Harrison, R. (1969b, July 29). [Memo. R. Harrison to File: RE Vail-Aspen trip report, July 29, 1969]. Brooks Scanlon Papers, Oregon Historical Society, On Loan to the Des Chutes County Historical Society, Bend, OR.

Hays, S. P. (1987). *Beauty, health, and permanence: Environmental politics in the United States, 1955–1985.* Cambridge, MA: Cambridge University Press.

Hines, D. J. (2010). In pursuit of experience: The postindustrial gentrification of the rural American West. *Ethnography*, 11(2), 285-308.

Hollern, M. (1968a, January 22, 1968). [Memo Michael Hollern to Conley Brooks RE: Mt. Bachelor Bill Healy Telephone Conversation]. Brooks Scanlon Records: Oregon Historical Society, On Loan to the Des Chutes County Historical Society, Bend, OR.

Hollern, M. (1968b, January 12). [Memo to The Record RE: Mt. Bachelor Inc. and Mt. Bachelor Lodging Inc, January 12, 1968]. Brooks Scanlon Records: Oregon Historical Society, On Loan to the Des Chutes County Historical Society, Bend, OR.

Hollern, M. (1971, August 30). [Memo. Michael Hollern to the Record. RE: Boise Cascade Recreational Land Business: August 30, 1971]. Brooks Scanlon Records: Oregon Historical Society, On Loan to the Des Chutes County Historical Society, Bend, OR.

Hollern, M. (1973, February 14). [Memo. Michael Hollern to The Record RE: Brooks- Scanlon - Brooks Resources Involvement in Bend Central Business District Feb 14, 1973]. Brooks Scanlon Records: Oregon Historical Society, On Loan to the Des Chutes County Historical Society, Bend, OR.

Hollern, M. (2008). Personal Interview with author, March 3, 2008, Bend, OR.

Hurley, P., & Walker, P. (2004). Whose vision? Conspiracy theory and land-use planning in Nevada County, California. *Environment and Planning A, 36*(9), 1529–1547. doi:10.1068/a36186

Kleinsasser, W. (1969). *A report of the environmental study of Bend, Oregon, which lasted from January until July 1968.* Private papers of William Kleinsasser, Department of Architecture, University of Oregon, Eugene, OR.

Klingle, M. (2007). *Emerald city: An environmental history of Seattle.* New Haven, CT: Yale University Press.

Larsen, S., & Hutton, C. (2012). Community discourse and the emerging amenity landscapes of the rural American West. *GeoJournal, 77*(5), 651–665. doi:10.1007/s10708-011-9410-1.

Moorman, J. (1994). *Sierra Club history committee oral history series: James W. Moorman interview with Mike McCloskey*. San Francisco: The Sierra Club.

Nave, J. (2002, June 9). *Black Butte Ranch: A brief history. Sisters County Historical Society. Sisters*. Retrieved from http://www.sisterscountryhistoricalsociety.org/OV-BBR.htm.

Nicholas, L., Bapis, E. M., & Harvey, T. J. (2003). *Imagining the big open: Nature, identity, and play in the new West*. Salt Lake City, UT: University of Utah Press.

Olson, B. (2010). Paper trails: The Outdoor Recreation Resource Review Commission and the rationalization of recreational resources. *Geoforum, 41*(3), 447–456. doi:10.1016/j.geoforum.2009.11.014.

Post, C. W. (2013). Heritage, amenity, and the changing landscape of the rural American West. *Journal of Cultural Geography, 30*(3), 328–355. doi:10.1080/08873631.2013.828484.

Proctor, J. D., & Pincetl, S. (1996). Nature and the reproduction of endangered space: The spotted owl in the Pacific Northwest and southern California. *Environment and Planning D, 14*, 683–708. doi:10.1068/d140683.

Redclift, M. (2006). *Frontiers: Histories of civil society and nature*. Amherst, MA: MIT Press.

Robbins, W. (2004). *Landscapes of conflict: The Oregon story, 1940–2000*. Seattle, WA: University of Washington Press.

Robbins, P., Meehan, K., Gosnell, H., & Gilbertz, S. (2009). Writing the New West: A critical review. *Rural Sociology, 74*(3), 356–382. doi:10.1526/003601109789037240.

Rothman, H. (1998). *Devil's bargains: Tourism in the twentieth-century American West*. Lawrence, KS: University Press of Kansas.

Rothman, H. (2000). *Saving the planet: The American response to the environment in the twentieth century*. Chicago: Ivan R Dee.

Rothman, H., & Nash, G. (1998). *The greening of a nation?: Environmentalism in the United States since 1945*. New York: Wadsworth Publishing Company.

Smith, B. (1969, August 28). [Project Agreement: Bill Smith to Marchia Stille. August 28, 1969]. Brooks Scanlon Papers, Oregon Historical Society, On Loan to the Des Chutes County Historical Society, Bend, OR.

Smith, B. (1972, June 13). [Letter. Bill Smith to Michael Hollern. June 13, 1972]. Brooks Scanlon Papers, Oregon Historical Society, On Loan to the Des Chutes County Historical Society, Bend, OR.

Smith, B. (1975, February 8). [Letter. Bill Smith to Jim Crowell, Feb 8, 1975]. Brooks Scanlon Papers, Oregon Historical Society, On Loan to the Des Chutes County Historical Society, Bend, OR.

Taylor, J. E., III. (2004). The many lives of the New West. *The Western Historical Quarterly, 35*(2), 141–165. doi:10.2307/25442968.

Travis, W. R. (2007). *New geographies of the American West: Land use and the changing patterns of place*. Washington, DC: Island Press.

Travis, W. R., & Robb, J. J. (Eds.). (1997). *Atlas of the new West: Portrait of a changing region*. New York: W.W. Norton.

United States Department of Agriculture. (1967). *The yearbook of agriculture, 1967. Outdoors USA*. Washington, DC: 90th Congress, House Document No. 29.

Walker, P., & Hurley, P. (2011). *Planning paradise: Politics and visioning of land use in Oregon*. Tucson, AZ: University of Arizona Press.

Wilder, R. (1970a). *Selected references for recreation and tourism: Special report 295*. Corvallis, OR: PROFIT, Agriculture Extension Service.

Wilder, R. (1970b). *Inventory of income producing recreation opportunities: Special report 300*. Corvallis, OR: PROFIT, Agriculture Extension Service.

Chapter 7
Death by a Thousand Cuts? The Moral Terrain of Neoliberal Environmental Governance in the South Carolina Lowcountry, USA

Annette Watson and Kate Skaggs

7.1 Introduction

The South Carolina "Lowcountry" is often romanticized for its marshy landscape, antebellum history, and acres of live oaks dripping with Spanish moss (Fig. 7.1). But the Lowcountry has been rapidly transforming from an agricultural and natural resource economy to a real estate economy—an exurbanization that people who live across that landscape passionately contest. But long-term residents are participating in planning processes that are reshaping not only these landscapes but also social futures. In this chapter we explore the role of specific actors in materializing the exurban vision through the planning process; specifically, we examine the formation of for-profit and nonprofit neoliberal environmental governance regimes in the ecologically sensitive landscapes of East Edisto and the Savannah River Preserve of South Carolina.

Both primary actors—the for-profit timber company MeadWestvaco[1] and the nonprofit Savannah River Preserve—consider exurbanization in their respective locations to be a triumph of neoliberal environmental governance. MeadWestvaco has enacted a master-planned development project in East Edisto, a tract of land over 78,000 acres, 30 min west of Charleston, SC. The development's website states, "We will continue to 'listen to the land.' Let it tell us what to do, what is appropriate, what fits, what preserves its natural character and beauty"

[1] MeadWestvaco merged with RockTenn in January 2015 to form the company West Rock.

A. Watson (✉)
Department of Political Science, Faculty in the Master of Environmental Studies Program, College of Charleston, 66 George St, Charleston, SC 29424, USA
e-mail: watsonam@cofc.edu

K. Skaggs
Coastal Communities Planner, Chesapeake & Coastal Program, Maryland Department of Natural Resources, 580 Taylor Ave, Annapolis, MD 21401, USA
e-mail: kate.skaggs@maryland.gov

L.E. Taylor, P.T. Hurley (eds.), *A Comparative Political Ecology of Exurbia*,
DOI 10.1007/978-3-319-29462-9_7

Fig. 7.1 An idealized landscape of the Lowcountry: the plantation estate. *Photo credit*: A. Watson

(MeadWestvaco 2010a). MeadWestvaco touts the four cornerstones of its East Edisto development as: "rural character, environmental responsibility, sustainable towns and villages, and education and employment opportunities." Currently, the Master Plan for East Edisto maintains three-quarters of its property as conserved land, parks, lakes, and density-restricted rural areas. Similarly, the Savannah River Preserve, located along the southern South Carolina/Georgia border, includes more than 650,000 acres of conserved land. This land is managed for the purpose of preserving a "way of life" for "future generations." The cornerstones of the Savannah River Preserve are educating landowners about "responsible" land stewardship, promoting interaction between neighbors, combining "traditional" and "rural" land uses, and encouraging landowners to support conservation through a land trust arrangement (Savannah River Preserve 2008). Taken together, these contrasting land development efforts suggest a concerted movement toward environmental sustainability in the Lowcountry.

Yet political ecologists have typically been skeptical of the ability of "neoliberal" governance approaches like these to achieve environmental sustainability. "Neoliberalism" is the term for a new kind of politics and political economy that emerged after the fall of the "Golden Age," the 25 prosperous years in the United States following World War II. As capitalist economic instability came about in the 1970s, businesses led a movement to increase influential power within unregulated, global markets and for private interests. This movement favored privatization by taking regulatory control out of the hands of the state: a move toward free trade and less government intervention in diverse market interactions and across decision-making processes. The hope for neoliberalism was originally to instigate capitalist economic productivity in both developed and developing nations, while providing solutions to social and environmental problems. Privatization, "reregulation" to facilitate free trade, and liberalization are the primary strategies of neoliberal policies. Privatization proponents argue that an unregulated, global economy with the least amount of federal government interference will result in the most productive (and efficient) flow of capital resources. This flow will, in turn, create perfectly competitive markets that lead to economic efficiency, high employment rates, raised GDP, increased investment rates, and an overall stable financial system (Crotty 2000; Liverman and Vilas 2006; McCarthy 2006).

Privatization as a solution for managing environmental resources is bolstered by the theory of the "tragedy of the commons" (Hardin 1968). Humans, biologist Garret Hardin famously argues, are morally greedy when presented with opportunities to "free ride." Hardin uses the analogy of herders who kept sneaking sheep onto a common pasture to illustrate his point. Hardin argues that either authoritarianism or regulation via private ownership may be used to successfully conserve the "commons." Given the Cold War context of Hardin's work, Hardin favored private ownership over authoritarian rule, the former of which would allocate more power to the individual. Individual landholders would be motivated to cease taking advantage of the commons and "free-riding," and thus through privatization, these individuals would assume "personal responsibility" for conservation (Hardin 1968; Lockie and Higgins 2007). Many policy-makers advocate the privatization of resources as the best solution for conservation; their neoliberal environmental policies aim to decrease the power of governments by relying upon businesses and local civil society organizations made up of these individuals to negotiate market-based solutions to environmental problems (Castree 2008a, b; McCarthy and Prudham 2004; Peck and Tickell 2002). The cases of East Edisto and Savannah River Preserve both represent different takes on the implementation of neoliberal, market-based solutions to environmental governance. East Edisto represents a for-profit conservation scheme, while Savannah River Preserve by contrast is an example of nonprofit conservation.

But what if neoliberal policies assume incorrectly that humans are at heart morally greedy? In recent years the moral constitution (and thus political capabilities) of human society has been up for theoretical debate, due to empirical research on small-scale social systems. Critics of Hardin's assumptions point to Elinor Ostrom's (1990) or James Scott's (1977, 1999) work, arguing that for many societies, there are already informal institutions and cultural practices that operate at small scales, effectively "governing" the commons without resorting to either authoritarian politics or privatization (e.g., Robbins 2004; St. Martin 2001). These critical literatures argue that indigenous and other rural communities are not immoral; the absence of privatization is not "lawlessness." Political ecologists have therefore described how these societies operate through "moral economies," political economies that result from the norms of a "subsistence ethic" based on a sense of reciprocity and justice rather than exclusively cash values (Scott 1977, 1999). But these researchers also record how the moral economies of subsistence-based social systems are often disrupted by the transition to neoliberal capitalism, and neoliberal policies have been charged with marginalizing indigenous peoples and the poor (Robbins 2004).

Thus, although promoted as a method of incentivizing moral environmental behavior, critics informed by Marxian analyses of political economy argue that neoliberalism's "free-market" system nevertheless maintains structures of power that promote exploitation, of both labor and natural resources. Scholars have found that market-based approaches encourage unsustainable rates of resource consumption and that environmental degradation often accelerates in places with weak regulatory oversight. As the implementation of neoliberal policies increases, so does the number of neoliberalism's critics (Büscher 2008; Liverman and Vilas 2006; McCarthy 2005, 2006; McCarthy and Prudham 2004; McCauley 2006).

But we hesitate to paint a picture of neoliberal economies as inherently immoral, taking a warning from geographer James McCarthy, who documented a trend wherein political ecologists more often than not celebrate the virtues of rural peoples in the developing world, while vilifying rural people within the First World (McCarthy 2002). Rather than engage in analysis that characterizes one or another economic system as "moral" (or not), we ask instead how actors in all environmental governance conceptualize moral action. In the first place, pro-neoliberal capitalists agree that, as consumption levels of natural resources rise globally and our human footprint grows, conservation efforts become ever more essential for the preservation and sustaining of resources for future generations (Dietz et al. 2003). In the second place, neoliberalism is considered a long-term process of managing relationships (rather than a static end goal for policy), and scholars have noted that processes of gentrification often predate the roll out of neoliberal policies (Sandberg and Wekerle 2010). We contend that if political ecologists want to sustain their critique of neoliberal environmental policy, we need to develop a more sophisticated critique, one that does not assume that in the absence of traditional economies, neoliberal economies are bereft of morality. Yet few political ecologists have examined the moral assumptions of rural elites in an advanced capitalist context. By "rural elite" in the context of South Carolina, we refer to predominantly white, upper- and middle-class rural residents, especially those whose families have long-term ties to the Lowcountry. These are the property owners who participate in local politics and businesses.

Specific concerns about gentrification related to neoliberal policies exist around land-use conflicts between existing and new landowners. A consistent theme throughout this literature of exurban amenity migration is a concern for the preservation of the natural amenities, which are drawing new generations of middle-class "come-heres" (Hiner 2014) from more urban areas, at the expense of the land-use practices of historic residents. Middle-class values for nature are often shown to supersede other visions for the landscape, and scholars have indicated that these new visions are detrimental to environmental sustainability (Hurley 2013; Hurley et al. 2008; Logan and Wekerle 2008; Robbins et al. 2011; Zimmerman 2001). At the same time, this gentrification research also suggests that local natural resource uses may be intentionally or organically included into the design and management of newly developed urban landscapes (Hurley 2013; Grabbatin et al. 2011).

Developers, government officials, and even many conservation organizations increasingly advocate approaches to achieve conservation goals that we identify as neoliberal in principle. Some see neoliberal practices for conservation goals as a "win-win" strategy, where supply and demand will eventually provide sufficient regulation of resource use for environmental protection (Büscher 2008; McCauley 2006; Sandberg and Wekerle 2010). There is cautious optimism for neoliberal policies from evidence in a few case studies, which have shown potential success through a robust planning process that paid special attention to the neoliberal process and the involvement of all "key" stakeholders (Reed 2007). In one example, Gilbert et al. (2009) argue that even in the neoliberal context, place attachment motivates residents to resist sprawl, saying this resistance comes from a sense of "bioregional citizenship." These authors note how contingent these successes are and how dependent the outcomes are upon local situations, including the local political, economic, and social landscapes (Reed 2007; Sandberg and Wekerle 2010).

Thus some scholars suggest there might be some "techniques of governance" to achieve environmental and social goals, even within neoliberal regimes (Ferguson 2010). By not treating neoliberalism as inherently evil, and perpetuated by immoral people and businesses, this opens up our analysis to explore possible techniques of governance to manage landscapes within an advanced capitalist context. Indeed, and despite the scholarship documenting neoliberalism's various injustices, our case studies of East Edisto and the Savannah River Preserve suggest that neoliberal environmental governance is guided by moral beliefs. Our cases explore how neoliberal actors in environmental planning—both within nonprofit and for-profit entities—develop a moral discourse that shapes the economy and how people will live on the landscape. People are motivated through all sorts of material and value-based commitments—they are always navigating a "moral terrain" that they themselves also shape. Our research documents what norms are functioning within the neoliberal political economy, and we argue that scholars can articulate "moral economies" in the "First World."

Our research interrogates what "moral terrain" is being produced in these advanced capitalist contexts. By "moral terrain" we mean the material, aesthetic, and ecological effects of an economy as physically inscribed upon the landscape—whether that economy be produced through subsistence practices or through capitalist relations. Moral terrains are landscapes that reflect a dominant culture's moral constitution, produced through the ways society conceptualizes and acts within a space—though others might argue that some terrains are also thus "immoral," we instead merely assess how people living within and actively managing these landscapes understand their actions morally. Political and cultural ecologists working in the developing world have often documented a direct relationship between human beliefs/morality and—not just environmental impacts—how the ecosystem is spatially and temporally influenced by human society and practices (and vice versa). For example, the classic work in cultural ecology by Rappaport, "Pigs for the Ancestors," describes, in effect, a "moral terrain" of specific Papua New Guinea tribes: their changing garden and pig densities, managed across space through rituals and beliefs, prompted both cycles of harvest and changes to the boundaries of human social territories (Rappaport 2000).

In this chapter we describe the moral vision of corporate and nonprofit planning efforts in a neoliberal context and show how these visions are physically shaping the exurbs in South Carolina's Lowcountry. This moral terrain has not been heretofore explored as a product of neoliberal discourse within an advanced capitalist context. Our analysis finds the following: (1) because neoliberalism promises control over local places, elite citizens seek to control their spaces through neoliberal strategies, and (2) these elites value subsistence activities and outdoor recreation in conjunction with valuing capitalist economic activities and (3) a material and spatial dimension to these neoliberal conservation practices. Stakeholders in the East Edisto development and the Savannah River Preserve not only talked lovingly about the Lowcountry and their historic uses of places in the landscape that they wished to preserve, but as discussed in other studies of exurbia in this book, they actively sought to translate these concerns into land-use plans that concentrated development and business ventures, while preserving critical habitats and many natural resources. We first describe our research methods and then discuss the East Edisto development and Savannah River Preserve in turn. We end this chapter with a discussion relating these findings to potential "techniques" of neoliberal environmental governance.

7.2 Methods

Within an overall political ecological approach, this comparative case study includes analyses of 2010 Census data within each development, participant observation of planning processes, and a discourse analysis of fieldwork, planning documents, and semi-structured interviews of stakeholders. Participant observation is a fieldwork method used to understand often-unspoken meanings of human experience (Watson and Till 2010), while discourse analysis is used to understand how words (text and spoken) and visual images reflect and create meaning (Dittmer 2010; Rose 2012). Cadieux and Taylor (2013) furthermore argue that ideologies of landscape are a large part of development discourses. Thus, to gather data for the discourse analysis, coauthor Skaggs conducted participant observation, further supported by in-depth interviews, including site visits and observations of planning practices between 2008 and 2009, during the formation of the two developments in this case study: East Edisto and the Savannah River Preserve. Watson returned to these developments in 2015 to photograph how the plans became translated into the built environment.

For the East Edisto development, the opportunities to observe planning processes were quite extensive. The landowner, paper company MeadWestvaco, conducted a series of public planning meetings (2007–2008) within these rural communities to introduce the idea of the East Edisto development and learn from public reaction. MeadWestvaco used these meetings to help design a "Preliminary Master Plan." After completing the plan, the company held four more public planning meetings to further gauge reaction. At each meeting, the economic, historic, environmental, transportation, and planning consultants spoke to the audience, and coauthor Skaggs not only took fieldnotes to understand the meanings of landscapes being promoted and debated but also used these opportunities to gather documents and visual data about these projects for analysis. Additionally, Skaggs received a guided windshield tour of the Savannah River Preserve by a local forester intimately familiar with the landscape and took fieldnotes on informal conversations from the tour and other such interactions with Savannah River Preserve stakeholders.

Skaggs conducted 12 semi-structured interviews with stakeholders from both projects, recruiting through contacts generated from the participant observation and through snowball sampling. Stakeholders include residents of the affected communities, conservation organizations, real estate agents, land managers, county and city planners, partners of the Savannah River Preserve, and representatives of MeadWestvaco. A majority of respondents were residents who felt either East Edisto or the Savannah River Preserve would be in some way impact them. In our interviewee data set, two respondents were female and ten were male. All interviewees were Caucasian and almost all hold bachelor degrees. Although almost all were employed, all participants had the luxury to take time out of their work day to be interviewed. Each had access to a car for personal mobility—which means that they could easily get to any of the planning meetings held by MeadWestvaco or the Savannah River Preserve. Our interviewees were given the choice of interview location, so that ideally each participant would be in a setting where they would feel most comfortable in expressing their views (Elwood and Martin 2000). Each chose their place of work (with the exception of one who was not employed), and the interviews were digitally recorded and transcribed from the digital recordings.

This sampling method yielded those most involved in the planning process—representative of the rural elite, a group that has been shown to sometimes partake in asymmetrical power relations with their neighboring African-American livelihood users (Grabbatin et al. 2011; Halfacre 2013; Johnson et al. 2008). We recognize the limitations of this research, where we sought to understand the nature of the changes being proposed and the motivations of those involved. Drawing upon this foundation, future research is needed to understand how elites may have acted to reinforce their privilege and suppress the voices of African-Americans who may have seen themselves as further marginalized by the process and landscape outcomes.

Analysis of the data was completed first through open and "grounded" coding for reoccurring themes related to a participant's expressions of morality or ethics (Crang 1997; Strauss and Corbin 1998; Watson and Till 2010). Coded words "grounded" in the ways the interviewees spoke included "good," "love," "smart," "right," and "responsible development." Thematic quotes that pertained to an interviewee's economic and recreational practices on the land were also coded. This coding process yielded a greater understanding of the meanings generated by these statements and informed the discourse analysis. Since discourse is understood to shape human ideas about what is possible (Foucault 1980), it is perhaps not surprising that we found strong moral content in this discourse about land, land-use, and environmental policy.

7.3 Results: Amoral or a Moral Neoliberalism in the Lowcountry?

7.3.1 East Edisto: The Moral Terrain of For-Profit Conservation

MeadWestvaco has developed a method for processing timber so that less pulp is needed to make the same amount of paper products. This new methodology, requiring fewer trees, meant that MeadWestvaco had about 80,000 acres of forestland available for use other than paper production. The company will now instead be gaining a profit as pieces of this property are sold off for private residential homes. As shown in Fig. 7.2, the current landscape where East Edisto is planned consists primarily of over 78,000 acres of pine timberland in Dorchester, Charleston, and

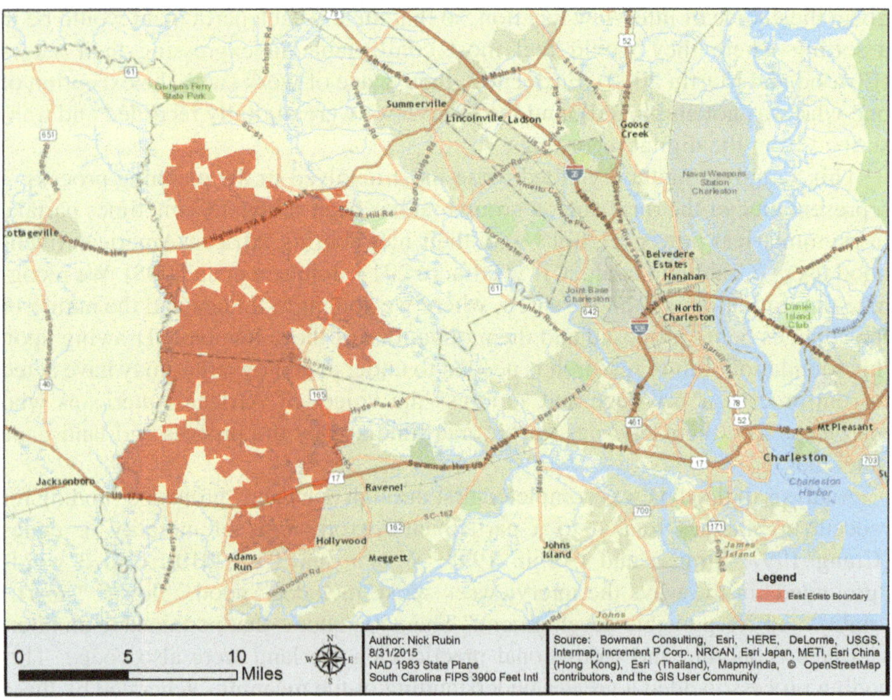

Fig. 7.2 The East Edisto boundary, outside of Charleston, SC. *Source*: East Edisto boundary estimated from maps by Bowman Consulting, 2015. Produced by N. Rubin using QGIS software

Table 7.1 2010 Census data for towns in and around East Edisto

Census data around East Edisto			
	Gettysville	Ridgeville	Summerville
Population	1610	1979	44,719
Median income	35,265	40,000	43,392
% Caucasian	39.5	33	72.1
% African-American	55.5	63.8	21.4
% who drive alone to work	84.2	70.2	82.2
Average travel time (min)	44	32	27.1

Berkeley counties in South Carolina. The northern edge of East Edisto is just a few minutes from Summerville and 20 min from downtown Charleston (MeadWestvaco 2010a, b). MeadWestvaco has owned this property for approximately 80 years for its timber supply, but before the company purchased the land, it was primarily used for farming. Thus, there are small communities and rural houses interspersed throughout this area. Census data for the proposed East Edisto development shows that the median household income is between $35,000 and $54,900; between 11.4 and 17.7 % of households, depending on locale, are therefore below the poverty line (U.S. Census Bureau 2010).

Table 7.1 shows statistics from three of the communities surrounding East Edisto; Gettysville is a community of 1610 people and Ridgeville is a community of 1979, located on the outskirts of Summerville and toward the top of the East Edisto tract. The population can range from about 40 % Caucasian and 55.5 % African-American (in Gettysville) to 33 % Caucasian and 64 % African-American (in Ridgeville). In Gettysville, 85 % of people drive alone to work and have a mean travel time of 44 min. In Ridgeville, 70 % of people drive alone to work and have a mean travel time of 32 min (U.S. Census Bureau 2010). These low-density exurbs are primarily bedroom communities, with little industry.

MeadWestvaco emphasizes its ties to surrounding communities and its concern for the future of the area. They claim to have a history of investing in the communities surrounding their timberlands throughout the United States. They also proclaim that they have a fiduciary responsibility to environmental stewardship (MeadWestvaco 2010a, b). Indeed, in 2005 MeadWestvaco was named to the Dow Jones Sustainability Index (DJSI). The DJSI was created in 1999 to track "the financial performance of the leading sustainability-driven companies worldwide" (Dow Jones Sustainability Indices 2013). The DJSI evaluates performance based on an assessment of economic, environmental, and social criteria to determine corporate sustainability. DJSI defines corporate sustainability as:

> a business approach that creates long-term shareholder value by embracing opportunities and managing risks deriving from economic, environmental and social developments. Corporate sustainability leaders achieve long-term shareholder value by gearing their strategies and management to harness the market's potential for sustainability products and services while at the same time successfully reducing and avoiding sustainability costs and risks. (Dow Jones Sustainability Indexes n.d.)

MeadWestvaco contends that, while they are extremely concerned with accountability to their shareholders, it is equally important that their business practices be conducted in an environmentally and socially "responsible" manner. MeadWestvaco dedicated part of their website to their "stewardship and sustainability" interests, stating:

> For MeadWestvaco, sustainability is both a business strategy and an ethical imperative. It is the motivating force behind our innovative products and environmentally responsible manufacturing processes….We create value for stakeholders by providing the sustainable packaging and business solutions people want and need to improve their lives, and use sustainable business practices to fulfill environmental, social and economic responsibilities. (MeadWestvaco 2010a, b)

MeadWestvaco proclaims there is a moral dimension to the company's business practices. But is this merely "greenwashing"? How does this morality actually play out on the landscape, on lands that are owned by MeadWestvaco?

MeadWestvaco's initial plans for East Edisto incorporated approximately "72,000 homes on 72,000 acres," as described by an interviewee. The company felt that this was a sustainable use of the landscape because the density of development would not be high, with one acre per home. Yet this idea was met with tremendous outcry from the local, elite community. One stakeholder described this initial idea and the sentiment that it was met with:

> I think they've just assumed that we were just like the rest of the South that you know, anything and everything is all what you want, more like what's going on in [some of] Dorchester and Berkeley County. They didn't realize the passion people have here for not only the built environment but the natural environment, and the awakening that has been taking place over the last twenty years….

As recalled by interviewees in our study, the resident community's understanding of sustainability was different than that reflected in MeadWestvaco's initial plans for the development. An interviewee quoted the then-South Carolina Republican Governor, Mark Sanford, "Please tell me when you start [construction] so…I know when to go out and stand in front of the bulldozers." The then-governor was one of the first of many to publicly express opposition to the tract home development scheme. Developers and many other interviewees expressed similar "passions" for the area. They communicated they would have fought initial East Edisto plans as initially envisioned by the company and strongly resisted necessary permitting approvals and other support for the project. The original plans for East Edisto by MeadWestvaco did not include discussion with the public, and their initial plan did not feature what interviewees considered a "unique," "conservation-oriented" design.

Rural landscapes are currently experiencing an immense restructuring (Cadieux and Hurley 2011; Gosnell and Abrams 2011). People are being drawn to rural areas because of their natural and cultural amenities; the Lowcountry in particular has mild winters, hot summers, beaches, unique critical habitat such as wetlands, as well as cities, with many creative activities such as theater and other arts. While downtown Charleston provides a historic, urban experience, its exurbs are a short drive away. These amenities have enticed population migration for decades, and it is predicted the Lowcountry's population will triple over the next 30 years (Allen and Lu 2003). Additionally, since Boeing's move to North Charleston and predicted increases in tourism, the South Carolina Lowcountry is expected to be among the top ten growing economies in the United States (Wenger 2009).

Demographic change has been acknowledged by conservation communities, residents, and developers alike. Outcry against the sprawl and tract home development style, such as seen in the East Edisto area, is only getting more noticeable in the Lowcountry. A 2006 *Post & Courier* article noted that Summerville was "the fastest-growing municipality in the Charleston area." Between 2000 and 2004 the population of Summerville grew 23 % compared to 18 % in Mount Pleasant, on the coast. At the time the article was written, citizen groups were outraged at the dramatic increase in approved building permits; by 2006 the town was on pace to approve more than 1200 residential housing permits, over a hundred more than the year before (Munday 2006a). Citizens expressed concern not only with increased housing units but also with the increased pressure on infrastructure such as schools, roads, available water, and public services (Munday 2006a, b). Residents who claim to care about the integrity and quality of life in the area are taking a stand—that is, residents like the then-governor, now senator: those residents who have enjoyed elite status in post-Civil War (the last 150+ years) history in the region. In 2009, another longtime resident won chairman of County Council on the platform of opposing the way Summerville has been developing, with a promise of promoting the long-term goals of economic stability and land preservation. The opposition's elite status, such as manifested in the appointment of the new anti-sprawl chairman of County Council, is being used to influence development plans.

Significantly, and as shown in our interviews, the potentially dramatic economic and population growth of East Edisto has these long-term residents wary of who the incoming population will be and if they will have similar values and beliefs. The elite community here and elsewhere (e.g., see Sandberg et al. 2013) is therefore trying to influence what types of industry and economic growth come to the area and define this area as "distinct" from the rest of the South.

In our interviews, stakeholders often used the words "nausea" and "distrust" when describing how they initially felt when learning of the East Edisto project. Consequently, although overall accepting the idea of development, stakeholders had been largely skeptical of those "from outside" the Lowcountry, because of possible agendas and values that do not reflect their own, or those "traditionally" found in the area. For example, long-term residents think that "outsiders" may not have the same "responsibility" that they feel toward the Lowcountry landscape. As expressed by one long-term resident and real estate developer:

> How many people can live here without making serious compromises? And while it sounds like a NIMBY [Not-In-My-Backyard] response, I'm perfectly happy to say that I think everybody deserves a place to live, not everybody deserves to live in Charleston.

This resident views people not from the Lowcountry as "everybody" else and expects that those who move there will not have the essential appreciation and "love for" the land that long-term residents have. Interviewees expressed that this tie to the land is unique in the Lowcountry and were skeptical that these newcomers would be as "respectful" of the land and historic land-use patterns. Another resident stated:

> It's a blessing and a curse that this place is so attractive to people .... And I think that there are people who come here that really don't have that deep understanding and appreciation for the coast ... that's, to me, very upsetting and I guess I'm feeling that there are people here that do not have that ethic.

Our interviews show that there is a specific set of values that these "traditional" residents refer to when contrasting the Lowcountry landscape from other places. Respondents were most anxious about potential changes they perceive may be brought about by the "outsider": specifically, they were anxious about potential changes to the subsistence-based and resource-dependent aspects of their lives. Although often unspoken, the imagined "outsider" was largely from the Northeastern Seaboard of the United States; this "outsider" is also imagined to only enjoy recreating in the nature that landscape provides—a threat to those who value nature through hunting and fishing.[2] The "long-term" residents of our interviews convey disdain and apprehension over the fact that the landscape they identify with may not be available for their grandchildren to experience in the same ways: not just for hiking, an activity valued in environmental management in the northeast, but also for subsistence-based activities, which provision their families along with the capitalist economy. These rural elites are satisfied with what they already enjoy of the land and see little need for changing their individual circumstances. A stakeholder who hunts deer put it plainly: "I think we have a good quality of life here and I want to try to protect that." Clearly, this is not the same land user that Hardin imagined bringing the commons to tragedy, unsatisfied and unthinking of the ecological consequences of human economies.

Our research shows that rural elites in the exurban Lowcountry are expressing an intense tie to the land *through its use* in *both* a subsistence and capitalist economy. Our results are supported by the work of Gibson-Graham (1996), who documented how capitalism is neither monolithic nor completely hegemonic in the ways people make their livelihoods. Our interviews, performed in an advanced capitalist context, echo concerns found in local communities across the developing world; political ecologists working in rural contexts have often found that "local" residents in developing contexts have a "conservation-oriented attitude" because of their history and use of the local landscape in subsistence-based economies (Robbins 2004). Like the work of Gibson-Graham (1996), our work documents how, even in the United States, the rural elite desires to preserve expanses for hunting and fishing: subsistence-based livelihoods and recreation, but subsistence livelihoods enmeshed in a complicated hybridity with capitalist economic practices. Subsistence economies persist even in the most advanced capitalist contexts (Emery and Pierce 2005).

[2] The fear of "northerners" is based on assumptions, as there is currently insufficient data to determine exactly where the majority of exurbanites relocating to the Lowcountry come from (Bartelme 2009).

We found in our interviews that a "moral economy" also persists, no matter the profession or institutional position of the interviewee, which included real estate developers, foresters, government officials, and general members of the conservation community. Even real estate developers involved in the MeadWestvaco project, who are often conceptualized as "greedy shepherds" when it comes to land management, articulated that they did not want more people to move to the area: they wanted their kids to have the same "outdoor experiences" that they had as children. One interviewee recounted:

> I'm just obsessed with what I'm going to leave my children....I grew up on a piece of land and I love the land. I love the land....I love that piece of property. I love everything on that piece of property. I love nature. I want to make sure that it's not compromised....I think that's very important to me.

A long-term resident and real estate developer recounted the evolution of this moral terrain among the community:

> All that's going on right now with protection of the natural environment, this time it's a bunch of guys, mostly hunters who said, "No, you don't have a right to go put condos on the property, ... it has a context and has a history and you can use and you do own the land, but you have to use it within that context of what we want to see, envision this thing to be." And ... every once and a while people ... wake up and go, "They were right!" And you would never get rid of the historic district and you would never ... get rid of the Ace Basin, and the conservation movement is never going to stop, because it's right...How do you sell [the land] when it's gone?

Scholars have documented similar findings of distrust by local communities toward "outside" entities (Hurley and Walker 2004; Liverman and Vilas 2006; McCarthy 2002, 2005; McCauley 2006; Nesbitt and Weiner 2001). Any outsider, whether conceived as a large corporation, NGO, government, or newer residents, may bring their own agenda to development, paying little respect for existing local politics or social relations—to the detriment of both the people and the environment. In one of many examples that populate the literature on political ecology in developing-world contexts, Ostrom et al. (1999) found that when the state government in Nepal tried to improve existing rural, dirt irrigation systems with modern, steel and concrete systems, the new systems proved less effective than the older systems. By instituting a system that did not take into consideration local norms and rules, markedly less water flowed to villages at the end of the irrigation system, which resulted in a decrease in overall agricultural productivity. The lesson was to learn from existing social relations that already managed resources.

Interviewees involved with the MeadWestvaco project expressed deep skepticism of not only individual "newcomers" to the area but of corporations that might not share the rural elite's subsistence-based values. For example, a conservation community representative noted that a large company seeking capital gains, with a primary responsibility to its shareholders, might not have the capacity to truly do "good." This interviewee defined "good" as taking the time to identify and acknowledge embedded historical, political, economic, and social relationships. One of our interviewees reiterated this concern about an "outside" big business wanting to develop in the Lowcountry:

> You still miss that...emotional responsibility, [the] tie of 'this is my land', [and] this is really important....My perception of the East Edisto project is that... they [MeadWestvaco] were absolutely floundering, looking for an answer that brought enough to the bottom line for them to be able to justify whatever they want to do....It's all market driven and it's all understandable...

Interviewees talked about how integrating conservation efforts in development plans might not be in the company's best interest, perceiving the company as yet another actor without any concrete ties or moral obligations to the local land. Our interviewees, although representing a variety of land-based professions, nevertheless expressed the same worries as those interviewees claiming to be from the conservation community.

However, while all the rural elite interviewees involved in the East Edisto project expressed distinct pro-conservation views, and most expressed particular limits on development, they nevertheless refused to diametrically oppose conservation with capitalist interests. For example, one local real estate developer explained:

> I would also say that I consider it a false premise, entirely false premise that conservation in any way is negative to economic development. The only thing we have to create wealth with is the natural resources that we have, and if you squander those resources, how in the world can you stand to make wealth?

The developer said that in order for a community to prosper, resources must be managed sustainably with no overexploitation.

Significantly, residents did not merely reject the East Edisto proposal; given the neoliberal context, they saw it as an opportunity. Because they are already elite, they saw the potential of working with a single landowner to shape what over 78,000 acres of the Lowcountry may look like for the next 50 years.

Even among those with lingering distrust of MeadWestvaco, residents expressed that they would prefer the East Edisto project instead of the uncertainty of what other developments could potentially bring to the area. As one representative of the conservation community said:

> Growth is coming to Dorchester County regardless of whether or not East Edisto is planned....Where it goes, and what it is, [that is] the big question mark. And if, if East Edisto answers that question then we know where it's going to go and we know what it's going to look like...

It was widely acknowledged that "more growth is coming" to the area and that the East Edisto project presented a "unique opportunity" for residents to see and help plan what that development might look like. Interviewees said that "if done right," and MeadWestvaco is "held accountable," this could be a design that sets this area of the Lowcountry "apart" from the rest of the southeast and could act as a model for future development by incorporating large areas for conservation into the design. Another resident was overtly positive about the planning process, saying "if [East Edisto's] what happens, it's the conservation—what a great thing, because we saw to what's going to happen on 70,000 acres, forever now."

In this case, neoliberal environmental governance allowed a relationship to form between the corporate landowner and these elite residents. MeadWestvaco recognized the initial skepticism and changed their approach early on for planning East Edisto. A developer stated that MeadWestvaco "pretty quickly knew [their plans would be rejected]…they pushed it for a while and they realized 'this ain't going to happen.'" Reflecting upon how the company responded to the feedback, another resident said, "I think they're somewhat sensitive to the community and I think they're realistic; if they weren't, they'd be run out of town." Another stated that "initially I was concerned that it was just purely money making.…But, the more I have read about their approach the more I think it's probably a reasonable approach."

These elites demanded and received a more transparent planning process for their shared future. Following the intense negative reaction to the initial plan, MeadWestvaco completely reorganized their planning and design for East Edisto. As a result, a representative of the conservation community said of MeadWestvaco after the planning process overhaul, "I think their heart's in the right place." This quote is especially important because it recognizes a human, moral characteristic to a for-profit, private company.

MeadWestvaco began to hold public planning meetings in 2007, meet with the conservation community regularly, and include public input in the final plans. Through the approval process, the company changed their original ideas to acknowledge local politics and local cultural values to achieve greater acceptance from these elites. MeadWestvaco asked for recommendations from the local conservation community and started regularly meeting with the local environmental organization, the Coastal Conservation League. MeadWestvaco acted not solely according to their responsibility to their shareholders; they also showed responsibility to the community, engaging in governance at a local scale.

At the end of each meeting held in the different communities, MeadWestvaco administered a survey via a third party, which asked the audience to react to various pictures that were displayed on a large screen in front of the room. At the end of the survey, the results were displayed for the audience to review. In one specific example of a community meeting, the audience was broken up into focus groups to discuss Watson Hill and MeadWestvaco's intent on annexing the property into East Edisto. Watson Hill is a 6600-acre property adjacent to East Edisto, owned previously by MeadWestvaco. MeadWestvaco sold the tract of land to a company who planned to develop 5000 homes on the property; the company went bankrupt and MeadWestvaco was able to repurchase Watson Hill, which has since been incorporated into the design for East Edisto. In the breakout sessions at this public meeting, each group worked with a consultant to address concerns or ideas residents have about Watson Hill.

All this research was used in the creation of the Final Master Plan for East Edisto, which was presented to the public in December 2009. The results of the consultant work consistently showed that people were concerned most about employment opportunities, increased traffic congestion during rush hours, and additional pressures on schools and police and fire services. People were also adamant that the aesthetics of the houses, office buildings, and industry parks fit with the "traditional Charleston style" architecture and that there be sufficient parks and recreational areas for children. People called for increasing the amount of land conserved, and downtowns with mixed-use spaces, and nature and biking trails.

While each of our interviewees saw their role in the future planning of the area as to "protect" the land they identify with, the discourse analysis revealed an explicit spatial dimension to this moral terrain. In one of the interviews, a representative of the conservation community stated:

> Our organization looks at growth as a reality and one of the roles we see ourselves [...] playing in that is that the development is done in a responsible way *in the right place*...The development and growth is coming anyway, it's going to be done *in a good spot*, it's going to be done *in a good pattern* development style...

They hope that development "is done ... in the right place." But what does this mean? Our analysis shows that residents of the area want to see future development follows what they recognize to be *spatially moral*: this means not only that the architecture of the houses and buildings mirror "traditional Charleston architecture" but that there are similar street designs, parks and open spaces, live oaks and palmetto trees, and yards for children to play in. But most important: concentrated development, so that spaces could be conserved for subsistence and recreational activities in the landscape.

Through its more transparent planning process, MeadWestvaco designed the Final Master Plan to reflect what we call the elite community's moral terrain. This 50-year Master Plan for East Edisto presents a three-phase, market-based plan. By the end of Phase One in 20 years, the Master Plan schedules the building of five towns: Summers Corner, Pine Hill, and Ashley Ridge, close to Summerville, and Greenwood and Good Hope, closer to the bottom of the tract near route 17. By Phase Three, to take place 50 years from now, the Plans call for 75 % of the property to be conserved. From the public meetings and meetings with stakeholders, the Master Plan reflects the desires of the elites for open green spaces, nature and bike trails, Charleston-style architecture, and dense, mixed-use, walkable town centers that are designed as "cluster development," as shown in Fig. 7.3 (Parker 2009; Theobald et al. 2005). That is, development will be clustered together in denser areas so that more open lands have the potential to be conserved. These "Corner Communities" are located on the outskirts of the denser downtowns. These are more rural, less dense communities with large homes but still not large "plantation" size. The architecture and trees resemble those found already in the Lowcountry: stately homes and live oaks. In areas where density will be approximately one home per 100 acres, these are labeled "plantation-style" homes (Fig. 7.4).

The East Edisto plan also focuses on the connectivity from one new town to the next; the Master Plan details bike paths and nature trails, which would link the towns. These paths and trails would be available to the public, not just East Edisto residents. And as depicted in their Master Plan and website, the developed area of the new town of "Summer's Corner," opened in September 2015, features traditional, Charleston-style houses and businesses (Figs. 7.5 and 7.6). The Master Plan's illustration shows tree-lined streets, surrounded by green fields and streams, and at Summer's Corner clusters these homes together to maximize open spaces (Fig. 7.7).

The images illustrated in the Master Plan and realized in Summer's Corner aim to show that the company incorporated local elite wishes for a conservation-driven

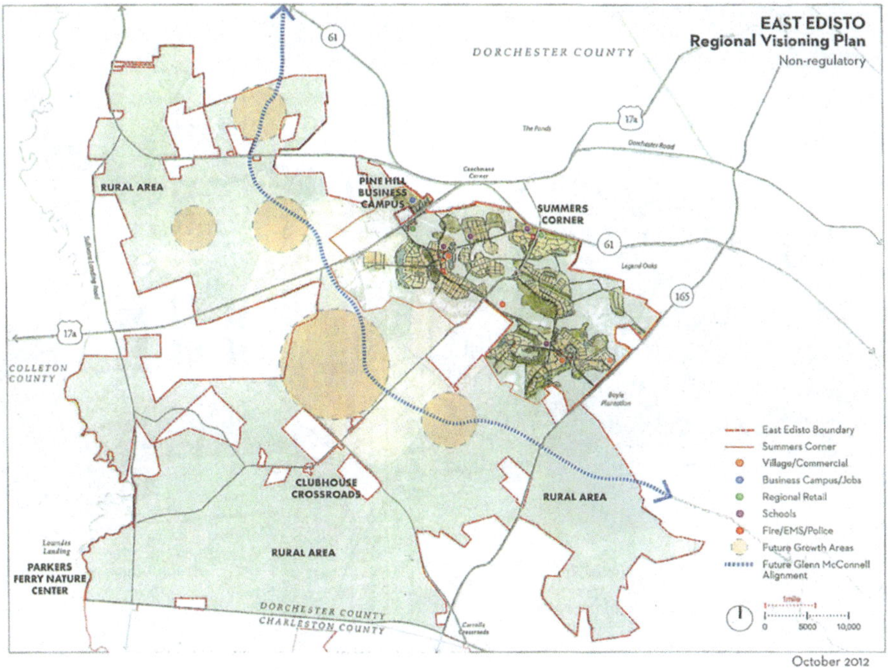

Fig. 7.3 Cluster-style development. *Source*: East Edisto District Master Plan, Ordinance 12-19, Dorchester County, S.C., 2012

Fig. 7.4 The size of a new "plantation-style" home. *Photo credit*: A. Watson

Fig. 7.5 A street in Summer's Corner. *Photo credit*: A. Watson

development plan. There will continue to be large areas of conserved land, bike and nature trails, and dense, mixed-use downtowns. This design plan resembles the "New Urbanist" paradigm through its "pedestrian orientation, accessible public spaces and community institutions, and celebration of unique local elements" (Meredith 2003). The architecture of the downtowns will resemble an architecture in which elite communities already identify with.

During one interview, a resident said that "people…want to be next to other property that is like their property and neighbors who have the same values that they have." Although this discourse of "similar values" sometimes takes racist and classist tones (and practices) with many elite communities whose discourse is examined, our interviewees largely referenced "outsiders" from the "northeast," rather than those families who have also lived in the area for hundreds of years. For example, one of our interviewees emphasized that "We don't want a gated community"—referencing a popular form of development in the Lowcountry that literally separates that development from surrounding poor (and often African-American) communities that had traditionally accessed the land for their own subsistence (Hurley et al. 2008). Instead, this interviewee clarified, "we want a community…people who would come to East Edisto to live, have to feel in their hearts that they're contributing to the community that attracted them there in the first place."

Sentiments such as this have been incorporated by MeadWestvaco into the design of East Edisto. Residents can live, work, and play in close proximity to each loca-

Fig. 7.6 Business area in Summer's Corner, where there is also a large commons space dedicated to communal use such as farmer's markets. *Photo credit*: A. Watson

tion, and images in the plan show people, couples, and families walking, shopping, and working in the same area. The Census data for this area shows that most people drive to work, with at least a 35 min commute; the Master Plan, with the mixed-use downtown where people both work and live, shows a desire for less time commuting and more time spent together as a family. It is a vision that is far from morally bereft, even though this project is neoliberal and profit motivated.

In *Lawn People*, Robbins (2007) notes how a similar moral goal promoted the lawn aesthetic; he reveals how images of the lawn as the center of family life were used to promote monoculture lawns as fundamental to creating a healthy family and community values. But he then shows that people sought the lawn aesthetic at the expense of both ecological and human health. It remains to be seen whether the East Edisto Master Plan will realize its promise to either the rural elite or the other existing families in this exurb; patterns of gentrification in other areas of the Lowcountry indicate that working class (and often African-American) individuals and communities become displaced due to increasing home values (Johnson et al. 2008), and these new homes are largely unaffordable to many existing African-American families. At Summer's Corner, for example, prices start at $250,000 and reach as high as $450,000—largely out of reach for the existing families of median income. But in the meantime, the development is largely being celebrated by this elite and its corporate partner as a success of exurban planning.

Fig. 7.7 Homes clustered in Summer's Corner. *Photo credit*: A. Watson

7.3.2 The Savannah River Preserve: The "Emotional Responsibility" of Private Landowners

Like in MeadWestvaco's East Edisto project, the Savannah River Preserve developed a plan that revolved around residents' "emotional responsibility" to the land, a term used by more than one of our interviewees from both case studies. Unlike East Edisto, however, the Savannah River Preserve is a nonprofit: an example of civil society being relied upon for neoliberal environmental conservation. In this case, private landowners organized to conserve land for the preservation of historical landmarks, watershed quality, long-standing agricultural practices, aesthetic value, and outdoor recreation (Fig. 7.8).

The Savannah River Preserve consists of primarily rural, low-density landscapes (Figs. 7.9 and 7.10). Collaborative organizations and agencies governing the Savannah River Preserve include The Nature Conservancy, Ducks Unlimited, Lowcountry Open Land Trust, South Carolina Department of Natural Resources, and the United States Fish and Wildlife Service. To enter one's land into the Savannah River Preserve, landowners submit an application for a conservation easement to the South Lowcountry (SOLO) Task Force. Landowners were granted $200 an acre by the South Carolina Conservation Bank for including their land in a

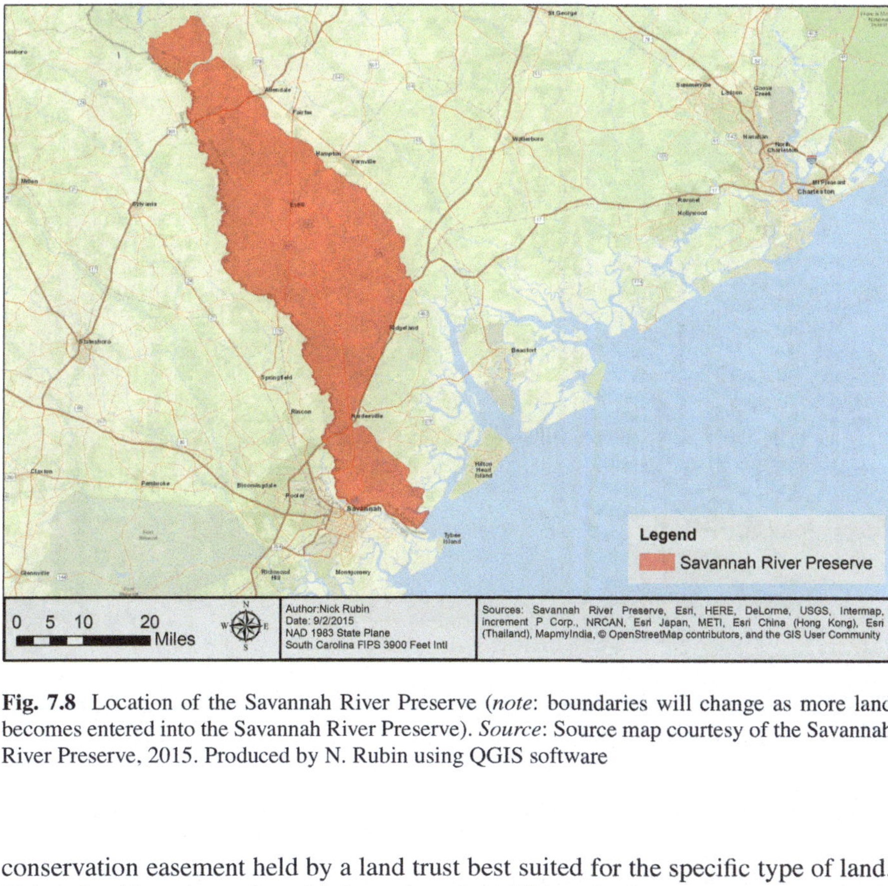

Fig. 7.8 Location of the Savannah River Preserve (*note*: boundaries will change as more land becomes entered into the Savannah River Preserve). *Source*: Source map courtesy of the Savannah River Preserve, 2015. Produced by N. Rubin using QGIS software

conservation easement held by a land trust best suited for the specific type of land. If their land is estimated at a higher value, the additional value may count as a charitable donation, and the landowner would receive tax credits. Through the collaboration of these stakeholders, the Savannah River Preserve currently consists of over 650,000 acres on the South Carolina side of the Savannah River.

Allendale County and the Town of Estill, two of the larger communities in the Savannah River Preserve, showed the effects of a difficult capitalist economy long before the recent downturn, as shown in Table 7.2. Estill has a population of 2040 (U.S. Census Bureau 2010), with 18 % of the population Caucasian and 78 % African-American. The median household income is $35,000 and 37 % of individuals are below the poverty line. In Estill, 81 % of those who work drive alone, with a mean travel time of 30 min (U.S. Census Bureau 2010). Most people in Estill work in educational service and health care (21 %), manufacturing (16 %), and arts entertainment, recreation, and accommodation/food services (15 %). According to the 2010 Census, Allendale, SC has a population of 10,419 with the median household income at $25,966; 37 % of individuals are below the poverty line. Most people in Allendale work in manufacturing (20 %), educational services and health care (19 %), and retail trade (12 %). As in the small, rural towns scattered throughout

Fig. 7.9 Mature pines in an exurban settlement within the Savannah River Preserve. *Photo credit*: A. Watson

East Edisto lands, there is little industry in the Savannah River Preserve and a lot of open space.

One stakeholder explains the beginnings of the Savannah River Preserve:

> It's a quiet, behind-the-scenes type of work.…It was mostly private landowners responding to development pressure in their region. The Savannah River region is fairly remote and it's still untouched from a development standpoint, and the landowners felt urgency to make sure that nothing's really going to change that…. Jasper County started approving all these high-density developments; so they came together and we sort of fashioned this program.… It was driven by major landowners, not by conservation groups; by the landowners wanting to keep their land the way it has been in their families for hundreds of years. That's the beauty of all of this: it's landowners making the decisions, not conservation groups. We're using conservation as a way to accomplish something like this.

This stakeholder emphasized that local landowners began the Savannah River Preserve and did so "quietly," with the inclusion of specific landowners and other stakeholders. This is drastically different from East Edisto, where MeadWestvaco used public meetings as part of their planning process. By emphasizing that this project has not been widely advertised, the interviewee highlights the workings of elite community members. One of the most repeated concerns of stakeholders involved in creating the Savannah River Preserve was to conserve large, contiguous

Fig. 7.10 Shows the edge of a field used primarily to grow turf grass in the Savannah River Preserve. *Photo credit*: K. Skaggs

Table 7.2 2010 Census data for towns in and around Savannah River Preserve

Census data around Savannah River Preserve		
	Allendale county	Estill
Population	10,419	2040
Median income	$22,946	$35,000
% Caucasian	25.5	17.7
% African-American	72.9	78.1
% who drive alone to work	82.7	81
Average travel time (min)	23.5	30

tracts of land. A long-term resident and representative of the conservation community states in one of our interviews:

> I think the biggest thing is just dividing, you know, making smaller tracts of land. And that affects the character and integrity of what's left. It's like death by a thousand cuts; the more it gets divided up, at some point you really start to lose the environmental benefits that you want to keep.

In a recent edition of Rachel Carson's *Silent Spring*, E. O. Wilson wrote in an afterword that, although the environmental movement has come a long way since the original publication in 1962, and "developers and policymakers come up with fewer spectacularly bad large projects, they continue to chip, saw, and drill away at the remains of the American natural environment…we just need a little more here and there." Wilson writes, "nature is" thus "dying the torture death of a thousand cuts" (Wilson 2002, p. 362). This term itself comes from the practice of Chinese torment, dating back to the tenth century, referring to the act of killing someone by slicing, an execution accomplished slowly, one cut at a time (Brook et al. 2008). This phrase has become translated into the contemporary environmentalist context in reference to something that happens in small increments but eventually leads to great destruction.

Stakeholders used the saying "death by a thousand cuts" to describe the slow demise of the land they love and know. A forester we interviewed stated, "I think the biggest thing is just dividing…that affects…the character and integrity of what's left. It's like death by a thousand cuts." One local forester claims:

> [The landowners] either grew up on these [Savannah River Preserve] lands or grew up within fifty miles of them and they visited them as young people; and they hunted on them and they traipsed across the fields and they have very fond memories of them. Like I did when I used to go hunting on Saturdays with my grandfather and great uncles and their friends, and it was like a very unorganized club of community effort, but it was just wonderful memories for me, and all that [land] was [now] under pressure, and it was threatened by this need for places for people to live.

The discourse about the Savannah River Preserve therefore echoes the discourse found in the for-profit example: subsistence activities are deeply valued by the local community. Additionally, these interviewees also deployed a discourse of local/outsider. This same interviewee recounts how he convinced "outsiders" of the "uniqueness" of his landscape:

> People would hear me talk and they didn't understand how I could be so in love with an area that I grew up with. And, I mean most people were just like, "Yeah [my birthplace] it's just where I grew up but I don't have any real deep connection to it." And my friends come visit me in Charleston during spring break and I took them out to my family's place in Beaufort, and I think they started to understand why I love this place so much.

These quotes mirror the thoughts of many stakeholders in the Savannah River Preserve and illustrate their connection and commitment to the land and preserving it for their community.

Many spoke of this connection specifically in terms of their "emotional responsibility" to the land. Coded transcriptions revealed stakeholders felt responsible to the stability of the ecology where they grew up and to preserve this ecosystem for their future generations. Stakeholders came together to design the Savannah River Preserve as the morally responsible thing to do. For example, one interviewee recounted:

> I think there have been more changes in my lifetime than in previous generations with the influx of residents into South Carolina and the ensuing development….I think for me, it's nostalgic, to see some of these rural places protected that remind me of places that I would have seen on John's Island or in Mount Pleasant growing up. So, it's in my blood, it's just something that's sort of fundamental to my, my whole being….And it's just so unique, unique cultural features to this area, and ecological, so now that I have children I really want them to see elements of my childhood demonstrated through the Ace Basin and the Savannah River Preserve. They'd be able to see wild places without them being spoiled.

"Spoiled" in this quote is later referenced again to indicate the division of large tracts of rural areas into subdivisions: death by a thousand cuts. Whether for-profit or nonprofit, stakeholders that we interviewed are concerned about how future development will change the landscapes they identify with and love. One stakeholder states:

> I think that South Carolina's uniquely poised for conservation because of the historic land use. And that is definitely in contrast to other states in the southeast, with the plantation being in the same family for generations and really, truly loving that land…there are a lot of public lands conservation that intertwine with private lands conservation so you have very large blocks of conservation, landscape scale conservation along the coast. So, it makes sense. You know, it's protecting our rural heritage which is so important for our economy, for our children and grandchildren…

Stakeholders are also concerned about what effects future development and population growth will have on existing infrastructure already at capacity, but as in the for-profit example of East Edisto, the initial efforts of the Savannah River Preserve were met with some skepticism. Residents in the Savannah River Preserve often clarified that the Preserve was not antidevelopment. A forester in the area stated, "There's a perception by the public that we're anti-development, and we're not. We're for planned development in the right places."

Indeed, similar to the East Edisto project, the Savannah River Preserve interviewees spoke about this "responsible" development being done in the "right places." For the Savannah River Preserve as well as in East Edisto, the "right places" reference a moral terrain. The opening paragraph in Savannah River Preserve's Executive Summary reflects what stakeholders see as a moral terrain:

> Only significant private conservation action on the part of all stakeholders can stabilize the region and ensure the continued rural economic traditions, the continued supply of clean water, and the continued quality of life this region offers… (The Nature Conservancy 2008).

Yet where the Savannah River Preserve is located in southeastern South Carolina, county governments heretofore have instituted little to no planning regulations or zoning ordinances, since populations in these areas have traditionally been low and not required zoning. That stasis has led to rapid change today, and the local community has therefore taken initiative and led private companies and NGOs to introduce land-use governance through conservation easements. Conservation easements in the Preserve and around local communities will help to establish zoning or at least to encourage zoning in already established towns. The founding document of the Savannah River Preserve clearly demonstrates the citizens' displeasure with "keyboarded" or tract home development and encourages instead infill of the existing infrastructure with the planned result of denser city centers and conservation easements that would act as a greenbelt around them. Infill in these areas could provide the greatest benefit to the greatest amount of people, such as encouraging a supermarket to the area.

The design of the Savannah River Preserve is also supposed to preserve the "rural traditions" they identify with, by conserving for future generations the large tracts of rural land that they feel emotionally responsible toward. "Rural traditions" not only translates to conserving land but also preserving "traditional" agricultural and subsistence economic practices in the region. The conservation easement placed on each piece of land entered into the Savannah River Preserve means that economic and recreational activities currently being conducted on the property might remain. The Savannah River Preserve Executive Summary states:

> A concerned group of South Lowcountry landowners now sees the trends of land sales in their back yard as the signs of imminent, unprecedented and unwelcome change. But they also see the current environment for change as an opportunity for equally unprecedented public and private conservation action. Their concerns range from the immediate impacts on their livelihoods and homes to much broader impacts across this five-county region that affect all South Carolinians. These broader impacts include a strong economic tradition in agriculture, forestry, hunting and outdoor recreation. Because of these concerns, the private landowners are now taking the initiative and providing leadership to meet the challenge for change....If we collectively fail to realize that opportunity, that which is now special and unique—the Savannah River Preserve landscape—may be gone forever. (The Nature Conservancy 2008)

People have migrated to the Lowcountry for generations because of the amenities offered in this area, including mild winters, beaches, and antebellum history and culture. This growth is expected to continue down the I-95 corridor to the rural areas in the Savannah River Preserve. The Savannah River Preserve may even act as a key amenity that encourages people with similar morals and values, like in East Edisto.

The Savannah River Preserve stakeholders we interviewed have acknowledged that the south Lowcountry expects population growth but also say that the area would "greatly benefit" from economic development. Participants in the Savannah River Preserve, similar to the elites in the East Edisto project, refuse to understand "economy" as antithetical to conservation practices. Locals hope the Preserve thereby fosters an even greater connection to the land through economic activity, creating an environment people can enjoy, hike in, but also hunt and fish and use in a variety of ways. A representative of the conservation community says:

> Conservation in a vacuum only works when you've got dedicated landowners of large tracks of land that can afford conservation ... 95 % of the counties' land might be tied up in timber and 5 % of the population have an economic connection to logging and timber, and that's not going to work. They [therefore] don't care what happens to that 95 % ...[that could become] whatever, factories, trailer parks, towns, subdivisions, anything to make money. They just have to drive by it, doesn't mean anything to them. We have to start ... [having] more diverse uses of the timber that may bring in clean industry...so that the wealth is shared by a broader base, then those people will say "don't mess with those woods." Without that, it doesn't mean anything to them.

While some residents may have expressed skepticism toward the intentions of those who established the Savannah River Preserve, all partners of the Preserve articulated that there could be a successful marriage between appropriate conservation governance and encouraging capitalist development.

7.4 Discussion: Effective Techniques of Neoliberal Environmental Governance?

Critics of the commodification of nature argue that neoliberal planning will nevertheless encourage a tragedy of the commons. People who may be attracted to each of these projects will identify in some way with the moral terrain of their designs; consequently, these conservation-driven governance plans may actually induce development in the area. Thus some scholars contend that the privatization of nature and the subjection of nature to the influences of supply and demand will actually lead to environmental degradation and unsustainable rates of resource consumption (Büscher 2008; McCauley 2006). There is a fine balance between using market solutions to help with conservation concerns and assigning market value to environmental resources for economic benefit or environmental protection (Liverman and Vilas 2006).

Nevertheless, East Edisto and the Savannah River Preserve are examples of corporations and civil society filling a gap where state governance has been diminished in the neoliberal context. Where the state has not or cannot act to govern resources, individuals and organizations will and have done so—and in the advanced capitalist context, local elites are able to effectively inform what moral terrain neoliberal environmental governance produces. Like their counterparts in developing world contexts, in the advanced capitalist context, there is a discourse of local landowners feeling an "emotional responsibility" to the land. Our interviewees, long-term residents of the exurban Lowcountry, also expressed the importance of subsistence activities like hunting, in addition to deploying a discourse of local/nonlocal in justifying their control of resources. We found that leaders within this Lowcountry context focus their moral concerns for the environment in ways that seek to reshape local policies and business practices, affected by a "moral economy" that persists in the "First World."

In an area where significant development and population growth is projected for the next 30 years, this localized governance becomes evermore important (Allen and Lu 2003). The overall moral discourse articulated through all interviews translates to physical transformations of the landscape and plans for continued "traditional" land tenure practices practiced by the elite. Because of a local public process, the design of East Edisto largely reflects what elite residents see as "good" development in the "right" places; although accomplished through different techniques, the conservation easements of the Savannah River Preserve direct development in the same way.

Because our analysis does not accept a binary between developing and advanced capitalist contexts, we are able to contribute to the literature on those "techniques of governance" that might be effective for neoliberal environmental regimes. As our evidence shows, stakeholders in both cases exude love, appreciation, and gratitude for these coastal landscapes, as well as demonstrate a passion for the land's preservation. This is something to pay attention to, as it demonstrates that the desire for a sustainable relationship with natural resources not only exists in subsistence cultures in developing countries but can also be found in the United States and across political constituencies. As Sandberg et al. (2013) point out, these sentiments will have implications for local development and conservation initiatives moving forward with state and policy initiatives. We thus argue that our case studies reveal pathways toward planning strategies that encourage sustainable use of resources.

7.5 Conclusion

Our case studies show how the advanced capitalist context presents challenges to traditional critiques of neoliberal environmental governance. Our cases present currently successful examples of neoliberal conservation governance for those elite stakeholders who played a key role in the design of both East Edisto and the Savannah River Preserve. Through each of their planning processes, actors have been able to help shape how common resources are managed and have significant input into how the land they identify with is shaped for future residential and economic development. But the literature critical of neoliberalism has often criticized its leaders and elites. For example, McCarthy (2005) notes that the ongoing neoliberalization of state regulations is being made by "elected officials who have histories of stark disregard for the environment, for the poor, and for sustainability by most definitions…," but this characterization does not easily fit the rural elite we interviewed for our two case studies.

Although in theory private ownership by civil society of natural resources would appear to lead to effective and sustainable conservation governance of resources, questions remain about what actually happens on the ground. Neoliberalism in the free market often favors pure capital and assumes that all humans value independence and economic rationality, leaving little room for environmental, intrinsic, or aesthetic values. This literature claims that, if the goal is environmental sustainability for future generations, neoliberalism may fall short. And indeed, when asked about overall thoughts on the East Edisto project, for example, a representative of the conservation community stated that "there are still devils in the details"; unease especially remains regarding whether MeadWestvaco will be accountable for the 75 % of conserved land they have promised. The concerns are related to the for-profit nature of MeadWestvaco and the skepticism that a company primarily concerned with capital gain might actually continue to allow local communities as much control as they have over the future of local resources.

Additionally, these instances of neoliberal management may be successful because of the largely non-diverse public engagement; the moral terrain that the rural elite generates might be serving to leave out other publics (Johnson et al. 2008; Sandberg et al. 2013). In our cases, there may be a tragedy of these commons for those who were not heard in the design of the plans, such as the poorest in these exurbs, including African-Americans, who in some cases represent an overwhelming majority of residents of affected areas. Thus, just because something is claimed to be moral does not mean that it is *right*—a point first made by geographer Paul Robbins about those *Lawn People* who ironically continue to pollute because of their love of their community (Robbins 2007). In other words, just because people claim to be "doing good" does not mean they *are*.

In that sense, neoliberal conservation may produce seemingly moral terrains that are uneven in their effect on local stakeholders. In the planning processes we document, our case studies suggest that neoliberal conservation can create new developments that reflect at least *some* local moral concerns about nature. Both the for-profit and nonprofit cases, like many neoliberal governance models, involve complex social, economic, and political relationships within the communities. It remains to be seen whether all current residents will be taken advantaged by these neoliberal governance strategies and what all the trade-offs are of putting a price on nature.

What we have demonstrated is that political ecological analysis of an advanced capitalist context yields a similar picture of morality, and of struggle and resistance, as do those analyses completed in the developing world. Rather than depict all neoliberal projects as inherently bereft of moral content, political ecologists are hereby challenged to articulate which "local" practices and beliefs have merit in the governing of environmental resources. Our cases show that some of these "techniques of governance" (Ferguson 2010) include tapping into the "emotional responsibility" that elites have about the landscape and generating participation in public as well as private local-scale plans that aim to control how "the future" changes their landscapes.

Acknowledgment The authors would especially like to acknowledge Patrick Hurley for his guiding role in formulating this research project.

References

Allen, J., & Lu, K. (2003). Modeling and prediction of future urban growth in the Charleston region of South Carolina: A GIS-based integrated approach. *Conservation Ecology, 8*(2). Retrieved from http://www.consecol.org/vol8/iss2/art2/.

Bartelme, T. (2009, October 25). Moving in from nearby. *Post & Courier*. Retrieved from http://www.postandcourier.com/news/2009/oct/25/moving-in-from-nearby/.

Brook, T., Bourgon, J., & Blue, G. (2008). *Death by a thousand cuts*. Cambridge, MA: Harvard University Press.

Büscher, B. (2008). Conservation, neoliberalism, and social science: A critical reflection on the SCB 2007 annual meeting in South Africa. *Conservation Biology, 22*(2), 229–231. doi:10.1111/j.1523-1739.2008.00894.x.

Cadieux, K. V., & Hurley, P. T. (2011). Amenity migration, exurbia, and emerging rural landscapes: Global natural amenity as place and as process. *GeoJournal, 76*(4), 297–302. doi:10.1007/s10708-009-9335-0.

Cadieux, K. V., & Taylor, L. (Eds.). (2013). *Landscape and the ideology of nature in exurbia: Green sprawl*. London: Routledge.

Castree, N. (2008a). Neo-liberalising nature I: The logics of de- and re-regulation. *Environment and Planning A, 40*(1), 131–152. doi:10.1068/a3999.

Castree, N. (2008b). Neo-liberalising nature II: Processes, outcomes and effects. *Environment and Planning A, 40*(1), 153–173. doi:10.1068/a39100.

Crang, M. (1997). Analyzing qualitative materials. In R. Flowerdew & D. Martin (Eds.), *Methods in human geography: A guide for students doing a research project* (pp. 183–196). Harlow, England: Longman.

Crotty, J. (2000). Slow growth, destructive competition, and low road labor relations: A Keynes-Marx-Schumpeter analysis of neoliberal globalization. *Political Economy Research Institute*. Working Paper Series: No. 6. University of Massachusetts Amherst. Retrieved from http://scholarworks.umass.edu/cgi/viewcontent.cgi?article=1002&context=peri_workingpapers.

Dietz, T., Ostrom, E., & Stern, P. C. (2003). The struggle to govern the commons. *Science, 302*(5652), 1907–1912. doi:10.1126/science.1091015.

Dittmer, J. (2010). Textual and discourse analysis. In D. DeLyser, S. Herbert, S. Aitken, M. Crang, & L. McDowell (Eds.), *The Sage handbook of qualitative methods in human geography* (pp. 274–286). London: Sage.

Dow Jones Sustainability Indexes. (n.d.). *Corporate sustainability*. Retrieved from http://www.mcilvainecompany.com/SURS/subscriber/Text/reference/CORPORATE_SUSTAINABILITY_OVERVIEW.pdf.

Dow Jones Sustainability Indices. (2013). *Results announced for 2013 Dow Jones Sustainability Indices Review; 24 sustainability industry group leaders named*. Retrieved from http://www.sustainability-indices.com/images/130912-djsi-review-2013-en-vdef.pdf.

Elwood, S. A., & Martin, D. G. (2000). Placing interviews: Location and scales of power in qualitative research. *Professional Geographer, 52*(4), 649–657. doi:10.1111/0033-0124.00253.

Emery, M. R., & Pierce, A. R. (2005). Interrupting the telos: Locating subsistence in contemporary US forests. *Environment and Planning A, 37*(6), 981–993.

Ferguson, J. (2010). The uses of neoliberalism. *Antipode, 41*(S1), 166–184. doi:10.1111/j.1467-8330.2009.00721.x.

Foucault, M. (1980). *Power/knowledge—Selected interviews and other writings 1972–1977*. In C. Gordeon (Ed.). Brighton, England: Harvester Press.

Gibson-Graham, J.-K. (1996). *The end of capitalism (as we knew it)*. Minneapolis, MN: University of Minnesota Press.

Gilbert, L., Sandberg, L. A., & Wekerle, G. R. (2009). Building bioregional citizenship: The case of the Oak Ridges Moraine, Ontario, Canada. *Local Environment, 14*(5), 387–401. doi:10.1080/13549830902903674.

Gosnell, H., & Abrams, J. (2011). Amenity migration: Diverse conceptualizations of drivers, socioeconomic dimensions, and emerging challenges. *GeoJournal, 76*(4), 303–322. doi:10.1007/s10708-009-9295-4.

Grabbatin, B., Hurley, P. T., & Halfacre, A. (2011). "I Still Have the Old Tradition": The co- production of sweetgrass basketry and coastal development. *Geoforum, 42*(6), 638–649.

Halfacre, A. C. (2013). *A delicate balance: constructing a conservation culture in the South Carolina low country*. Columbia: University of South Carolina Press.

Hardin, G. (1968). The tragedy of the commons. *Science, 162*(3859), 1243–1248. doi:10.1126/science.162.3859.1243.

Hiner, C. C. (2014). 'Been-heres vs. come-heres' and other identities and ideologies along the rural–urban interface: A comparative case study in Calaveras County, California. *Land Use Policy, 41*, 70–83. doi:10.1016/j.landusepol.2014.05.001.

Hurley, P. T. (2013). Whose sense of place? A political ecology of amenity development. In W. P. Stewart, D. R. Williams, & L. E. Kruger (Eds.), *Place-based conservation: Perspectives from the social sciences* (pp. 165–180). Dordrecht, The Netherlands: Springer Netherlands. doi:10.1007/978-94-007-5802-5.

Hurley, P. T., Halfacre, A. C., Levine, N. S., & Burke, M. K. (2008). Finding a 'disappearing' non-timber resource: Using grounded visualization to explore urbanization impacts on sweetgrass basketmaking in Greater Mount Pleasant, South Carolina. *Professional Geographer, 60*(4), 1–23. doi:10.1080/00330120802288941.

Hurley, P. T., & Walker, P. A. (2004). Whose vision? Conspiracy theory and land-use planning in Nevada County, California. *Environment & Planning A, 36*(9), 1529–1547. doi:10.1068/a36186.

Johnson, C. Y., Halfacre, A. C., & Hurley, P. T. (2008). Resistant place identities in rural Charleston County, South Carolina: Cultural, environmental, and racial politics in the Sewee to Santee area. *Human Ecology Review, 16*(1), 1–16.

Liverman, D., & Vilas, S. (2006). Neoliberalism and the environment in Latin America. *Annual Review of Environmental Resources, 31*, 327–363. doi:10.1146/annurev.energy.29.102403.140729.
Lockie, S., & Higgins, V. (2007). Roll-out neoliberalism and hybrid practices of regulation in Australian agri-environmental governance. *Journal of Rural Studies, 23*, 1–11. doi:10.1016/j.jrurstud.2006.09.011.
Logan, S., & Wekerle, G. R. (2008). Neoliberalizing environmental governance? Land trusts, private conservation and nature on the Oak Ridges Moraine. *Geoforum, 39*(6), 2097–2108.
McCarthy, J. (2002). First world political ecology: Lessons from the wise use movement. *Environment and Planning A, 34*(7), 1281–1302. doi:10.1068/a3526.
McCarthy, J. (2005). Devolution in the woods: Community forestry as hybrid neoliberalism. *Environment and Planning A, 37*(6), 995–1014. doi:10.1068/a36266.
McCarthy, J. (2006). Neoliberalism and the politics of alternatives: Community forestry in British Columbia and the United States. *Annals of the Association of American Geographers, 96*(1), 84–104. doi:10.1111/j.1467-8306.2006.00500.x.
McCarthy, J., & Prudham, S. (2004). Neoliberal nature and the nature of neoliberalism. *Geoforum, 35*, 275–283. doi:10.1016/j.geoforum.2003.07.003.
McCauley, D. J. (2006). Selling out nature. *Nature, 443*(7), 27–28. doi:10.1038/443027a.
MeadWestvaco. (2010a, January 27). Retrieved April 2, 2010, from http://newsroom.mwv.com/press-release/corporate/mwv-reports-strong-fourth-quarter-and-full-year-2009-profit-growth.
MeadWestvaco. (2010b). Maps—East Edisto regional context. Retrieved from http://www.eastedisto.com/maps/regional-context/.
Meredith, J. R. (2003). Sprawl and the new urbanist solution. *Virginia Law Review, 89*(2), 447–503.
Munday, D. (2006a, April 4). Summerville setting growth records. *The Post & Courier*. 4 April. Retrieved from http://nl.newsbank.com/nl-search/we/Archives?p_product=CPCB&p_theme=cpcb&p_action=search&p_maxdocs=200&p_field_label-0=Section&s_dispstring=summerville%20setting%20growth%20records%20AND%20section(all)%20AND%20date(all)&p_field_advanced-0=&p_text_advanced-.
Munday, D. (2006b, April 29). New housing blamed for well water woes. *The Post & Courier*. Retrieved from http://nl.newsbank.com/nl-search/we/Archives?p_action=doc&p_docid=111F25939ACFEFF8&p_docnum=1.
Nesbitt, J. T., & Weiner, D. (2001). Conflicting environmental imaginaries and the politics of nature in Central Appalachia. *Geoforum, 32*(3), 333–349. http://www.ingentaconnect.com/content/els/00167185/2001/00000032/00000003/art00047.
Ostrom, E. (1990). *Governing the commons: The evolution of institutions for collective action*. Cambridge: Cambridge University Press.
Ostrom, E., Burger, J., Field, C. B., Norgaard, R. B., & Policansky, D. (1999). Revisiting the commons: Local lessons, global challenges. *Science, 284*(5412), 278–282. doi:10.1126/science.284.5412.278.
Parker, M. (2009, December 9). MeadWestvaco plan transforms East Edisto land into antebellum neighborhoods. *Charleston Regional Business Journal*. Retrieved from http://www.charlestonbusiness.com/news/32058-meadwestvaco-plan-transforms-timberland-into-antebellum-neighborhoods.
Peck, J. A., & Tickell, A. (2002). Neoliberalizing space. *Antipode, 34*(3), 380–404. doi:10.1111/1467-8330.00247.
Rappaport, R. A. (2000). *Pigs for the ancestors: Ritual in the ecology of a New Guinea people*. Reissued Long Grove, IL: Waveland Press.
Reed, M. G. (2007). Uneven environmental management: A Canadian comparative political ecology. *Environment and Planning A, 39*(2), 320–338. doi:10.1068/a38217.
Robbins, P. (2004). *Political ecology: Critical introductions to geography*. Malden, MA: Blackwell Publishing.
Robbins, P. (2007). *Lawn people*. Philadelphia: Temple University Press.
Robbins, P., Martin, S., & Gilbertz, S. (2011). Developing the commons: The contradictions of growth in exurban Montana. *The Professional Geographer, 64*, 1–15. doi:10.1080/00330124.2011.601193.
Rose, G. (2012). *Visual methodologies: An introduction to researching with visual materials*. London: Sage.

Sandberg, L. A., & Wekerle, G. R. (2010). Reaping nature's dividends: The neoliberalization and gentrification of nature on the oak ridges moraine. *Journal of Environmental Policy & Planning, 12*(1), 41–57. doi:10.1080/15239080903371915.

Sandberg, L. A., Wekerle, G. R., & Gilbert, L. (2013). *The Oak Ridges Moraine battles: Development, sprawl, and nature conservation in the Toronto region*. Toronto, ON, Canada: University of Toronto Press.

Scott, J. C. (1977). *The moral economy of the peasant: Rebellion and subsistence in Southeast Asia*. New Haven, CT: Yale University Press.

Scott, J. (1999). *Seeing like a state: How certain schemes to improve the human condition have failed*. New Haven, CT: Yale University Press.

St. Martin, K. (2001). Making space for community resource management in fisheries. *Annals of the Association of American Geographers, 91*(1), 122–142. doi:10.1111/0004-5608.00236.

Strauss, A., & Corbin, J. (1998). *Basics of qualitative research: Techniques and procedures for developing grounded theory*. Thousand Oaks, CA: Sage.

The Nature Conservancy. (2008). *Savannah River Preserve: Executive Summary* [Pamphlet]. Charleston, SC: The Nature Conservancy.

Theobald, D. M., Spies, T., Kline, J., Maxwell, B., Hobbs, N. T., & Dale, V. H. (2005). Ecological support for rural land-use planning. *Ecological Applications, 15*(6), 1906–1914. doi:10.1890/03-5331.

U.S. Census Bureau. (2010). Quick Facts: Allendale County, SC. Online at https://www.census.gov/quickfacts/table/PST045215/45005,00

Watson, A., & Till, K. E. (2010). Ethnography and participant observation. In D. DeLyser, S. Herbert, S. Aitken, M. Crang, & L. McDowell (Eds.), *The sage handbook of qualitative geography* (pp. 121–137). London: Sage.

Wenger, Y. (2009, December 3). Welcome to boomtown. *Post & Courier*. Retrieved from http://www.postandcourier.com/article/20091203/PC1602/312039875.

Wilson, E. O. (2002). Afterword. In R. Carson (Ed.), *Silent spring* (pp. 357–363). New York: Houghton Mifflin Company.

Zimmerman, J. (2001). The "nature" of urbanism on the new urbanist frontier: Sustainable development, or defense of the suburban dream? *Urban Geography, 22*(3), 249–267. doi:10.2747/0272-3638.22.3.249.

Chapter 8
"If That Would Have Happened": The Moral Imperative of Environmental History

Michael H. Finewood and Lou Martin

8.1 Chemical Plants and Moral Narratives

Bordering the northwest corner of Southern Beaufort County, the Okatie, Colleton, and Chechessee Rivers flow into each other, respectively, and then into the Port Royal Sound. It was in this system where, in 1969, South Carolina state officials celebrated the industrial company BASF's announced intent to develop a petrochemical factory on 1800 acres of the state-owned Victoria Bluff, a largely undeveloped escarpment located along the first bend of the Colleton River. The tract of land's location near the sound was ideal for the production and distribution of plastics and chemicals. In return for the anticipated regional economic boost, state and county officials promised significant incentives, including land at cut rates, tax exemptions, abundant labor, and water from the Savannah River. Public funds would also support infrastructure and 10 miles of channel dredging for a deep-water port. State officials expected little push back from their public-private deal. Indeed it was a *cause célèbre* for growth proponents throughout the county and state as a means to end endemic regional poverty via industrial-based economic development (Bryan 2011). The new plant meant 600–1000 new skilled jobs, with predictions of 5000–7000 more ancillary jobs, a windfall for one of the poorest parts of South Carolina.

At the time of the proposal, Hilton Head Island (approximately 3 miles down-river from the site) was headlong into its own transformation (Danielson 1995;

M.H. Finewood (✉)
Environmental Studies and Science, Pace University,
861 Bedford Road, Pleasantville, NY 10570, USA
e-mail: mfinewood@pace.edu

L. Martin
Department of History, Political Science, & International Studies,
Chatham University, 1 Woodland Road, Pittsburgh, PA 15232, USA
e-mail: LMartin3@Chatham.edu

L.E. Taylor, P.T. Hurley (eds.), *A Comparative Political Ecology of Exurbia*,
DOI 10.1007/978-3-319-29462-9_8

Shannon and Taylor 2003). Since the Civil War, Hilton Head was the location of vast, privately owned timber stands as well as the home to largely poor, isolated (by water) African American farming communities. But in the 1960s, spurred by the development efforts of Charles Fraser and the Sea Pines Company, the island's timber stands were being converted into the amenity-based development complex many people are familiar with today. Key to this development has been an emphasis on the integration of the barrier island's "natural" landscape (e.g., green space, trees, beaches) with retirement and resort lifestyle offerings (e.g., housing, amenities, services). For many residents at the time—including some local African American—the BASF plant threatened to degrade both the integrity of the region's natural resource amenities (what attracted many new residents to Hilton Head in the first place) and the growing property values that coastal exurban property was garnering (Bryan 2011).

Both opponents and proponents of the plant framed their arguments in moral terms. Proponents argued that new skilled industrial jobs were the necessary boon for one of the poorest regions in the country. Indeed, several regional and national African American organizations rallied behind pro-industry forces and the potential economic growth that would come with the plant (Bryan 2011; Danielson 1995, p. 156). Specifically, the BASF plant promised jobs for a region plagued with poverty and racism, particularly acute in local African American communities who subsisted on agriculture and fishing. Industries such as oyster canning and sugar production, which had historically employed African Americans, had come and gone, leaving few other opportunities for work. However, Hilton Head's newest, white residents aligned with African American fisherman to mobilize a campaign against BASF built on the importance of a healthy environment for local communities. Leveraging considerable social capital, BASF opponents allied with federal politicians to put pressure on state decision-makers, who eventually scrapped the plans.

Over time, the narrative of BASF's defeat has become an important part of the region's dominant historiography, often presented as both a symbolic victory of local stakeholders defeating the special interests of government and corporate power, as well as the unique and strong character of local environmental concerns. Importantly, too, BASF's defeat shifted the regional political economy away from heavy industry toward the kind of exurban development that was beginning on Hilton Head in the 1960s. In this chapter, we discuss this event and the place in which it occurred—Southern Beaufort County, South Carolina—to consider the role of historiography and narrative in exurban politics. Specifically, we explore discourses that have emerged from the 1969 defeat of the BASF[1] chemical plant in one of the last "undeveloped" east coast estuaries.[2] The BASF narrative is an important story for understanding the way past events have shaped today's landscape, particularly as it normalizes the vision and materialization of today's amenity-based development. Thus we argue that the often taken-for-granted BASF *moral narrative* is mobilized

[1] BASF, or Badische Anilin and Soda Fabrik, is a German chemical firm.

[2] I use the phrasing "undeveloped" deliberately to represent the politics of land-use decision-making. Here, Victoria Bluff is seen as undeveloped when it is not producing for human needs.

as part of a broader discourse that legitimates a pattern of unequal geography on the development landscape of Southern Beaufort County, South Carolina.

The defeat of the BASF plant is an interesting story in the light of South Carolina political ecology and its growth machine politics. The South Carolina coastal zone, colloquially referred to as the Lowcountry,[3] boasts a great diversity of places that span the rural-urban spectrum, including dense urban centers, comparatively small fishing villages, and tourism hot spots. The 187 miles of coastline[4] also touch distinct language communities, a long history of resource dependence, and two National Estuarine Research Reserves.[5] South Carolina's coastal zone also offers some of the most fantastic, relatively undeveloped environs on the east coast. Despite their diversity and significance, virtually all of these places are experiencing socio-ecological change resulting from development-based land-use planning and associated population growth (Allen and Lu 2006; Kleppel et al. 2006). The BASF narrative persists today as part of a broader discourse around this regional development and environmentalism.

In this light, the BASF narrative undergirds a planning/management discourse that legitimizes a particular form of amenity-based development, largely built around new residential neighborhoods and commercial retail for amenity migrants (Finewood 2012). Thus, when the BASF story is mobilized in more recent conversations about exurban environments, it is often intended to strengthen the value of environmental stewardship via amenity migration-based development (Gill et al. 2001), in other words, engendering environmental health through development. Thus, while the BASF story is clearly about historic environmental activism, we suggest it is also about how memory and historiography serve a particular purpose (Ayers 1995).

Importantly, the BASF narrative represents something different when race and power are included in the analysis. In other words, the argument above comes into a clearer focus when considering the ways some local African Americans perceive this historiography and the related outcomes of the BASF defeat (Faulkenberry et al. 2000; Johnson and Floyd 2006). The sanitized version of BASF represents a moral narrative of environmental activists and locals as *winners* and industry as *losers*, obscuring the complicated roles that race and privilege have played in land-use decision-making during the last five decades. More specifically, this narrative contributes to a powerful conceptualization of the appropriate "form and function" of the regional amenity landscape (Larsen and Hutton 2011). In this case, the most appropriate form and function serve the interests of developers and amenity migrants, not necessarily long-standing African American communities.

We approach this argument from a political ecology perspective, a field that encompasses a wide range of ways to think about and research the human dimensions of environmental change (Peet and Watts 1996; Robbins 2004; Rocheleau et al. 1996; Biersack 2006). In this case, political ecology is utilized as a framework

---

[3] Also referred to as the "low country" or "lowcountry," the Lowcountry is both a geographic and cultural marker for South Carolina's coastal socio-ecology.

[4] There is also 2876 miles of tidal shoreline.

[5] South Carolina National Estuarine Research Reserves are North Inlet-Winyah Bay and the ACE Basin.

to interrogate who benefits and who pays in the context of environmental change and how those costs and benefits cleave across race and class lines and emerge in the exurban experience. We also consider how environmental history and political ecology can be synthesized to think critically about memory and historiography (Brannstrom 2004; Offen 2004). Political ecologists have been critical of the ways that narratives normalize the landscape, often under the guise of apolitical stories about the environment (Schein 1997; Duncan and Duncan 2004). Likewise, remembering—and telling—a story can help to legitimize current planning/development efforts by discursively linking to, and moralizing, events in the past (Alderman 2008; Daniels and Lorimer 2012). By merging political ecology and environmental history, we can position the stories we tell about ourselves as cogs in a broader discourse about current land-use decision-making that exacerbate and normalize inequality. In the case of Southern Beaufort County, the BASF narrative contributes to the reinforcement of uneven participation and outcomes between historic African American communities, wealthy white developers, and recent white amenity migrants (Schein 2006; Mercer 2002).

Here we encourage researchers, planners, and other decision-makers to consider the subtle but powerful influence of narratives in the land-use planning process and their implications for the socio-ecological landscape. By focusing in this chapter on a story and the way it is mobilized in exurban politics, we hope to contribute to the ongoing project of deconstructing the way particular conceptualizations of nature produce uneven geographies through development and environmental change, "obstructing better discussion, rather than facilitating it" (Cadieux 2011, p. 341). Thus, contributing to a text that explores the multiple pathways through which power is exercised in exurban landscapes, our goal is to develop a more just land-use planning process by providing an analysis of alternative perspectives.

In the following section, we further elucidate the value of political ecology and environmental history when interpreting the BASF narrative, which helps to think critically about tensions around the impacts of coastal development in South Carolina. We focus specifically on divergent discourses and how they contribute to place making. We then follow with a discussion of *moral narrative* and how discourses about BASF are operationalized in Southern Beaufort County. This development is key to understanding regional environmental change, as coastal populations grow and where over 60% of the population increase comes from retiring baby boomers who are moving to the region for the amenities it provides (Allen and Lu 2006). We conclude the chapter by synthesizing the literature and considering the powerful role of narrative in exurban development in Southern Beaufort County.

8.2 Exurban Political Ecology and Environmental History

As a research framework, political ecology's focus on equity and environmental change brings a necessary critical analysis to the South Carolina Lowcountry (Gallardo and Stein 2007). While Southern Beaufort County has historically gone through several, dramatic socio-ecological transformations, key to the argument here is recent amenity-based migration and exurban development. *Amenity migration*— a growing subfield of research in political ecology (Cadieux and Hurley 2011; Taylor 2011; Abrams et al. 2012)—investigates the processes of social-ecological transformation that occur in destination communities, whereby "the movement of people based on the draw of natural and/or cultural amenities, can be thought of as both driver and outcome of this transition, resulting in significant changes in the ownership, use, and governance of rural lands, as well as in the composition and socioeconomic dynamics of rural communities" (Gosnell and Abrams 2011, p. 303). Thus, as amenity migrants move to exurbia—those places outside of, but connected to, the city—they demand idealized natures that produce particular kinds of exurban spaces (Cadieux and Taylor 2013).

These exurban landscapes become "postproductivist," whereby once-rural communities transition from extractive, production-based economies to service, retail-based economies designed to attract a growing demographic of wealthy, urban elite (Murdoch 2003; Halfacree 1997; McCarthy 2008). Thus, despite often being characterized as *rural*, exurban landscapes are distinct because of their connection to urban networks and the cultural subjectivities of the urban elite who migrate from there (or back and forth on vacation) (Cadieux and Taylor 2013; Walker and Hurley 2011; Zimmerman 2001). Although these communities are diverse, they often share reputations for providing a respite from the vagaries of the city through amenities such as green space, golf, clean air, and other outdoor activities. Planners attract retiring/vacationing urban elites to exurbia by providing features such as ideal(ized) places to live (i.e., conservation subdivisions) alongside retail and a broad array of services similar to metropolitan regions (Starnes 2003). In Southern Beaufort County, this has meant converting vast silviculture and agriculture tracts into new housing, shopping centers, and health services in exurban communities that capitalize more broadly on regional, coastal amenities.

Unsurprisingly, as exurban communities grow, tensions abound over competing visions for the landscape (Boucquey et al. 2012; Campbell and Meletis 2011; Hurley and Walker 2004; Larsen and Hutton 2012; Nesbitt and Weiner 2001; Walker and Fortmann 2003) and the socio-ecological changes that are engendered as differing interests vie to influence how the landscape is shaped (Ghose 2004; Kondo et al. 2012; Robbins et al. 2012). These changing material landscapes increasingly mirror racialized urban social and political inequalities, whereby participation in planning for the landscape is unequal and/or the negative outcomes are felt unevenly (Bullard 2007). In exurban communities, powerful real estate developers marshal economic and political resources to realize a particular vision of land use that often marginalizes historic local communities and under-resourced people of color. Here again political ecology compels researchers to investigate such spatial organization in exurbia, where predominant ways of framing and consuming the environment create/exacerbate inequality (Cadieux 2011).

Regional discourses about balancing development and conservation often mask power inequalities that are embedded in land-use planning and policy. For Southern Beaufort County, specifically, we draw on Hurley and Halfacre (2011), who have written extensively about South Carolina's coastal political ecology (see also Halfacre et al. 2010). Their focus on race and the environment emphasizes that *how we talk about the environment* matters. Indeed, our analysis of multiple discourses surrounding development in Beaufort County reveals that the way we talk often facilitates elite control over landscape development strategies, normalizing outcomes that unevenly distribute costs and benefits across race and class lines (Finewood 2012; Johnson et al. 2009; see also Lee 2009).

In Southern Beaufort County, two important and often contradictory discourses have emerged around the historic defeat of the BASF plant. The first discourse is a *public memory* of the defeat, which takes the form of a moral narrative that simplifies the past. The concept of public memory helps to understand the important role of narrative in exurban landscapes. French (1995) suggests public memory reflects the "social contexts in which people shape their group identities and debate their conflicting perceptions of the past" (p. 9). Writing about the Vietnam Veteran's Memorial, Bodnar (1992) points out that these memories are not just a subject of "significant contention and debate," but they also present "a triumph of one set of interests over another" (p. 6). Narratives do similar work by conveying particular messages or meanings in specific places (Hoelscher and Alderman 2004; Price 2010). However, despite their intended meaning, narratives are a "covert exercise of power [that] inevitably sanctions some voices while silencing others" (Cronon 1992, p. 1350). Thus, a narrative should be conceived of as an exercise of power, invoking cultural constructs of history that can generate advantages in regional political economies (Cronon 1992; Kenny and Zimmerman 2003).

Narratives no doubt shape our way of seeing, but it is important to consider how they also *make* places (Schein 1997). In other words, stories serve a purpose in places like Southern Beaufort County, because they moralize particular land-use strategies (Mercer 2002). The region's economy is largely driven by new commercial and residential development and tourism with a wide range of costs and benefits to socio-ecological systems. Nonetheless, negative externalities and the costs of amenity-based development are apoliticized by moral narratives—like the one about BASF—whereby past environmental activism helps to normalize today's environmental decision-making.

A second discourse is revealed through community members as well as the work of scholars who embrace complexities and eschew linear explanations of events in the past. In particular, environmental historians focus on the roles that culture and power play in human/environment relationships and how those change over time. McNeill (2010) posits *environmental history* as the political documentation of "the history of self-conscious human efforts to regulate the relationship between society and nature as well as between social groups in matters concerning nature," whereby, "the outcome of struggle carries major implications for the land itself as well as the people involved" (p. 347). A key component of this effort is to cast "suspicion on straightforward linear explanations of environmental change" (McNeill 2010, p. 362). White (2004), writing about the cultural turn in environmental history, identifies two key emphases: a greater attention to "discourse, story, and narrative" as both "reflecting and shaping human relations to nature" and a stronger consideration of consumption (pp. 558–559).

Furthermore, some environmental historians like Sutter (2009) urge particular attention to the "social and racial histories of environmental thought and practice" (p. 4). Almost universally, race and class infuse US southern socio-ecologies (Alderman and Graves 2011; Hurley and Carr 2010; Inwood 2011), and environmental history can offer a fuller account of the various agents "enmeshed in the natural world" (Sutter 2009, p. 4). Here we can focus on how BASF stories provide a historical narrative that justifies events in the present and the inequalities that are tied closely to them (Offen 2004, p. 4).

This intersection of political ecology and environmental history is what Offen (2004) calls *historical political ecology*: "a field-informed interpretation of society-nature relations in the past…and the significance of those interpretations for improving social justice and nature conservation today" (p. 21). The intersection of these two fields (Brannstrom 2004), therefore, examines these multiple histories, how they are mobilized in current discourses, and how they might reinforce or exacerbate uneven geographies. And it is here that we can see narrative doing work (Hoskins 2010). The BASF narrative plays an influential—albeit subtle—role in regional land-use decision-making and practice. Racial politics are coded within land-use decision-making via conceptions of suitable and unsuitable land-use practices (Mollett 2010; Leib 2002; Brabec and Richardson 2007). This not only marks older African-American neighborhoods as inconsistent—or out of place—with new forms of amenity-based development (Cloke 2006; Duncan and Duncan 2004; Schein 2009) but also represents a particular type of nature in regional politics that reinforces white privilege (Dwyer and Jones 2000; Pulido 2000).

8.3 Analytical Approach

We argue that the BASF *moral narrative* is mobilized as part of a broader discourse that legitimates a pattern of unequal geography on the development landscape in Southern Beaufort County. We explore the political ecology of the BASF narrative by teasing out its public memory and complicating it via an environmental history framework. Historians of public or collective memory typically analyze constructions of the past that resonate widely with the public. These constructions often take the form of memorials and monuments as well as unofficial remembrances and widely accepted narratives of the past, or what Bodnar would call "vernacular culture" (Bodnar 1992, pp. 13–14). Although some historians call for a more sophisticated theoretical foundation for this subdiscipline, many nevertheless see great value in analyzing popular constructions of the past to reveal their discursive role in present-day recounting (Confino 1997).

Historians deconstruct official and vernacular cultural representations of the past by exploring both their apparent significance as well as underlying or hidden meanings. Our understanding of the public memory of the defeat of the BASF plant was developed from research in Southern Beaufort County from 2006 to 2009, largely based on 75 interviews with planners, politicians, activists, and other stakeholders (see Finewood 2012; Finewood and Porter 2010). These interviews revealed a common narrative of historical land-use patterns, and some further hinted at the complexities that such a narrative obscured. This understanding of the popular discourse and the role it has played allows for a reexamination of seemingly banal public exchanges, resort advertisements, environmental activism, and declarations about the environmental benefits of exurban development.

We paid particular attention to the discursive dissonances within the topics of land-use decision-making, race, and power. The suggestion by some interviewees that past complexities were being obscured by narratives like BASF led to a careful examination of scholarly histories of the region to help locate those complexities. Our analysis of the scholarly discourse was based on an environmental history of the proposed BASF plant, environmental histories of the South Carolina coast more generally, histories of recent land development and prominent individuals, and histories of African Americans on the coast. These studies challenge, often implicitly, the moral narrative that many Southern Beaufort County residents take for granted and remind us of the importance of critically analyzing such narratives in future land-use decision-making processes. Scholarship is rooted in primary sources that include oral history and present-day memories but also written records that must be interrogated for biases.

Increasingly historians have recognized the multiplicity of human experience and reject the idea of a single, knowable past, which only adds to the complexity of discourse overall. Contrasting the complexities of the scholarly discourse with the simplicity of the more popularly repeated BASF narrative reveals the intentionality of a usually hidden moral tale. Here we add to the scholarly discourse by embracing a more complex understanding of past events and how they contribute to the present landscape of coastal South Carolina.

8.4 The Role of the BASF Moral Narrative in Exurban Development

In discourses about exurban change in Southern Beaufort County, the BASF narrative is often mobilized to strengthen arguments for environmental stewardship via amenity migration-based development. Since the initial development of Hilton Head Island, this has resulted in vast tracts of rural property in the coastal zone being committed or converted to residential, commercial, or retail development. In this framing, the defeat of the BASF plant is linked to these development outcomes, legitimizing ongoing exurban growth. For example, a local developer suggested that the defeat of BASF led to a positive regional environmental outcome. The moral imperative is discursively written into the narrative *in opposition to* "what could have happened":

> We are actually fortunate here, to have the [economy] we have…Versus, in 1969, BASF wanted to build a deep water port at Victoria Bluff on the Colleton River, fabricating liquid natural gas storage tanks to ship overseas. I mean if that would have happened, I mean a

> multi-billion dollar industry, the complexion of this whole area would have been different. It would have been an industrial area…in comparison to [an industrial area], we are fortunate to have what we have now. Our business is tourism, retirees, and the support folks—white collar and blue collar—who serve the retirees and the tourists. That is what we have. So it is not a polluting industry by any means, in comparison to other things that you could have (Personal conversation, with permission, 2009).

The moral narrative is mobilized to normalize specific regional development and land-use outcomes. In a similar vein, an Associated Press article describes the 1991 congressional approval of $1 million to purchase Victoria Bluff: "A 155-acre pristine area along the Colleton River that repeatedly has been threatened by industrial development may become home to a park, nature preserve, fish hatchery and education center" (Associated Press 1991). Typical of how the battle over Victoria Bluff is remembered, this proclamation builds on a binary between industrial development and the value of nature as amenity. The article further explains that a BASF chemical plant and a Chicago Bridge and Iron Works had been proposed for the site in the past, but "environmentalists and others have feared such projects could harm the Colleton River, which is among the most pristine rivers in South Carolina" (Associated Press 1991).

In the above sense, BASF's defeat is characterized as a symbolic, moral victory for environmental amenities that simultaneously undergirds support for exurban amenity-based development. Regional tourist industries—and similar groups—are able to capitalize on the "nature" that has been protected through BASF's defeat. For example, *USA Today* covered Hilton Head's 50th anniversary in a 2006 article titled "As Resort Turns 50, Its Founder's Eco-Friendly Development Philosophy Is Hugely Influential." In the article, a former resort executive notes that Hilton Head founder Charles Fraser changed the design of an entire marina to "save that one tree." One resident originally from New York noted that Hilton Head had done a "great job of maintaining the natural beauty and health of the ecosystem." Further, a Sea Pines Plantation executive recalled that profit was not Fraser's motive: "Environment was way up there. But profit? No" (Sloan 2006). This construction recovers the potential negative impacts of regional development by producing a nature that was somehow protected through that very same development.

In other words, just as the BASF narrative tells the story of development as a choice between a chemical plant and an amenity destination, it also creates the space for idealized environments that are engendered by amenity-based development. The plant's defeat set the stage for today's development, and the moral narrative is subtly written onto the landscape (Schein 1997). Promotional materials, for example, capitalize on the narrative by reinforcing Hilton Head Island's nonhuman environment:

> Hilton Head Island has carefully nurtured and protected its lush and tropical environment. As a result, outdoor adventures in Hilton Head allow visitors to view the same waters where bottle nose dolphins play, catch glimpses of a loggerhead sea turtle on guided turtle tours, walk or horseback the woodlands where bobcats prowl, or watch a flock of seabirds swoop over a sandy shore. Explore the natural wonder of nature and the outdoors. The Hilton Head Island outdoors experience is one like no other. ("Flora and Fauna n.d.")

With a different purpose in mind, a local politician, business owner, and sportfisher penned a newspaper article (widely referenced in stakeholder interviews), lamenting declining water quality due to increased development. The author opens his article with the memory of BASF and uses it as a stepping-stone to critique today's development strategies:

> Its ironic that this site was also the one that was destined to be a BASF chemical plant in the 70's which would have killed all sea life for 30 miles around. Instead of a quick merciful death at the hands of the chemical giant we chose instead to drag it on with a slower, more insidious one with the May and the Okatie Rivers. All we need are a few more dominoes to fall and we can stage our last Chesapeake Bay style futile exercise in ecosystem recovery at Calibogue and Port Royal Sounds. (Harter 2009)

Despite the different didactic strategy from above, the author moralizes the BASF narrative as a broader story about regional environmentalism and stewardship. The point, however, when considered through a political ecological lens, is that the narrative—in either form—belies important links between regional development and race.

In this telling, the BASF defeat was not a victory against development altogether but a victory against a particular type of development (i.e., heavy industry). Major economic support for the opposition to BASF came from local real estate interests who had an eye for the undeveloped tracts of land spanning Highway 278, as well as property values on Hilton Head. So the seemingly less apparent irony of BASF's defeat (as a trope of environmental triumph) is that by preventing heavy industry, it likely facilitated the residential development pattern we see in the region today (commercial and retail development around planned neighborhoods), reshaping the landscape and opening up territory for occupation by exurbanites. Characterizing BASF as an environmental victory aligns those exurbanite interests with moral tales of environmental stewardship.

8.5 Complicating the Narrative

An alternative discourse about the Lowcountry, however, eschews simple, moral narratives and presents a deeper understanding of the region's historical political ecology. Seen this way, the BASF moral narrative conceals the rich history of local African Americans and their complex relationship to powerful decision-makers in the Lowcountry. In fact, the BASF episode exacerbated an already tenuous relationship between regional African Americans and newly arriving whites in the region. In particular, the opponents of BASF were largely made up of Hilton Head's elite community of wealthy white retirees and development interests. BASF opponents emerged from the island as it was being developed into an exurban, amenity-destination community, framing their arguments in terms of environmentalism (i.e., good stewardship) and hyping the type of jobs an exurban economy would produce (i.e., a growing industry of construction and services was emerging around new exurban amenity migrant needs).

Backed by the political economic connections of Charles Fraser and other Hilton Head-based developers, the relatively small but growing contingent of Hilton Head raised national awareness of their opposition to BASF and its pollution potential (this was no small political economic feat in and of itself). The contingent of BASF opposition was able to tap into the national sentiment of environmentalism popular in the late 1960s (Bryan 2011), when the prospect of a German firm with one of the worst pollution records in the world developing one of the last undeveloped estuaries on the east coast (including a projected daily discharge of 2.5 million gallons of waste) eventually drew the concern of federal senators. The opposition to the plant resulted in BASF abandoning their plans to develop on Victoria Bluff.

This narrative belies the role of African Americans in shaping the Lowcountry and its environmental history. West African slaves were critical to the cultivation of rice along the coast of South Carolina and Georgia. Those who survived the harrowing Middle Passage brought with them a wealth of experience and knowledge of rice cultivation that planters exploited in the creation of their famous "rice aristocracies" in the Lowcountry. The construction of elaborate systems of dikes and canals represented the marriage of West African agriculture and British technology. Possibly because of their exclusive knowledge, the slaves of the rice plantations insisted on the task system (rather than the gang labor system) because the task system afforded them more independence and flexibility. After the Civil War, emancipated slaves seized abandoned plantations, especially on the Sea Islands, and began a marvelous experiment in self-government that has drawn the attention of several historians (Foner 1983).

These communities focused on food production and fishing to achieve self-sufficiency, but their efforts were short-lived. During the late 1860s, the federal government restored white deed holders' rights to the land. African Americans were forced to accept sharecropping contracts that severely constrained their ability to choose which crops to plant and destined many to a lower standard of living for decades to come. During reconstruction between 1865 and 1877, the rice aristocrats watched their empires crumble as they could not profitably maintain the dikes and canals with wage labor, and the African Americans continued to exhibit considerable independence. White elites in South Carolina reasserted their dominance through extreme violence in the 1870s and 1880s, constructing barriers to African American access to political office and education that persists today (Carney 2002; Foner 1983; Saville 1994; Stewart 1996).

Amid the racial and economic turmoil of the late 1800s, African Americans in coastal South Carolina and Georgia pieced together an economy that rested on the turpentine plantations of the longleaf forests of the pine belt, industrial logging, small-scale fishing along the coast, truck farming, and service work in the emerging tourism sector. As early as the 1880s, resort hotels constructed on the Sea Islands were attracting visitors from the city to enjoy "nature's restorative powers." Developers drained marshes, drilled artesian wells, and constructed railroads to Charleston and Savannah. African-Americans secured jobs as day laborers, kitchen help, caddies, bellhops, and housekeepers at the resorts (Stewart 1996). Because of their distinctive dialect, the Gullah communities have garnered writers' attention since the early twentieth century. Ambrose Elliott Gonzales published numerous Gullah stories, written in their dialect, during the 1920s, and Duke University professor of religion Mason Crum published his study of the Gullah community in

1940 with the hopes that a deeper understanding of "Negro progress and the Negro's contribution to American life in economics, in letters, and in art" might ease racial tensions in the South (Crum 1940, p. ix).

In the mid-twentieth century, Sea Island African American communities were an important source of activism during the Civil Rights Movement. Born in Charleston, South Carolina, Septima Clark, one of the Movement's most influential leaders, experimented with the concept of a Freedom School in the Sea Islands. The Freedom School was a gathering place for African Americans to organize, to learn how to register to vote, to learn to read, and to learn job skills. Clark's idea for the Freedom Schools would spread from the Sea Islands across the South to Mississippi where the Student Nonviolent Coordinating Committee adopted the idea as a way to empower local African Americans. This cemented one of the most enduring legacies of the Movement throughout the South (Payne 1995).

Nonetheless, several accounts of BASF—which recognize that it was not specifically about developing Victoria Bluff but rather about what kind of development would be acceptable there—obscure the role of race both in that pivotal moment and over time. Retirees, second-home owners, and others who were interested in Southern Beaufort County as an amenity destination did not want heavy industry despoiling the political ecological niche they had carved out in the region. Many new residents did, however, want development that created housing, infrastructure, and services, as Danielson (1995) suggests:

> At the heart of the controversy over BASF were competing economic visions, two completely different mind-sets about economic development and the road to prosperity, mutually incompatible and involving very different environmental assessments. The BASF fight was fundamentally about what kind of an economy would prevail in lower Beaufort County and what its impact would be on the environment. Hilton Head was firmly committed to resort development with a strong emphasis on environmental amenities, which precluded heavy industry. The state was equally devoted to industrial development, which would surely undermine Hilton Head's cachet. Though each of these kinds of developments would change the environment, heavy industry would be more destructive than condominiums, golf courses, and marinas. (p. 151)

The defeat of BASF was, in this sense, a white exurban narrative. Many African American residents today find more complexity in this pivotal decision over development. Even in 1970, proponents of BASF noted that not building the plant would mean that 160 unskilled African Americans would lose an opportunity for employment, while opponents observed that constructing the plant would harm the fishing industry on which many African Americans depended. Therefore, African Americans stood to lose out regardless of whether the plant was constructed or not (Webster 1970).

Recent interviews with local African Americans further complicate the moral narrative. A local councilwoman observed that the economic development of recent decades in Bluffton had resulted in an improved standard of living that included sidewalks, new sewer systems, and resort jobs that were closer than Hilton Head. She also noted that factory work was not particularly desirable anymore and that the younger generation can go off to college and return to good jobs in the service industry. Another African- American interviewee sees recent land development as more complex, noting that some land developers unscrupulously bought land for less than market value and that there had been a growing sense of powerlessness among the African American community prior to Barack Obama's election. While there have been some net positive outcomes from the BASF defeat, there were several issues related to race that continue to be buried under the dominant narrative.

A narrative of BASF that occludes race and power is much simpler. The contestation over the plant is often characterized as two "sides" of a development debate. And an alliance of activism and development won the debate. Despite continued attempts to develop industrial plants on Victoria Bluff before 1991, it is clear today that the amenity-based development vision has persisted. A bird's eye view of the Colleton River reveals a string of private docks that line the southern edge of the river. The bluff itself is the site of a putting green for an exclusive golf course. The efforts of early Hilton Head residents and developers to defeat BASF put a stamp on the region: Southern Beaufort County would not have an industrial economy but instead would have an economy based on construction, retail, and services. Thus the economic development pattern/strategy that began on the southern tip of Hilton Head (Sea Pines) in the 1960s has spread along the east/west running Highway 278 to US-95, filling in nearly every developable tract of land between the May and Colleton Rivers with housing and commercial retail development. It is quite clear that this particular vision for regional development was legitimized and pursued, opening up territories for the occupation of amenity landscapes by exurbanites.

Today, Southern Beaufort County's development continues to mirror (to a large degree) what started on Hilton Head Island. Vast tracts of exurban land between Hilton Head Island and I-95 have been converted to new housing developments surrounded by commercial and retail centers. Amenities such as climate, golf, retail, gated conservation communities, and vast estuarine systems are key to the local economy (50 % of the region is water), particularly as selling points for the housing market. Exacerbating a global trend of coastal migration, Southern Beaufort County has experienced explosive population growth and development. As mentioned earlier, demographics have become profoundly older and whiter, with over 60 % of the population increase comes from immigrating, retiring baby boomers (Allen and Lu 2006).[6]

8.6 Conclusion

Today's debates about Southern Beaufort County's environmental change and development have unfolded in a similar way, framed by moral arguments between economic growth and environmental health. In other words, debates about the region's development are generally polarized between the environmental impacts of

[6] A major portion of the research that inspired this paper took place in Bluffton, South Carolina, located a few miles inland from Hilton Head Island. Bluffton's growth is representative of regional development. Since the late 1990s, Bluffton has grown from 1 sq. mile to 53 sq. miles, with a population increase from 1000 people (1985) to 10,000 people (2005). If you add the adjacent county population, population numbers jump to nearly 30,000 people.

rapid development and the politics of economic growth and job creation. And it is in this context that the BASF narrative has (re)emerged.

Earlier in this chapter, we suggested that planners and other stakeholders think critically about narrative and its discursive role in exurban politics. In this sense, the taken-for-granted BASF moral narrative is mobilized as part of a broader discourse that opens up territories for the occupation of amenity landscapes by exurbanites while legitimizing a pattern of unequal geography in Southern Beaufort County. A key aspect when considering this argument is that, until the 1960s, people of color outnumbered whites at a ratio of 10:1. After the Civil War, most of Southern Beaufort County was rural and African American, with pockets of wealthy white communities in places like Bluffton and Beaufort. Regional African American communities have been historically poor and marginalized. And although the defeat of BASF depended on, in part, an alliance between local African Americans and whites, regional growth has had, at best, a mixed outcome for many African American communities. Traditional livelihood strategies are all but gone, and today's economy is largely built around services. Likewise, long-standing African American communities are fragmented and diminished. Positive stories about development coincide with white narratives about the region.

Most of the African Americans we spoke with (for this project) were relatively indifferent about the defeat of BASF, particularly as it represented events nearly 40 years in the past. As most people do, they also recognized the complexity and contradictions that reside in the need for jobs and the desire to protect the environment (as they often knew people who made their living with a diverse livelihood strategy that included both industry and resource-based work). However, the popular way of telling the BASF narrative did represent an omission of how many African Americans felt about the BASF defeat. As one participant suggested, both the defeat and the way of telling the story were just one in a long line of micro-injustices that were chipping away at the African American community "brick by brick" (personal conversation, with permission, 2009). The story posed yet another example of outsiders (usually white) who were shaping the landscape in a way that marginalized the interests/needs (even the presence) of local African American communities.

Thinking of stories in this frame therefore implores environmental decision-makers to pay special attention to the role the BASF narrative plays in exurban landscapes. A historical political ecology (Offen 2004) framework suggests the BASF narrative is one of the several possible stories about the past that provide a moral valuation for exurban development, and we should cast suspicion on such linear explanations of environmental change (McNeill 2010, p. 362). In other words, stories like BASF are often "celebratory and serve to inculcate residents with traditional values" (Duncan and Duncan 2004, p. 5). These values coincide with the interests of exurban elites (Taylor 2011). Importantly, by binding moral imperatives to the narrative's outcome (Cronon 1992), the story of the BASF's defeat is reflective of good environmental stewardship via amenity-based development (Gill et al. 2001). When stakeholders invoke moral narratives about the past, they are trying to tell stories about ourselves and our relationships with nonhuman nature. Here we can see that, far from being benign, the BASF moral narrative is meant to convey value to exurban development, particularly in opposition to *what could have happened*.

More specifically, we must see the discursive nature of the BASF narrative. Here, the narrative not only moralizes good and bad development but also legitimizes the uneven geographies that result from exurban development. When implying that amenity-based development is an appropriate outcome from the BASF narrative, several other moral themes are implicit but discursively masked in the practice of exurban development (Hurley and Carr 2010). The ongoing loss of control and displacement of African American communities via exurban development suggest that the moral imperative of BASF is not benign.

In this view, conveying moral narratives *matters* as they facilitate elite control over landscape development strategies (Lee 2009), normalizing outcomes that unevenly distribute costs and benefits across race and class lines (Finewood 2012; Johnson et al. 2009). Thus, power within the broader exurban landscape is reinforced via how we talk about it. The merger of political ecology and environmental history, therefore, provides critical insight into the ways stories about the past normalize control over the landscape, demonstrating the role of the exurban vision of nature in state and local politics and how that vision opens up territories for the occupation of amenity landscapes by exurbanites.

References

Abrams, J. B., Gosnell, H., Gill, N. J., & Klepeis, P. J. (2012). Re-creating the rural, reconstructing nature: An international literature review of the environmental implications of amenity migration. *Conservation and Society, 10*(3), 270–284. doi:10.4103/0972-4923.101837.

Alderman, D. H. (2008). Place, naming and the interpretation of cultural landscapes. In B. J. Graham & P. Howard (Eds.), *The Ashgate research companion to heritage and identity* (pp. 195–213). Burlington, VT: Ashgate.

Alderman, D. H., & Graves, W. (2011). Innovations in southern studies within geography. *Southeastern Geographer, 51*(4), 505–512.

Allen, J. S., & Lu, K. S. (2006). Predicting trajectories of urban growth in the Coastal Southeast. In G. S. Kleppel, M. Richard DeVoe, & M. V. Rawson (Eds.), *Changing land use patterns in the coastal zone: Managing environmental quality in rapidly developing regions* (pp. 47–67). New York: Springer.

Associated Press. (1991, November 10). S.C. prefers nature reserve at Victoria bluff. *Spartanburg Herald-Journal,* B4.

Ayers, E. L. (1995). Memory and the south. *Southern Cultures, 2*(1), 5–8. doi:10.1353/scu.1995.0037.

Biersack, A. (2006). Reimagining political ecology: Culture/power/history/nature. In A. Biersack & J. B. Greenberg (Eds.), *Reimagining political ecology* (pp. 3–40). Durham, NC: Duke University Press.

Bodnar, J. (1992). *Remaking America: Public memory, commemoration, and patriotism in the twentieth century*. Princeton, NJ: Princeton University Press.

Boucquey, N., Campbell, L. M., Cumming, G., Meletis, Z. A., Norwood, C., & Stoll, J. (2012). Interpreting amenities, envisioning the future: Common ground and conflict in North Carolina's rural coastal communities. *Geojournal, 77*(1), 83–101. doi:10.1007/s10708-010-9387-1.

Brabec, E., & Richardson, S. (2007). A clash of cultures: The landscape of the Sea Island Gullah. *Landscape Journal, 26*(1), 151–167.

Brannstrom, C. (2004). What kind of history for what kind of political ecology? *Historical Geography, 32*, 71–87.

Bryan, W. D. (2011). Poverty, industry, and environmental quality: Weighing paths to economic development at the dawn of the environmental era. *Environmental History, 16*(3), 1–31. doi:10.1093/envhis/emr050.
Bullard, R. D. (Ed.). (2007). *The black metropolis in the twenty-first century: Race, power, and politics of place*. New York: Rowman and Littlefield.
Cadieux, K. V. (2011). Competing discourses of nature in exurbia. *Geojournal, 76*(4), 341–363. doi:10.1007/s10708-009-9299-0.
Cadieux, K. V., & Hurley, P. T. (2011). Amenity migration, exurbia, and emerging rural landscapes: Global natural amenity as place and as process. *Geojournal, 76*(4), 297–302. doi:10.1007/s10708-009-9335-0.
Cadieux, K. V., & Taylor, L. (Eds.). (2013). *Landscape and the ideology of nature in exurbia: Green sprawl*. New York: Routledge.
Campbell, L. M., & Meletis, Z. A. (2011). Agreement on water and a watered-down agreement: The political ecology of contested coastal development in Down East, North Carolina. *Journal of Rural Studies, 27*(3), 308–321. doi:10.1016/j.jrurstud.2011.06.002.
Carney, J. (2002). *Black rice: The African origins of rice cultivation in the Americas*. Cambridge, MA: Harvard University Press.
Cloke, P. (2006). Rurality and racialized others: Out of place in the countryside? In P. Cloke, T. Marsden, & P. Mooney (Eds.), *Handbook of rural studies* (pp. 379–387). London: Sage.
Confino, A. (1997). Collective memory and cultural history: Problems of method. *The American Historical Review, 102*(5), 1386–1403. doi:10.2307/2171069.
Cronon, W. (1992). A place for stories: Nature, history, and narrative. *The Journal of American History, 78*(4), 1347–1376.
Crum, M. (1940). *Gullah: Negro life in the Carolina Sea Islands*. Durham, NC: Duke University Press.
Daniels, S., & Lorimer, H. (2012). Until the end of days: Narrating landscape and environment. *Cultural Geographies, 19*(1), 3–9. doi:10.1177/1474474011432520.
Danielson, M. N. (1995). *Profits and paradise: The development of Hilton Head Island*. Columbia, SC: The University of South Carolina Press.
Duncan, J. S., & Duncan, N. G. (2004). *Landscapes of privilege: The politics of the aesthetic in an American suburb*. New York: Routledge.
Dwyer, O., & Jones, J. P., III. (2000). White socio-spatial epistemology. *Social and Cultural Geography, 1*(2), 209–222. doi:10.1080/14649360020010211.
Faulkenberry, L. V., Coggeshall, J. M., Backman, K., & Backman, S. (2000). A culture of servitude: The impact of tourism and development on South Carolina's coast. *Human Organization, 59*(1), 86–95.
Finewood, M. H. (2012). Feeling the squeeze: A political ecology of race and amenity-based development in coastal Bluffton, South Carolina. *Local Environment, 17*(9), 991–1011. doi:10.1080/13549839.2012.688733.
Finewood, M. H., & Porter, D. E. (2010). Theorizing an alternative understanding of "disconnects" between science and management. *Southeastern Geographer, 50*(1), 130–146. doi:10.1353/sgo.0.0068.
Flora and Fauna. (n.d.). Retrieved from http://hiltonheadsunrise.com/flora-and-fauna/.
Foner, E. (1983). *Nothing but freedom: Emancipation and its legacy*. Baton Rouge, LA: Louisiana State University Press.
French, S. A. (1995). What is social memory? *Southern Cultures, 2*(1), 9–18. doi:10.1353/scu.1995.0049.
Gallardo, J. H., & Stein, T. V. (2007). Participation, power, and racial representation: Negotiating nature-based and heritage tourism development in the rural south. *Society and Natural Resources, 20*(7), 597–611. doi:10.1080/08941920701216545.
Ghose, R. (2004). Big sky or big sprawl? Rural gentrification and the changing cultural landscape of Missoula, Montana. *Urban Geography, 25*(1), 528–549. doi:10.2747/0272-3638.25.6.528.
Gill, N., Klepeis, P., & Chisholm, L. (2001). Stewardship among lifestyle oriented rural landowners. *Journal of Environmental Planning and Management, 53*(3), 317–334. doi:10.1080/09640561003612890.

Gosnell, H., & Abrams, J. (2011). Amenity migration: Diverse conceptualizations of drivers, socioeconomic dimensions, and emerging challenges. *Geojournal, 76*(4), 303–322. doi:10.1007/s10708-009-9295-4.

Halfacre, A., Hurley, P. T., & Grabbatin, B. (2010). Sewing environmental justice into African-American sweetgrass basket-making in the South Carolina Lowcountry. *Southeastern Geographer, 50*(1), 147–168. doi:10.1353/sgo.0.0064.

Halfacree, K. (1997). Contrasting role for the post-productivist countryside: A postmodern perspective on counterurbanisation. In P. Cloke & J. Little (Eds.), *Contested countryside cultures: Otherness, marginalisation and rurality* (pp. 70–93). London: Routledge.

Harter, D. (2009, March 11). We're writing the autopsy of an estuary. *Bluffton Today,* A19.

Hoelscher, S., & Alderman, D. H. (2004). Memory and place: Geographies of a critical relationship. *Social & Cultural Geography, 5*(3), 347–355. doi:10.1080/1464936042000252769.

Hoskins, G. (2010). A secret reservoir of values: The narrative economy of Angel Island immigration station. *Cultural Geographies, 17*, 259–275.

Hurley, P. T., & Carr, E. R. (2010). Forum on political ecology of the U.S. South: Introduction: Why a political ecology of the U.S. South? *Southeastern Geographer, 50*(1), 99–109.

Hurley, P. T., & Halfacre, A. C. (2011). Dodging alligators, rattlesnakes, and backyard docks: A political ecology of sweetgrass basket-making and conservation in the South Carolina Lowcountry, USA. *Geojournal 76*(4), 383–399.

Hurley, P. T., & Walker, P. A. (2004). Whose vision? Conspiracy theory and land-use planning in Nevada County, California. *Environment and Planning A, 36*(9), 1529–1547. doi:10.1068/a36186.

Inwood, J. F. (2011). Geographies of race in the American South: The continuing legacies of Jim Crow segregation. *Southeastern Geographer, 51*(4), 564–577.

Johnson, C. Y., & Floyd, M. F. (2006). A tale of two towns: Black and white municipalities respond to urban growth in the South Carolina Lowcountry. *Human Ecology Review, 13*(1), 23–38.

Johnson, C. Y., Halfacre, A. C., & Hurley, P. T. (2009). Resistant place identities in rural Charleston county, South Carolina: Cultural, environmental, and racial politics in the Sewee to Santee Area. *Human Ecology Review, 16*(1), 1–16.

Kenny, J. T., & Zimmerman, J. (2003). Constructing the 'genuine American city': Neo-traditionalism, new urbanism and neo-liberalism in the remaking of downtown Milwaukee. *Cultural Geographies, 11*, 74–98. doi:10.1191/1474474003eu294oa.

Kleppel, G. S., DeVoe, R., & Rawson, M. V. (Eds.). (2006). *Changing land use patterns in the coastal zone: Managing environmental quality in rapidly developing regions*. New York: Springer.

Kondo, M. C., Rivera, R., & Rullman, S., Jr. (2012). Protecting the idyll but not the environment: Second homes, amenity migration and rural exclusion in Washington State. *Landscape and Urban Planning, 106*(2), 174–182. doi:10.1016/j.landurbplan.2012.03.003.

Larsen, S., & Hutton, C. (2012). Community discourse and the emerging amenity landscapes of the rural American West. *Geojournal, 77*(5), 651–665. doi:10.1007/s10708-011-9410-1.

Lee, C. W. (2009). Conservation as a territorial ideology. *City and Community, 8*(3), 301–328. doi:10.1111/j.1540-6040.2009.01290.x.

Leib, J. I. (2002). Separate times, shared spaces: Arthur Ashe, Monument Avenue and the politics of Richmond, Virginia's symbolic landscape. *Cultural Geographies, 9*, 286–312. doi:10.1191/1474474002eu250oa.

McCarthy, J. (2008). Rural geography: Globalizing the countryside. *Progress in Human Geography, 23*(1), 129–137. doi:10.1177/0309132507082559.

McNeill, J. R. (2010). The state of the field of environmental history. *Annual Review of Environmental Resources, 35*, 345–374. doi:10.1146/annurev-environ-040609-105431.

Mercer, D. (2002). Future-histories of Handford: The material and semiotic production of a landscape. *Cultural Geographies, 9*, 35–67. doi:10.1191/1474474002eu232oa.

Mollett, S. (2010). Racial narratives: Miskito and colono land struggles in the Honduran Mosquitia. *Cultural Geographies, 18*, 43–62. doi:10.1177/1474474010384928.

Murdoch, J. (2003). Co-constructing the countryside: Hybrid networks and the extensive self. In P. Cloke (Ed.), *Country visions* (pp. 263–282). London: Prentice-Hall.

Nesbitt, J. T., & Weiner, D. (2001). Conflicting environmental imaginaries and the politics of nature in central Appalachia. *Geoforum, 32*(3), 333–349. doi:10.1016/S0016-7185(00)00047-6.

Offen, K. H. (2004). Historical political ecology: An introduction. *Historical Geography, 32*, 19–42.

Payne, C. (1995). *I've got the light: The organizing tradition and the Mississippi freedom struggle*. Berkeley, CA: University of California Press.

Peet, R., & Watts, M. (Eds.). (1996). *Liberation ecologies: Environment, development, social movements*. New York: Routledge.

Price, P. L. (2010). Cultural geography and the stories we tell ourselves. *Cultural Geographies, 17*, 203–210. doi:10.1177/1474474010363851.

Pulido, L. (2000). Rethinking environmental racism: White privilege and urban development in Southern California. *Annals of the Association of American Geographers, 90*(1), 12–40.

Robbins, P. (2004). *Political ecology*. Malden, MA: Blackwell.

Robbins, P., Martin, S., & Gilbertz, S. (2012). Developing the commons: The contradictions of growth in exurban Montana. *The Professional Geographer, 64*(3), 317–331. doi:10.1080/00330124.2011.601193.

Rocheleau, D. E., Thomas-Slayter, B. P., & Wangari, E. (Eds.). (1996). *Feminist political ecology: Global issues and local experiences*. New York: Routledge.

Saville, J. (1994). *The work of reconstruction*. Cambridge, England: Press Syndicate of the University of Cambridge.

Schein, R. H. (1997). The place of landscape: A conceptual framework for interpreting an American scene. *Annals of the Association of American Geographers, 87*(4), 660–680.

Schein, R. (Ed.). (2006). *Landscape and race in the United States*. New York: Routledge.

Schein, R. (2009). Belonging through land/scape. *Environment and Planning A, 41*, 811–826. doi:10.1068/a41125.

Shannon, M. A., & Taylor, S. W. (2003). Astride the plantation gates: Tourism, racial politics, and the development of Hilton Head Island. In R. D. Starnes (Ed.), *Southern journeys: Tourism, history, and culture in the modern south* (pp. 177–194). Tuscaloosa, AL: University of Alabama Press.

Sloan, G. (2006, September 1). Hilton Head, way ahead of its time. *USA Today*. Retrieved from http://usatoday30.usatoday.com/travel/destinations/2006-08-31-hilton-head_x.htm.

Starnes, R. D. (2003). Introduction. In R. D. Starnes (Ed.), *Southern journeys: Tourism, history, and culture in the modern south* (pp. 1–14). Tuscaloosa, AL: University of Alabama Press.

Stewart, M. (1996). *"What nature suffers to groe": Life, labor, and landscape on the Georgia Coast, 1680–1920*. Athens, GA: University of Georgia Press.

Sutter, P. S. (2009). No more the backward region: Southern environmental history comes of age. In P. S. Sutter & C. J. Manganiello (Eds.), *Environmental history and the American South* (pp. 1–24). Athens, GA: University of Georgia Press.

Taylor, L. (2011). No boundaries: Exurbia and the study of contemporary urban dispersion. *Geojournal, 76*(4), 323–339. doi:10.1007/s10708-009-9300-y.

Walker, P., & Fortmann, L. (2003). Whose landscape? A political ecology of the "exurban" Sierra. *Cultural Geographies, 10*(4), 469–491. doi:10.1191/1474474003eu285oa.

Walker, P. A., & Hurley, P. T. (2011). *Planning paradise: Politics and visioning of land use in Oregon*. Tucson, AZ: University of Arizona Press.

Webster, B. (1970, February 1). Carolina plan to build a plant on unspoiled coast arouses protest. *New York Times*, 40.

White, R. (2004). From wilderness to hybrid landscapes: The cultural turn in environmental history. *The Historian, 66*(3), 557–564. doi:10.1002/9781444323610.ch10.

Zimmerman, J. (2001). The "nature" of urbanism on the new urbanist FRONTIER: Sustainable development, or defense of the suburban dream? *Urban Geography, 22*(3), 249–267. doi:10.2747/0272-3638.22.3.249.

Chapter 9
No (Back)Sliding: Amenity Migration, Viewsheds, and Contesting Steep Slope Ordinances in Western North Carolina

Jessica McCallum Breen, Patrick T. Hurley, and Laura E. Taylor

9.1 Introduction

The Appalachian Mountain region in North Carolina is a study in exurban transformation. Environmental politics in the region have always been strained. Fierce fights over timber clear-cutting, national forest and national wilderness park creation, and energy production were waged in the postwar era (Newfont 2012), but now these have given way to battles over the protection of scenic landscape features from exurban residential development. Development on steep slopes to some in the community threatens the "integrity" of these hillsides and creates safety hazards, while for others restrictions on development are unnecessary government interventions into land-use and property rights. These battles might be seen by geographers and other social scientists as extensions of familiar conflicts between environmentalist outsiders and long-time locals favoring continuing natural resource extraction, either through traditional resource commodity extraction or amenity real estate development. New land-use ordinances (in North Carolina these are local municipal laws, which direct comprehensive plan-making, zoning, and development review processes) were created to manage the exurban transition underway in the region,

J.M. Breen
Geography Department, University of Kentucky,
817 Patterson Office Tower, Lexington, KY 40506, USA
e-mail: jessica.breen@uky.edu

P.T. Hurley (✉)
Department of Environmental Studies, Ursinus College,
P.O. Box 1000, Collegeville, PA 19426, USA
e-mail: phurley@ursinus.edu

L.E. Taylor
Faculty of Environmental Studies, Planning Program, York University,
4700 Keele Street, Toronto, ON M3J 1P3, Canada
e-mail: taylorL9@yorku.ca

L.E. Taylor, P.T. Hurley (eds.), *A Comparative Political Ecology of Exurbia*,
DOI 10.1007/978-3-319-29462-9_9

represent a pro-environment approach to protecting natural heritage, but also enhance and create natural amenities for the benefit of real estate investment (see also Robbins et al. 2012).

Conflicts over environmental management practices and land-use priorities have long been a focus of geographic research in the American West and other areas of the country where amenity migration is a driver of socio-economic landscape change. Rather than representing a purely urban/rural or newcomer/long-time resident cultural and political clash, research on these conflicts demonstrates how they occur as the result of competing ideas of landscape stemming from opposing rural capitalisms (Walker and Fortmann 2003; Hurley and Walker 2004). By competing rural capitalisms, we mean the ways in which different industries and associated economic actors are tied to different ways of extracting value from landscapes and landscape features. Some long-term rural residents favor environmental regulatory practices generally reflective of their ties to extractive natural resource production with limited concern about environmental protection, while others embrace the adoption of new environmental management practices emphasizing the protection of scenic landscapes, recreational opportunities, or generally permitting consumptive uses of the landscape (Fortmann and Kusel 1990; Hurley and Walker 2004; Nesbitt and Weiner 2001; Reed 2007; Walker and Fortmann 2003). In so doing, both newcomers *and* long-term rural residents foster a transition of the local economy, from one focused on natural resource production to one focused on new markets in real estate and land development. This is the case in Jackson County, North Carolina, where amenity transformations like those found in the Sierra Nevada Mountains of California and other landscapes of the American West are underway in the southern Appalachians (Culbertson et al. 2008).

In Jackson County, a 2007 controversy surrounding new land-use regulation—specifically the creation and passage of subdivision and steep slope ordinances—reveals how an environmental management conflict fueled by amenity migration plays out within the land-use planning process in this region. This conflict takes place against the backdrop of Jackson County's transformation from an extractive economy based largely on timber to a natural amenity economy based on recreation and residential development. Given this economic restructuring, decisions about what landscapes are worthy of protection—including how and by whom they should be protected—are likely to become increasingly embattled. Our case study examines how particular communities in a county experiencing exurban dynamics emerge in favor of or in opposition to the creation and implementation of new landscape-related ordinances. In this study, we interviewed Jackson County residents about their opinions of the subdivision and steep slope ordinances of 2007 in an effort to better understand the extent to which discourse coalitions emerge in a place where exurban socio-demographics and corresponding ideas associated with competing rural capitalisms come to dominate these contests. Our results provide greater clarity on how opinions across land-use constituencies associated with competing rural capitalisms converge and diverge, with implications for how these coalitions may affect the County's future land-use planning and environmental management. These results have implications for planners and policymakers working on land-use regulation in exurbia.

9.2 Whose Slopes? Reterritorialization, Amenity Landscapes, and Competing Rural Capitalisms

With few studies of land-use conflict related to amenity migration in the eastern United States (see Nesbitt and Weiner 2001 for a notable exception), we turn to the existing literature on the American West for key insights about the drivers of this conflict and ways these conflicts are often centered on land-use regulation and planning arenas. In the American West, fights over protecting landscapes, natural heritage, conservation areas, and other natural amenities have come to define exurban land-use conflicts and land-use decision-making processes. Scholars increasingly recognize the important role that amenity migration plays in shaping this convergence of factors, including the complex ways in which rural economies shift from a focus on extracting natural resources to a focus on extracting real estate value from natural landscape features. Housing demand driven by individuals seeking second homes, a place to retire, and other forms associated with increased in-migration have meant an expanded demand for residential dwellings. Receiving communities increasingly feel they need to respond to new forces shaping landscape uses and aesthetics. Scholars highlight the ways these fights often center on processes that are transferring or may transfer control of resource access from one group to another (a process referred to as "reterritorialization" by Brogden and Greenberg 2003; see also Hurley and Walker 2004; Robbins 2006; Walker and Fortmann 2003). Key to these processes and the conflicts that result are questions about the following: the role science should play in formulating new rules over landscape and environmental management; the clash of landscape ideologies and visions about future development that shape the extent to which particular coalitions embrace particular forms of management; and whose vision of the future wins out within the decision-making processes that shape the material transformations of the landscapes in question (Hurley and Walker 2004; Walker and Hurley 2004).

Too often, as we have already noted, this convergence of opinions is framed in terms of "newcomer" vs. "long-time local," in which the former is seen as supportive of strong growth controls protecting environmental and amenity landscape values and the latter supportive of laissez-faire or "pro-growth" policies. Instead, we draw on the concept of competing rural capitalisms to provide an alternative framing. Competing rural capitalisms extends ideas about the changing forms of resource control that are part of the process of "rural societies shifting their focus, economies, and identities from one defined by extractive activities like forestry, mining, and agriculture into one defined by natural amenities such as scenic beauty, wildlife, and outdoor recreation" (Schewe et al. 2012, p. 4). Nature under traditional rural economies is understood in its commodity form, whose value under capitalism is realized as raw material ripped free of its landscape. Nature under the new amenity economy is valued in situ as part of a whole landscape, including biophysical patterns and features that are valued for their beauty, nature protection, and recreational opportunities. The shift is from one emphasizing production to consumption (Woods 2005). In our view, both represent different forms of extraction as both result in the environmental degradation of the landscape. Moreover, the shift from production to consumption is never smooth or complete, a common feature of exurbanization. The shift is uneven, producing spaces of extreme wealth and extreme poverty in different times and places (see Smith 2010 [1984]). Human–nature relations are marked by conflict at the local landscape level but changes come about because of global capital seeking to grow and find new means of exploiting nature. The concept of competing rural capitalisms considers the shift in ideologies of nature as necessary to understanding the shift in political-economic power relations that is the focus of reterritorialization. As described in the book's Introduction, this shift is often complex, uneven, and incomplete.

This complex, uneven, and incomplete shift is partly a result of the rise of land-use planning regimes in arenas where either they have largely never existed or where they were not needed. Thus, conflicts may center on whether land-use regulatory interventions are even necessary to protect continuing traditions of extracting economic value from local landscapes, whether those traditions center on natural resource extraction or forms of extraction associated with real estate development. Drawing on Reed (2007, discussed in the Introduction), we note the important ways in which changing drivers of amenity migration, when mixed with local politics, lead to "uneven environmental management" outcomes. That is, different coalitions within civil society come together—often with different planning and landscape priorities—to shape the ways iconic landscapes and vulnerable forms of nature become regulated or protected. Even when similar globally influenced ideas about landscape protection and ecological restoration come into play, local social and political relations mediate how those processes shape local landscapes.

Our chapter is a case study of a step in one community's transition from a resource- to amenity-based economy and is intended to reflect on the politics of how a particular land-use planning regime emerges. The focus is on the ways conflicts over land-use decision-making emerge and which regulatory approaches come to be supported by different discursive coalitions. Our case study from the southern Appalachia has a number of affinities with existing research from the American West. First, fights over residential units associated with resort-development outside of legislatively established urban growth boundaries in Central Oregon highlight efforts to avoid setting precedents about the forms of urbanization that may threaten natural resource use and livelihoods tied to the rural countryside (Walker and Hurley 2011). Second, struggles over land-use decision-making in Montana focus on conflicts about efforts to minimize the impact of unregulated growth on the amenity value of landscapes, including the ways in which developers and long-time residents come to support regulatory efforts (Ghose 2004; Robbins et al. 2012). Critically, these locals may do so in situations where there is no history of widespread support for—but rather a history of outright revolt against—efforts to develop formalized plans and planning procedures. Third, conflicts over planning centered on landscape protection in Nevada County, California highlight struggles over competing visions and the ways that technical scientific processes may heighten anxiety within the electorates that ultimately determine who will make land-use decisions (Walker and Fortmann 2003; Hurley and Walker 2004).

The Nevada County case, in particular, provides a number of important points of comparison for our case study. By attempting to impose their amenity view of landscape and their idea of "rural" on Nevada County, exurban in-migrants and the rural form of real estate capitalism challenged traditional ideals of the landscape as a source of commodity production and traditional political control of the county's government (Walker and Fortmann 2003). The political response by voters suggested that concerns over environmental change and landscape degradation were important, but that particular regulatory approaches to conserving or protecting landscape ideals might not be acceptable to a majority of voters within an area characterized by exurban dynamics (Hurley and Walker 2004). Despite their concerns about the environmental impacts of residential development, efforts that prioritized ecological science over local understandings—those of the people living in Nevada County—proved untenable. Importantly, this case raises serious questions about what types of landscape regulation are acceptable within an exurban transition. This chapter provides a comparable example, in which we further highlight the types of regulatory approaches contested by competing forms of rural capitalism, even when those competing forms are centered on different visions of residential development (and not just traditional forms of resource extraction vs. residential development).

In the remainder of the chapter, we first describe the County to give background context. We then discuss our methodological approach to surveying local actors about their opinions using the Q method (described below). Finally, we offer conclusions about the success of establishing a new planning regime based on local coalitions that completely disrupt assumptions about simple newcomer vs. old-timer coalitions.

9.3 Case Study Overview

The Jackson County Chamber of Commerce had good reason to choose mountainlovers.com as their Internet domain name. Nestled in the far western corner of North Carolina, but within an approximately 3-hour drive of both Atlanta, Georgia and Charlotte, North Carolina, Jackson County boasts scenic mountain lakes, waterfalls, and pristine forests. Indeed, the county's webpage intones, "Thanks for visiting Jackson County, a place of lofty peaks, rushing water and spectacular scenery" and then proceeds to highlight its proximity to both the Great Smoky Mountains National Park and the Blue Ridge Parkway as well as the plentiful opportunities for outdoor recreation. The county's rugged topography includes "some 185 named summits and approximately 250 named rivers and creeks," which provide "lofty vistas and fast-flowing water." Richland Balsam, the county's tallest peak, stands at 6140 ft and rushing waters fall some 411 ft at Whitewater Falls. The website further highlights that the county is "nationally known for the quality of our trout-fishing waters. Our Tuckaseegee River is viewed as a fly-fisherman's paradise, and beautiful mountain lakes along the river's East and West forks provide opportunities for boating and other outdoor recreation" (Jackson County 2014). Many of these tourist-orientated features can be found inside and outside of the Nantahala National Forest. Jackson County's abundance of natural amenities and public lands has not gone unnoticed by tourists, second homeowners, retirees, and other amenity migrants seeking these natural landscape characteristics.

Between 1990 and 2010, Jackson County saw a 24 % increase in population, growing from 26,846 to 33,121, 91 % of which was attributable to in-migration, as many former visitors to the area chose to make Jackson County their permanent home (Jackson County Planning Board 2005). Between 2000 and 2010, the county's population grew to 40,271, with recent estimates pointing to 40,919 in 2013, less than projected in the county's 2006 planning document, suggesting that in-migration may be slowing but that it still continues. Further, of the county's 19,291 total housing units in 2000, nearly 24 % of them (compared to just under 4 % for the State of North Carolina) were characterized by seasonal use. This large number reflects the County's significant number of tourist-oriented residential units and second homes (many of which are owned by people who have primary residences in Florida due to that state's absence of an income tax and lower property taxes). These numbers point to an appreciable majority of amenity migrants currently living in the county.

Jackson County was created in 1851 in response to the difficulty, because of distance and terrain, certain landowners had in being able to file legal paperwork related to land use and management (Williams 2001). In the years since, the county's economy has largely centered on natural resource extraction, particularly logging and mining augmented by agriculture, as well as manufacturing and education. While logging activity has diminished in Jackson County, lumber mills have persisted in the County and there is still a paper mill located in Sylva, the County seat. Mining operations in Jackson County wound down in the 1960s, but once included kaolin clay, mica, copper, corundum, olivine, gold, nickel, chrome, talc, feldspar, and vermiculite (Williams 2001). Agriculture in Jackson County has been steadily declining since the 1970s, with an increasingly smaller portion of suitable acreage in agricultural production each year. The decline in local agriculture has been sped along by development pressures that facilitate the conversion of agricultural land into residential home sites. Manufacturing has been declining in Jackson County as elsewhere in the United States as factories move overseas motivated by cheap labor and less stringent environmental rules. In large part, the economic declines of some sectors have been balanced by growth in others.

The service economy is one area where growth has increased. Education jobs have been on the rise as enrollments at Western Carolina University and Southwestern Community College have increased. Healthcare jobs have also increased as the County's population increases in age. Tourism has been a source of economic activity in Jackson County since the 1840s when the Town of Cashiers started hosting wealthy South Carolinians seeking to escape the summer heat (Williams 2001).

The boundary of the county generally runs along the ridgeline, so the mountains literally surround the historic settlements in the valleys of the County. The dedication of the Great Smoky Mountains National Park in 1934 and the Blue Ridge Parkway in 1935 helped to draw more tourists to the region. Improvements in infrastructure including the creation of NC Highway 107, the Great Smoky Mountains Expressway (US Hwy 74), and the Jackson County Municipal Airport have increased the accessibility of the area. The transition from rural to exurban may have been a long, slow process, but the acceleration of growth in recent years, including an increased demand for housing, infrastructure, and services, has brought competing rural capitalisms into sharp focus.

Early land planning efforts began in 1979 in response to a Congressional mandate in Title IV of the Housing and Community Development Act of 1973, which required recipients of Housing and Urban Development Agency (HUD) funds to conduct a comprehensive planning process. Together with the seven far western counties of North Carolina, Jackson County produced a joint land development guide. A complete land-use evaluation including socio-economic and biological data, the guide classified the then current land uses in each county and predicted future needs and concerns based on anticipated population growth. The shifting demographic landscape of Jackson County was notable even in 1979 and the guide notes the exceptional levels of in-migration and predicted that, without increased planning, conflict would arise between long-term rural residents and in-migrants (Ferell and Killian 1978).

It was not until 2006 that Jackson County would make a comprehensive update to its land-use plan. Noting the particularly steep and rugged terrain, the plan indicated that almost half of land in the County was unbuildable (Jackson County 2005). Moreover, early on in the document the authors pointed out the dramatic number of new subdivisions, some of which were built as gated communities and many of which were being constructed by companies from outside the county and state. The report called out the dramatic increase in building lots they represented—some 5454—and the projected additional 10,000 plus people that would result. This trend was in addition to the more common, piecemeal smaller-scale parcelization patterns. Meanwhile, several hazard events occurred in the surrounding area that brought into question the safety of the mountainside construction that had become commonplace, including several landslides; most notably was a hurricane-induced landslide in Macon County that killed five people (Martin 2007). There had also been a battle in the County to prevent the Smoke Rise Gun Club from siting a new shooting range for their members in first the Caney Fork and later the Tilley Creek communities (Kucharski 2005). Many in the County believed it was time to take action.

A perfect storm of concern among Jackson County residents concerning the environmental, economic, and related social changes that were going on in the County led to the election of two new County Commissioners. Like their predecessors and those on the board, they were male and were born and raised in the county. They would join one returning Commissioner in endorsing a platform of land-use planning in 2005, which would lead to the creation of the new plan. Moreover, the new Commissioners would eventually help pass specific ordinances preserving farmland, restricting light pollution, bolstering existing erosion control guidelines, and regulating subdivisions and steep slope development. But faced with ongoing development pressure, the newly elected commissioners also chose to re-evaluate the County's land-use regulations regarding land subdivision more broadly, leading to the creation of several controversial *new* ordinances.

Recognizing concerns about the potential impacts of continuing development, the Board of Commissioners worked with the county's newly expanded planning board and a new county planner to create a Subdivision Ordinance and the aptly named Mountaintop and Hillside Development Ordinance (generally referred to as the "steep slope" ordinance). The report supporting the subdivision ordinance specifically refers to the creation of "2369 new subdivision lots" between "January, 2005 and January, 2007"; it also states that, "growth being experienced by Jackson County is promoting sprawl, traffic congestion, air pollution, and the loss of prime agricultural and forest land due to inefficient and excessive land consumption" (Jackson County 2007, p. 1). Critically, the report recognized that "one of the goals [of the ordinance] is the preservation of the county's rural character" (Jackson County 2007, p. 2) and addressed landscape issues, pointing to the fact that "pasture lands are disappearing and once uninhabited mountain slopes and ridgetops are being developed" (Jackson County 2007, p. 1). The resulting ordinance sought to, among other things, ensure "economically [sic] and stable development," "provision of needed community open space in new land developments through dedication or reservation of land…or provision of funds in lieu of dedication," and "wise and timely development of new areas, in harmony with…official plans of the county" (Jackson County 2007, p. 3). In many ways, then, this ordinance addressed longstanding concerns with negative impacts of residential development (see Rome 2001) and specified planning approaches commonly used across the country.

Meanwhile, the "steep slope" ordinance specifically acknowledged that "the protection of mountain hillsides is an urgent matter given growing public concern about development on them." In particular, it states the possibility that improper development would increase erosion, create slope failures and other hazards, destroy "unique vegetation communities and wildlife habitats", and threaten aesthetic values for municipalities and the "community's sense of identity" as well as "natural beauty which is vital to the tourism and recreation industry" (Jackson County 2007, pp. 2–3). The resulting ordinance requires that, "prior to commencing any development or land disturbed activity and prior to making application for any permits and/or other approvals," (Jackson County 2014) a slope analysis must be conducted and approved by Jackson County Planning Director to ensure that a specified process of technical review for building plans would be followed and to set out methods to be used for construction. Perhaps most importantly from the perspective of development, the ordinance set new minimum lot sizes and density limits, which increased in size and decreased in density as "average slopes" got steeper. For slopes with gradients between 30 and 34 %, the minimum lot size would be 2 acres and a maximum density for clustered building sites of 0.5 lots per acre. By contrast, slopes of 45 % or more would require a minimum lot size of 10 acres or more or 0.1 lots per acre. In the process, the ordinance likely reduced perceived development entitlements on existing private parcels, but did so in keeping with limits on steep slope development elsewhere and in a way that still allowed for some types of development, such as clustered communities, that would provide diverse forms of housing in the county.

These two ordinances, which County Commissioners claimed were necessary to prevent the sort of natural disasters seen in other counties (Martin 2007), directly affected the building boom that Jackson County was experiencing at the time. The new provisions would ensure an overall reduction in the number of dwelling units that could be built as well as reducing the extent of development in particular areas. Conflict between the Commissioners and their supporters, and development and construction interests, was further heightened when, in an effort to prevent a land rush upon release of the plan, the County Commissioners announced their intention to set a 5-month moratorium on new subdivisions (Hotaling and Hall 2007). Jackson County erupted. Opponents took out full-page ads in the County's only weekly newspaper in the name of "Property Owners of Jackson County Organization," decrying the devastation to the County's economy they asserted would occur if the moratorium passed.

The moratorium and the ordinances were bitterly debated in the Board of Commissioners meetings. Typically unattended and generally ignored by the public, the meetings were swamped with concerned citizens wanting to opine on the newly proposed rules. In July 2007 more than 1300 people converged on the county seat of Sylva to participate in a public hearing on the matter (Martin 2007). The moratorium passed with a 4–1 vote, with the Chairman, an employee of one of the larger, gated developments, voting in the negative. The moratorium was lifted 5 months later as another 4–1 vote led to the passage of the subdivision and steep slope ordinances (Gobel 2007). For their work in passing what arguably became the most stringent land-use ordinances in the State of North Carolina, the Jackson County Board of Commissioners was honored with the Governor's Conservation Achievement Award for Municipal Conservationist of the Year by the North Carolina Wildlife Federation in conjunction with the North Carolina Governor's Office.

The ordinances are explicitly intended to regulate land use with the goal of protecting amenity landscapes. They specifically identify viewsheds and aesthetics as key elements to protect. Indeed, these are the very sort of planning policies one would expect to appease amenity migrants. At the same time, the kind of large-lot rural residential development these ordinances were attempting to restrict as well as the more desirable picturesque locations where development might occur are the very sites that amenity migrants seek out: mountaintop parcels with expansive views. On the one hand, the ordinances create a competition between the very aspects of rural landscapes that competing forms of rural capitalism depend upon: between the protection of landscape qualities and features versus the development of those same features. On the other hand, the ordinances, through protected area zoning, can leave material places in the landscape untransformed by development, in terms of a reduced number of newly carved out building sites, enhancing the natural landscape qualities that attract development.

Support for the new ordinances, then, might be interpreted as an attempt by amenity migrants (nearly all newcomers) to "close the door" behind them and prohibit other amenity migrants from coming into the area. In the process, long-time locals and newcomers agree that landscapes and the steep slopes that produce hazards are seen as a commons needing restrictive government regulation. At the same time, support for the ordinances by long-time residents in this rural area might be seen as protecting the very same landscapes and qualities that are part of longstanding traditions, having more to do with supporting livelihoods and less with scenic amenities.

Yet if these ordinances were genuinely without support from newcomers and long-term rural residents, suggesting a desire to continue traditional extractive uses and with little government implemented land-use regulation, we would expect to see the sort of political backlash that followed the failed Natural Heritage 2020 plan in Nevada County (Hurley and Walker 2004; Walker and Hurley 2004). This type of backlash was not forthcoming; indeed, all five members of the Jackson County Board of Commissioners retained their posts in the election following the adoption of the ordinances. Natural amenity and landscape protection were environmental imaginaries that prevailed in this case with the resulting reterritorialization of the landscape. Despite the controversy that surrounded the passing of these two ordinances, protective views among residents (not developer and builders interested in continued pro-growth perspectives) supported these regulations that would shape future growth and change in this Appalachian Mountain landscape. To be sure, long-term building would not stop, but its scope, extent, and location within the landscape would be very different. The political ecology of landscape has shifted where the capital transformation hinges on the shift in environmental ideology in the local communities that supports the regulation of land-use favoring amenity development.

9.4 Locating Land-Use Coalitions Within Exurban Communities

While the competing rural capitalisms concept helps to identify the key political-economic sources of conflict in an area experiencing amenity migration, it does not specifically provide a mechanism for better understanding how particular constituencies or coalition communities arise. Indeed, it leaves scholars in a similar position as the "newcomer" vs. "long-time local" binary it replaces (see Chap. 3, Hiner). Given our desire to better understand the dynamics of socio-cultural change in restructuring rural capitalisms, we chose to investigate how different actors align their landscape values (within which are embedded livelihoods, ancestral and contemporary family ties, and nature ideologies) with environmental regulation. To do so, we use a methodology designed specifically for the study of human subjectivity. Commonly known as the Q method (McKeown and Thomas 1988), it has been applied to a broad range of social science research over the past 75 years, including political ecology research on environmental management conflicts and other environmental studies (see Addams and Proops 2000; Dayton 2000; Niemeyer et al.

2005; Robbins 2006). An advantage of using Q methodology versus other discourse analysis techniques is that it provides for a consistent comparison of participants' responses, as each are reacting to the same set of stimuli (Webler et al. 2009). The Q method also reveals the trade-offs people make between competing ideas, something that can be lost in standard survey methodologies.

Unlike traditional survey methodology, the Q method does not require taking a representative sample of individuals from a population. Participants are selected for the intensity and breadth of their opinions rather than their demographic representation, thus minimal demographic data about participants is collected. However, there are some things about the participants in this case that we can say for certain. Participants were generally white and while some do possess Cherokee ancestry, none were registered members of the Eastern Band of Cherokee. Tenure in Jackson County among participants ranged from lifelong residence (for one 80 year-old participant) to less than 1 year. Participants ranged in age from their 30s to their 80s. None of the participants, as far as we know, were residents of gated communities. Livelihoods of participants included a real estate agent, student, professor, farmer, schoolteacher, logger, and more. Participants were fairly evenly split across gender lines.

Prior to beginning key informant interviews, informal meetings were held with local real estate agents, county employees, and citizens interested in the ordinances. A combination of convenience and snowball sampling methods was used to identify probable participants beginning with individuals known to the author, local government officials, and individuals who wrote letters to the editor or spoke at public meetings on the subdivision and steep slope ordinances. Participants were asked to sort pre-prepared printed statements into those that were "most like the way (they) think about land use in Jackson County" and those that were "least like" (see Fig. 9.1). The placements of statements were recorded and a factor analysis was computed. The statements provided to the participants were as follows:

- Our primary cash cows in this county are education and tourism.
- This county has nothing else to offer, it has to develop the property it has.
- Developers and real estate agents are clearly concerned for the future and the community's general welfare.
- Developers put these cutesy names on new subdivisions like preserve or reserve to make you think they are real conservation minded, but it's just marketing.
- Growth is going to happen here, but every square foot should not be allowed to be a potential house site.
- There are some places we're not allowed to go to anymore because they've been bought up and gates and signs put up around them for exclusive gated community type things and that's not right.
- The removal of thousands and thousands of acres of land from the public domain put in the hands of a few wealthy people for their own amusement and pleasure is wrong.
- People supported the steep slope and subdivision ordinances because they saw what the outside developers were doing and how it raises their personal property taxes and wanted to work to try to curtail that.

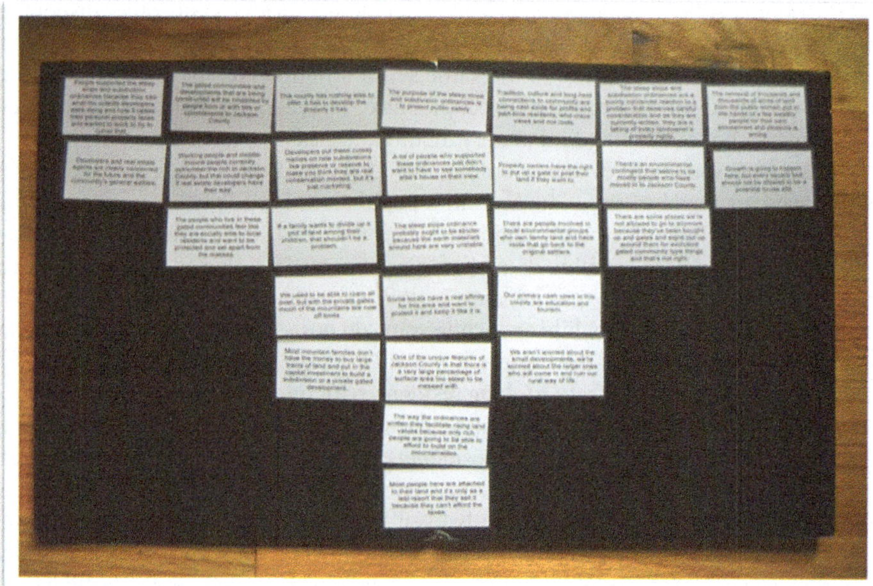

Fig. 9.1 The sorting board used by participants, shown with notecards in place

- Most people here are attached to their land and it's only as a last resort that they sell it because they can't afford the taxes.
- Most mountain families don't have the money to buy large tracts of land and put it in the capital investment to build a subdivision or a private gated development.
- Working people and middle-income people currently outnumber the rich in Jackson County, but that could change if real estate developers have their way.
- The people who live in these gated communities feel like they are socially elite to local residents and want to be protected and set apart from the masses.
- We aren't worried about the small developments, we're worried about the larger ones who will come in and ruin our rural way of life.
- Tradition, culture, and long-held connections to community are being cast aside for profits and part-time residents, who crave views and not roots.
- The gated communities and developments that are being constructed will be inhabited by people from or with ties or commitments to Jackson County.
- The steep slope ordinance probably ought to be stricter because the earth materials around here are very unstable.
- We used to be able to roam all over, but with the private gates, much of the mountains are now off limits.
- One of the unique features of Jackson County is that there is a very large percentage of surface area too steep to be messed with.

- If a family wants to divide up a plot of land among their children, that shouldn't be a problem.
- The steep slope and subdivision ordinances are a poorly conceived reaction to a problem that deserves careful consideration and as they are currently written, they are a taking of every landowner's property rights.
- The way the ordinances are written they facilitate rising land values because only rich people are going to be able to afford to build on the mountainsides.
- A lot of people who supported these ordinances just didn't want to have to see somebody else's house in their view.
- Property owners have the right to put up a gate or post their land if they want to.
- The purpose of the steep slope and subdivision ordinances is to protect public safety.
- There are people involved in local environmental groups who own family land and have roots that go back to the original settlers.
- Some locals have a real affinity for this area and want to protect it and keep it like it is. There's an environmental contingent that seems to be mostly people who have moved in to Jackson County.

The pre-prepared statements were collected from archival research of the Sylvia Herald, the county's only newspaper, and a series of 12 semi-structured interviews with key informants that took place in the Fall of 2010. Archival research determined that March 15, 2007 was the first mention of the proposed ordinances in the newspaper. Interviews with respondents were conducted from mid-January to mid-February of 2011 in locations participants requested to ensure their comfort. Some Q-sorts were conducted in participant's homes, others in their offices or a public venue like the local coffee shop. The findings were subject to content analysis and nine emergent themes were identified. Three representative statements from each of the themes were selected and then printed on notecards. Some statements were revised for length or general clarity from the original, but editing was minimized in order to preserve the voice and intent of the original speaker. The themes or "factors" captured the divergence of residents' values with respect to the proposed slope regulations. Based on the archival research and interviews, we found that people were generally concerned about the effect on landscape and culture, livelihood and the economic transition.

9.5 No (Back)Sliding? Convergent and Divergent Community Opinions about Steep Slope Ordinances

Our study confirms that land-use decision-making is a nexus for tension within emerging rural capitalisms, where aspects of reterritorialization are both encouraged and countered by people in the community. Still, we seek to demonstrate that coalitions emerge within the competing rural capitalism frame. Similarities in

Table 9.1 Factor characteristics

	F1	F2	F3
Eigenvalue	8.4567	1.8613	1.3015
No. of defining variables	8	5	5
Average rel. coef.	0.80	0.80	0.80
Composite reliability	0.970	0.952	0.952
S.E. of factor scores	0.174	0.218	0.218

responses between those who one might think would have very different aspirations for the County, in fact, have a high degree of consensus. Q method's consensus statements refer to points of agreement in the discourse; they can identify non-issues, opinions that no consensus group is particularly attached to, or they can identify rallying points. This approach allows us to better see how and why particular coalitions form. In doing so, they provide key evidence as to why a particular land-use regulatory approach was acceptable within the Jackson County exurban context.

There is a substantial amount of consensus among the three factors, or clusters of opinion, identified in the factor analysis. Respondents focus on the low impact of natural resource use and its relation to development, the role of real estate in extracting value from the landscape, and issues related to public safety. All three factors have responses that highlight strong negative opinions about developers and real estate agents. They all agree that there is a large percentage of land in Jackson County that is too steep to be developed and that there should be some limits on development. They also all agree that land development is not the only option for Jackson County's future. These are all things that could be the basis for greater cooperation between the three factors (Table 9.1).

Consensus also emerged around a group of issues that were not of strong importance to any of the three factors. These "non-issues" included the following: (1) the ability of mountain families to afford to build large developments themselves; (2) the perceived elitism of people living in gated communities; (3) the ability of a family to subdivide land for its children; and (4) support for the steep slope and subdivision ordinances being based in a desire to combat rising property taxes. Importantly, these results do not indicate a lack of concern about these topics by our informants, but rather that there are other topics about which our informants have stronger feelings. These consensus statements indicate issues that provide the opportunity for compromise, as they are not particularly salient to any one factor.

Distinguishing statements are those statements that were significantly different from one factor to another. These illustrate the conflict between and among factors. Below, we describe the distinguishing statements for each factor in terms of the themes they illustrate and the types of individuals who loaded heavily on these factors. We also use information gleaned from the recreation of the discourse to flesh out the rough skeleton of opinion provided by the factor analysis.

9.5.1 *Convergent Communities: Low Impact Resource Use and Amenity*

This factor focuses on traditional access, loss of community, and classism/elitism. Such factors speak to issues surrounding natural resource use and access on forested private land as well as the ways that changing social-economic demographics may be (or are perceived to be) challenging other shared social norms. These are areas that scholarship on social–political dynamics in exurbia often suggests as key dimensions of land-use conflict. Interestingly, the individuals who loaded heavily on this factor included amenity migrants and a *particular* sub-set of long-term rural residents. The combination of the two demographic groups is what makes this factor particularly interesting.

The long-term rural residents whose Q-sorts aligned with this factor are notable for their focus on the preservation of Appalachia's cultural heritage (Table 9.2). They are concerned with rising property values and their tax implications, because of the potential that residents and their children will be priced off their land and away from their heritage. Changes in the rural community concern them because of the impacts on the continuation of cultural traditions, such as specific foodways, dialects, religious observances, and livelihood practices. They are also concerned about the loss of traditional access to recreation areas and to places to gather non-timber forest products, including ginseng and ramps (wild leeks). These residents see large-scale gated developments as a threat to natural resource access, and support land-use planning as a means of curtailing new developments that will restrict access. These respondents draw parallels between the current real estate developers, many of whom are not from the region, and the logging companies that beset the area in the late 1800s. The logging companies bought up vast tracts of forestland and employed local residents in their operations, but ultimately when the trees were gone the logging companies left, along with the jobs, leaving residents to cope with the environmental devastation from the large-scale clear-cut logging operations. More recently, this cynicism about hit-and-run "outsider" development was

Table 9.2 Distinguishing statements for Factor 1—convergent communities: low impact resource use and amenity

Statement	Rank	Score
Tradition, culture, and long-held connections to community are being cast aside for profits and part-time residents, who crave views and not roots	3	1.91
We aren't worried about the small developments, we're worried about the larger ones who will come in and ruin our rural way of life	0	0.04
Working people and middle-income people currently outnumber the rich in Jackson County, but that could change if real estate developers have their way	0	–0.09
Most people here are attached to their land and it's only as a last resort that they sell it because they can't afford the taxes	0	–0.15

bolstered by the environmental damage left behind when the 2008 global economic downturn resulted in the failure of several real estate development companies operating in the County, who left the landscape scarred with partially cleared subdivisions. The resulting land clearance, including removal of vegetation and reworked earth, may or may not be temporary.

Amenity migrants in Jackson County are generally wealthier and more educated than the average Jackson County resident (Jackson County 2005). They are "classic" exurbanites, as they very often originate in more urban areas and bring with them urban-specific environmental ideas born of their experiences in city-oriented environments. Amenity migrants whose Q-sorts aligned with this factor also oppose large-scale development and support land-use planning initiatives that protect environmental quality and viewsheds. Given these concerns, their opinions align with preservation-minded long-term rural residents and are based in the desire of amenity migrants to maintain the qualities of the area that drew them to it in the first place. Though individually they may be more concerned for the biophysical environment than the preservation of Appalachian culture, amenity migrants benefit from the political clout of long-term rural residents and see this coalition as advantageous. During an interview, an amenity migrant stated that in public meetings, they make certain that the individual speaking for their position possesses a local accent. The interviewee expressed that they felt that this choice meant their position was more likely to be listened to by long-term rural residents.

The two groups that converge on this factor do make concessions to each other. These groups support planned development that respects the region's unique Appalachian heritage. While amenity migrants might prefer to severely limit development to retain their image of an ideal rural landscape, long-term rural residents fear that their children and grandchildren will have to leave the area to find work. The continuing loss of younger generations to the wider world because of a lack of jobs at home is a severe threat to Appalachian culture. For their part, the long-term rural residents understand that there are aspects of Appalachian culture that many amenity migrants find unappealing, such as the hunting of bears using dogs and snake-handling. While long-term rural residents do not disown those unappealing traditions, they do tend to focus much of their efforts on the preservation of activities and landscape uses more palatable to amenity migrants, thus ameliorating some of the potential exurban/rural culture clash that might otherwise occur.

Many people during the recreation of the discourse described gated communities in terms of classism and spoke of the residents who live in these mostly second-home developments as being elitist. This strategy of differentiation by year-round amenity migrants seemed like a strategy to differentiate themselves from a different class of exurbanite in order to be part of the community, thereby distancing themselves from the more undesirable form of amenity migrant: the disengaged seasonal, second homeowner.

9.5.2 *Divergent Communities: Extracting Real Estate Gold from Steep Slopes*

The cluster of opinions that define Factor 2 strongly reflect the area's ties to natural resource extraction (Table 9.3). With twice as many distinguishing statements as the other two factors, the major themes in this factor are property rights, traditional access, and elitism/classism. The unifying trait of all the individuals who loaded highly on this factor was a livelihood based in natural resource extraction, whether it was a traditional form, such as logging or farming, or the more modern incarnation of real estate development and construction. These individuals are primarily long-term rural residents, though some are more recent arrivals, and they regard land-use planning, particularly zoning, as an attack on their property rights and livelihoods. They are skeptical of the motives of people who support land-use planning, often viewing such support as self-serving and elitist. Those that have longstanding family ties to the area resent what they see as the implication that they are not good stewards of *their* land. They acknowledge the hazards of building on steep slopes, but believe that the extent of the hazard has been over-stated and that problems with landslides and erosion have been the result of landowners' unwillingness to spend the necessary money to do the work properly, and/or their tendency

Table 9.3 Distinguishing statements for Factor 2—divergent communities: extracting real estate gold from steep slopes

Statement	Rank	Score
The way the ordinances are written they facilitate rising land values because only rich people are going to be able to afford to build on the mountainsides	3	1.72
Property owners have the right to put up a gate or post their land if they want to	3	1.52
A lot of people who supported these ordinances just didn't want to have to see somebody else's house in their view	2	1.42
There's an environmental contingent that seems to be mostly people who have moved in to Jackson County	1	0.53
The removal of thousands and thousands of acres of land from the public domain put in the hands of a few wealthy people for their own amusement and pleasure is wrong	0	0.14
The steep slope and subdivision ordinances are a poorly conceived reaction to a problem that deserves careful consideration and as they are currently written, they are a taking of every landowner's property rights	0	0.09
Tradition, culture, and long-held connections to community are being cast aside for profits and part-time residents, who crave views and not roots	0	−0.07
There are some places we're not allowed to go to anymore because they've been bought up and gates and signs put up around them for exclusive gated community type things and that's not right	−2	−0.96
We used to be able to roam all over, but with the private gates, much of the mountains are now off limits	−3	−1.56

to hire workers from outside the County who are ignorant of how to do the work safely and effectively. They are concerned about rising land values and changes to their rural community, but do not see land-use regulations like the steep slope and subdivision ordinances as being the solution to the problem.

One of the key differences between Factor 2 and Factor 1 are the rankings of the statements regarding traditional access. Factor 1 indicates that people's freedom to traverse the mountains was being impeded by gates and posted land. In contrast, Factor 2 disagrees strongly with that same concept. It is not that individuals who loaded heavily on Factor 2 do not believe that there has been an increase in the number of gates or that they do not think that gates impede people's ability to move across the mountains, but rather they simply disagree that they were ever able to freely roam the mountains. Nearly everyone interviewed was queried about how access to privately owned land was traditionally gained and the response was universal: the standard practice was to ask the landowner for their permission to enter their property for hiking, hunting, fishing, or collecting forest resources and it was usually granted. On occasions when it was not granted, the custom was to abide by the landowner's decision and go elsewhere. Why the two factors diverge so strongly on this issue is not clear. Potentially, Factor 2 includes individuals who do not believe people were ever free to roam the mountains. Rather, these individuals acknowledge that one had to ask permission to access someone else's land. By contrast, individuals in Factor 1 may be basing their response on the change in frequency with which permission is granted today or indeed the degree of comfort with newcomers for access at all, where exurbanites value the seclusion and wilderness of their properties.

9.5.3 Community Awakenings: Protecting the General Public

Close correlation between factors is an indicator that the factors are saying similar things. The correlation between Factor 1 and Factor 3 is .77, which is fairly high. We might have chosen to omit Factor 3 from our analysis, because it is statically so similar to Factor 1. However, participants varied widely on their loadings between the two factors, an indicator that, though the factors are similar, there is some aspect that is distinct .

The distinguishing statements for Factor 3 are primarily concerned with public safety (Table 9.4). The individuals who loaded heavily on this factor were all in-migrants associated with Western Carolina University. A regional comprehensive college founded in 1889, the University is part of the local University of North Carolina-system and is located in unincorporated Cullowhee. This factor reflects concern over a series of landslides in neighboring counties that resulted in several deaths. In contrast to Factor 2, Factor 3 believes that the steep slope ordinance should be stricter to protect public safety. Individuals who loaded heavily on this factor demonstrate some concern for changes in the community caused by development, but lack the longstanding family ties to the region that might make these landscape changes more salient to their use of natural resources.

Table 9.4 Distinguishing statements for Factor 3—community awakenings: protecting the general public

Statement	Rank	Score
The steep slope ordinance probably ought to be stricter because the earth materials around here are very unstable	3	1.72
The purpose of the steep slope and subdivision ordinances is to protect public safety	1	0.67
Tradition, culture, and long-held connections to community are being cast aside for profits and part-time residents, who crave views and not roots	1	0.63
There are people involved in local environmental groups who own family land and have roots that go back to the original settlers	−1	−0.40

9.6 Impacts of Amenity Migration on Future Land-Use Planning in Jackson County

Appalachia possesses a famously insular culture. While in Nevada County, California amenity migrants were able to gain elected office and set the environmental management agenda (Hurley and Walker 2004), the story in Jackson County is not the same. Instead, the process of reterritorialization in Jackson County is playing out differently, with newcomers showing some success at reshaping natural resource access dimensions, particularly when it comes to questions of continued access for traditional landscape uses (e.g., access for hunting or gathering). There appears to be a clear correspondence of this reterritorialization process with the efforts to protect landscapes and their amenity values. But this push does not appear to come from the forceful interventions of amenity migrants or through the imposition of some type of outside values. Instead, the push for landscape protection and efforts to protect an "amenity commons" appear to come from a widespread coalition of individuals living within the community, which is grounded in widespread concern about the continued impacts of residential development and the resulting loss of access to lands for traditional recreational uses.

We do not mean to suggest that amenity migrants do not have any influence on local land-use planning, but to the extent that they do, their influence clearly comes through coalitions with long-term rural locals, particularly if exurbanites want to have any sort of control over the decision-making process that determines how landscapes are transformed. Interestingly, the extent to which amenity migrants can effect change may be constrained by the choices these landowners make about how to benefit from preferential income taxation out-of-state (therefore wishing to avoid registering as permanent residents of North Carolina) and, thus, whether they can directly influence voter outcomes.

In practice, those migrants who are concerned about protecting landscape amenity values have been fairly successful in combining forces with long-term rural residents motivated to preserve the area's Appalachian heritage. The success of this coalition is evidenced by the positive outcome of the effort to prevent the relocation of a private gun club to two rural neighborhoods (Kucharski 2005) and the election of three pro-planning Commissioners to the five-member County Board. But again, it should be noted that this is a tactic that only works for amenity migrants who do not live in gated communities, for reasons inherent to the goals of the long-term rural residents involved. For amenity migrants who do live in the large gated communities, their hope for power is contingent upon their rising numbers in the county, a process largely curtailed (at least for the moment) by the global housing bust. In this sense, the full implications of reterritorialization may not be clear. And the continued in-migration of amenity residents (and developers; see Hurley 2013) may mean further conflicts between the competing rural capitalisms involved, with new potential threats to and opportunities for shaping particular landscape features.

Though they are not specifically amenity migrants in the same sense as many second homeowners and retirees, the influence of in-migrants affiliated with Western Carolina University—some of whom are exurbanites themselves—should also be taken into account. In contrast to amenity migrants, many of whom retain residency in other states, this group of in-migrants are generally residents of Jackson County and thus able to vote in local elections. They are well-educated and as such tend toward political involvement. Their environmental management preferences are likely to be similar to those of amenity migrants and they too lack the longstanding family ties to the area possessed by long-term rural residents. However, they do not possess the income to reside in the large gated communities and their need to get to work is likely to keep them off the mountainsides favored by amenity migrants, where steep mountain roads are impassable in bad weather and even a little ice deters the most adventurous four-wheel drivers. At the same time, there is potential for this group to coalesce with the amenity migrants in support of more proactive environmental management schemes, but that does not yet appear to be the case.

9.7 Conclusions

Similar to efforts in Nevada County, California, the land-use ordinances passed by the Jackson County Board of Commissioners in 2007 were intended to guide development in the county toward a particular vision of the future (Hurley and Walker 2004). The vision was one in which new residential housing would be minimized in areas with iconic landscape features and one that sought to protect natural amenity values. Our research uncovered three distinct factors, or clusters of convergent and divergent opinions, among the population regarding land-use planning in Jackson County. The first sought to use land-use planning to stave off a loss of community and protect a unique culture. The second spoke to long-term rural residents who retain their ties to a history of natural resource extraction and its opposition to land-use regulations. The third factor supported land-use planning as a means of hazard prevention in light of a series of landslide events in neighboring counties.

In an area like Jackson County, with a long history of natural resource extraction and more recent amenity migration, we expect to see conflict between competing rural capitalisms. New in-migrants and long-term rural residents have an increasing interest in protecting landscape qualities. Being able to achieve landscape protection is derived from the shifts in political power over land-use decision-making. This is the process of reterriorialization, as the intertwined effects of changing economic and political power are exercised in exurban landscapes.

In the case of Jackson County, the embrace of new ordinances appears to emerge as the result of a coalition between amenity migrants and long-time locals who are concerned about elements of cultural landscape heritage and how these might be protected through land-use ordinances that privilege consumptive rather than extractive uses of the landscape. This coalition suggests that long-time locals are actively engaged in protecting the very landscape qualities that amenity migrants seek out. Both groups are, it would seem, fully engaged in valorizing the natural amenity commodity values of these landscapes.

By contrast, many long-term rural residents and some new in-migrants resist these new efforts to regulate land development. Indeed, one of the interesting things that this case study has illuminated is the extent to which in-migrants, even those who can specifically be defined as amenity migrants, are actively involved in supporting the continued extractive uses of the landscape for residential development. Reflecting trends elsewhere, the overwhelming majority of real estate agents and developers in Jackson County are themselves amenity migrants.

There is, in fact, historical precedent for this trend in Jackson County. The Blue Ridge Timber Company, the logging company that was responsible for the lion's share of timbering done in Jackson County in the late 1800s, was owned by in-migrants from Maine, drawn to the area by its vast timber stores (Williams 2001). The Blue Ridge Timber Company faced a number of difficulties as a result of attempting to transplant to North Carolina timbering techniques honed in the very different landscape of Maine (Williams 2001). In a similar fashion, the mistakes of recent development (i.e. landslides, poorly built roads, faulty drainage plans) have been blamed on the attempts of real estate developers, contractors, and even real estate buyers to apply building techniques and ideas about landscape use that originate in environments other than the Southern Highlands. In this sense, the Blue Ridge Timber Company suffered from its own newcomer challenges.

These findings, and our case generally, are consistent with research elsewhere, particularly in the extent to which conflict over environmental management has come to characterize the politics of land use in the area. To date, however, in contrast to the case in Nevada County, California (Hurley and Walker 2004), the pro-growth coalition has been unable to overcome the clear concern within the community over changing the way development occurs on steep slopes. Use of the Q method demonstrates that the demographic changes associated with reterritorialization may not immediately or automatically lead to power being shifted from one group to another when it comes to controlling development. Instead, the coalitions around landscape amenity value may arise from within existing community dynamics to limit development. Yet pro-growth coalitions are not likely to go down without a fight; future battles over the limits on building opportunities enshrined in the land-use ordinances are likely to be fought for sometime to come as these pro-growth coalitions seek to extract unfettered development rights from the land.

Recognizing that reterritorialization itself is an uneven process—complete with alternate pathways involving the same general cast of characters—has significant implications for thinking about the stability of changes in environmental management and the protection of landscape qualities within exurbanizing areas. Land development in areas experiencing amenity migration may be subject to fluctuating periods of protection and exploitation as the balance of decision-making power sways between competing rural capitalisms and the coalitions that support different ways of extracting value from the landscape. Some regulations and planning approaches are quite likely to be more acceptable than others, with real consequences for biophysical landscape and community transformations. Various approaches and protection strategies are likely to lead to new and shifting discursive alliances and coalitions, as residents with divergent opinions struggle to protect landscape qualities they value and/or seek to exploit particular landscape qualities associated with opportunities related to changing development trends.

References

Addams, H., & Proops, J. (2000). *Social discourse and environmental policy: An application of Q methodology*. Northampton, MA: Edward Elgar Publishing.

Brogden, M. J., & Greenberg, J. B. (2003). The fight for the West: A political ecology of land use conflicts in Arizona. *Human Organization, 62*(3), 289–298.

Culbertson, K., Case, D., Fowler, D., Morgan, H., & Schwellenbach, S. (2008). Moving to the mountains: Amenity migration in the Sierra and Southern Appalachian Mountains. In J. L. Wescoat Jr. & D. M. Johnston (Eds.), *Political economies of landscape change: Places of integrative power* (pp. 77–88). New York: Springer.

Dayton, B. W. (2000). Policy frames, policy making, and the global climate change discourse. In H. Addams & J. Proops (Eds.), *Social discourse and environmental policy: An application of Q methodology* (pp. 71–99). Northampton, MA: Edward Elgar.

Ferell, R. S., & Killian, C. D., Jr. (1978). *Land development guide for North Carolina state planning region A*. Cullowhee, NC: Western Carolina University.

Fortmann, L. P., & Kusel, J. (1990). New voices, old beliefs: Forest environmentalism among new and long-standing rural residents. *Rural Sociology, 55*(2), 214–232. doi:10.1111/j.1549-0831.1990.tb00681.x.

Ghose, R. (2004). Big sky or big sprawl? Rural gentrification and the changing cultural landscape of Missoula, Montana. *Urban Geography, 25*(6), 528–549. doi:10.2747/0272-3638.25.6.528.

Gobel, J. (2007, August 9). County leaders OK subdivision, steep-slope ordinances. *Sylva Herald & Ruralite*.

Hotaling, L., & Hall, A. (2007, March 15). Commissioners OK moratorium on new subdivisions. *Sylva Herald & Ruralite*.

Hurley, P. T. (2013). Whose sense of place? A political ecology of amenity development. In W. Stewart, D. Williams, & L. Kruger (Eds.), *Place-based conservation: Perspectives from the social sciences* (pp. 165–180). Dordrecht, The Netherlands: Springer.

Hurley, P. T., & Halfacre, A. C. (2011). Dodging alligators, rattlesnakes, and backyard docks: A political ecology of sweetgrass basket-making and conservation in the South Carolina Lowcountry, USA. *GeoJournal, 76*, 383–399. doi:10.1007/s10708-009-9276-7.

Hurley, P. T., & Walker, P. A. (2004). Whose vision? Conspiracy theory and land-use planning in Nevada County, California. *Environment and Planning A, 36*(9), 1529–1547. doi:10.1068/a36186.

Jackson County. (2007). *Jackson County ordinance*. Retrieved from http://www.jacksonnc.org/jackson-county-ordinance.html.

Jackson County. (2014). Retrieved from http://www.jacksonnc.org.

Jackson County Planning Board. (2005). *Jackson County land development plan*. Retrieved from http://planning.jacksonnc.org/html/jc_land_dev_plan.html.

Kurcharski, S. (2005). Shooting range argument continues with hackles raised. Smoky Mountain News.

Martin, E. (2007, July). *A slippery slope: Battle lines are drawn as mountain counties try to balance development with perserving the region's natural beauty*. Retrieved from www.businessnc.com.

McKeown, B., & Thomas, D. (1988). *Q methodology*. Sage University paper series on quantitative applications in the social sciences, series no. 07-066. Beverly Hills: Sage.

Nesbitt, J. T., & Weiner, D. (2001). Conflicting environmental imaginaries and the politics of nature in Central Appalachia. *Geoforum, 32*(3), 333–349. doi:10.1016/S0016-7185(00)00047-6.

Newfont, K. (2012). *Blue Ridge commons: Environmental activism and forest history in western North Carolina*. Athens, GA: University of Georgia Press.

Niemeyer, S., Petts, J., & Hobson, K. (2005). Rapid climate change and society: Assessing responses and thresholds. *Risk Analysis*, 25(6), 1443–1456. doi:10.1.1.468.6294.

Reed, M. G. (2007). Uneven environmental management: A Canadian comparative political ecology. *Environment and Planning A, 39*(2), 320–338. doi:10.1068/a38217.

Robbins, P. (2006). The politics of barstool biology: Environmental knowledge and power in greater Northern Yellowstone. *Geoforum, 37*(2), 185–199. doi:10.1016/j.geoforum.2004.11.011.

Robbins, P., Martin, S., & Gilbertz, S. (2012). Developing the commons: The contradictions of growth in exurban Montana. *The Professional Geographer, 64*(3), 317–331. doi:10.1080/00330124.2011.601193.

Rome, A. (2001). *The bulldozer in the countryside: Suburban sprawl and the rise of American environmentalism*. Cambridge: Cambridge University Press.

Schewe, R. L., Field, D. R., Frosch, D. J., Clendenning, G., & Jensen, D. (2012). *Condos in the woods: The growth of seasonal and retirement homes in Northern Wisconsin*. Madison, WI: University of Wisconsin Press.

Smith, N. (2010 [1984]). *Uneven development: Nature, capital, and the production of space* (3rd ed.). New York: Verso.

Walker, P. A., & Fortmann, L. P. (2003). Whose landscape? A political ecology of the 'exurban' Sierra. *Cultural Geographies, 10*(4), 469–491. doi:10.1191/1474474003eu285oa.

Walker, P. A., & Hurley, P. T. (2004). Collaboration derailed: The politics of "community-based" resource management in Nevada County. *Society & Natural Resources, 17*(8), 735–751. doi:10.1080/08941920490480723.

Walker, P. A., & Hurley, P. T. (2011). *Planning paradise: Politics and visioning of land use in Oregon*. Tucson, OR: University of Oregon Press.

Webler, T., Danielson, S., Tuler, S., Kalof, L., & Shockey, I. (2009). *Using Q method to reveal social perspectives in environmental research*. Greenfield, MA: Social and Environmental Research Institute.

Williams, M. R. (Ed.). (2001). *The history of Jackson County* (Sesquicentennialth ed.). Sylva, NC: Jackson County Historical Association.

Woods, M. (2005). *Rural geography: Processes, responses, and experiences in rural restructuring*. London: Sage.

Chapter 10
The Paradox of Engagement: Land Stewardship and Invasive Weeds in Amenity Landscapes

Peter Klepeis and Nicholas Gill

10.1 Introduction

Australia is an oft-used case study highlighting land transformation. The large-scale political and economic shifts initiated by colonialism led to vast socio-ecological changes, including modifications of fire regimes, the purposeful and inadvertent introduction of exotic invasive species, and the proliferation of attitudes about nature rooted in utilitarian frameworks and capitalism (Dunlap 1997; Flannery 1994; Head 2000). That landscapes of private property and production predominate in the post-colonial era suggests uniformity in ideologies of nature and land stewardship. Ideas about rural land use are heterogeneous and evolving, however (Gill 2014). In recent decades, rural to exurban transitions in Australia are diversifying attitudes and values among rural residents, creating a complex "management mosaic" (Epanchin-Niell et al. 2010; Gosnell et al. 2006) and leading to both intended and unintended socio-ecological consequences (Abrams et al. 2012; Gill et al. 2010; Mendham et al. 2012).

Amenity migration involves an influx of newcomers (exurbanites) to rural areas who do not necessarily seek to earn a living from the land but rather celebrate it for the lifestyle amenities it provides (Gosnell and Abrams 2011), such as "natural scenery, proximity to outdoor recreation, cultural richness, or a sense of rurality" (Abrams et al. 2012, p. 270). Landscapes once dominated by cultivated fields, extractive industries, and livestock are being transformed into areas that are multi-functional—mixed landscapes of production and consumption (Argent et al. 2007;

P. Klepeis (✉)
Department of Geography, Colgate University, 13 Oak Drive, Hamilton, NY 13346, USA
e-mail: pklepeis@colgate.edu

N. Gill
School of Geography and Sustainable Communities, University of Wollongong, Wollongong, NSW 2522, Australia
e-mail: ngill@uow.edu.au

L.E. Taylor, P.T. Hurley (eds.), *A Comparative Political Ecology of Exurbia*,
DOI 10.1007/978-3-319-29462-9_10

Gosnell and Abrams 2010; Holmes 2005; Wilson 2007). Both exurbanites and longtime rural residents (usually primary producers earning a living from the land and whom we call ranchers) are agents of change in these multifunctional landscapes. Each group represents diverse environmental ideologies and land-use goals. The interaction between ranchers and exurbanites leads to both a co-construction of social relations and also of land management regimes. Through this interaction, amenity migration is stimulating land-use and land-cover (land system) change across broad areas of Australia, Europe, North America, and parts of the Global South (Cadieux and Hurley 2011; Matarrita-Cascante and Stocks 2013).

Rural to exurban transitions result in what Breen et al. (2016, this volume) call "competing rural capitalisms." The term underscores ways in which both ranchers and exurbanites seek economic change and opportunity. While much of the amenity migration literature assumes ranchers want to maintain land use in some kind of time-honored way, ranchers often embrace modified management practices or, indeed, entirely new land uses. Many seek regulatory and technical assistance to control weeds, mitigate wildfire risk, and maintain fence lines, all examples of management that facilitates primary production. But they also seek opportunities to subdivide their lands and develop real estate. Similarly, exurbanites tend to favor opportunities for new real estate markets as well, although they generally seek to develop landscapes of consumption where activities such as recreation or hobby farming are embraced. In short, in addition to areas of distinct contrast, there is often overlap between what ranchers and exurbanites value (Breen et al. 2016).

Our analysis of the socio-ecological implications of rural to exurban transitions acknowledges shifting attitudes among ranchers and exurbanites alike. Reinforcing a core argument threaded throughout this edited book, and using the example of exotic invasive weeds, we demonstrate ways in which competing rural capitalisms represent shifting ideologies and shifting ecologies. We start by reviewing how the spread of invasive plants intersects with rural to exurban transitions. We then explore notions of land stewardship, social capital, and calls within the scholarly literature for increased knowledge sharing among ranchers and exurbanites. These two sections underpin the analysis of our three case studies, which are part of a long-term study of rural change in New South Wales (NSW), Australia. Results show that both ranchers and exurbanites value privacy and do not tend to interact regularly with members of the other group. But there is significant overlap in stewardship ideology, which suggests the potential for enhanced collaboration among ranchers and exurbanites in the management of transboundary invasive weeds.

10.2 Amenity Migration and Invasive Weeds

The spread of exotic invasive species constitutes an important component of human-induced global environment change (Vitousek et al. 1997). During the anthropocene, in particular, the scale of exotic plant introductions has led to the creation of novel ecosystems with impacts on socio-ecological conditions and processes (Ellis et al. 2012; McNeely 2004; Mooney and Cleland 2001). Given the implications for nature and society, there have been repeated calls to investigate the human dimensions of invasive species (e.g. Ellis et al. 2012; Head and Atchison 2009; Robbins 2004).

In addition to climate change and the translocation of species via trade or other travel, land-use change is one of the most important factors in the introduction and spread of invasive species (Decker et al. 2012; Polce et al. 2011). For millennia people have moved plants from place to place both knowingly—even if unaware that the species is invasive—and inadvertently (Butcher and Kelly 2011; Crosby 1986). In addition, as agents of landscape change people are able to create conditions favorable to invasives (Maestas et al. 2003; Robbins 2004). In a prominent example, land-use change connected to amenity migration holds the potential for both the direct introduction of invasives as well as the kind of landscape preparation that makes an area vulnerable to the establishment and spread of invasive species.

Many political ecologists deem the term "invasive species" to be problematic because it privileges a particular ideology of nature or a particular relationship that one group of people has with local ecology over another (e.g., see Cadieux and Taylor 2012). In the context of "invasive weeds," for example, some people may champion a plant for its aesthetic value whereas others may target it for control because of perceived negative impacts on biodiversity or the land's grazing value. We acknowledge the risk in using a term that may privilege particular ideas about nature as being "right" or "appropriate." To be clear, our use of the term invasive species (hereafter called "invasives") refers to the spread of plants in ways that lead to potential negative impacts on ecosystem services and primary production.

The types of consequences linked to rural change that attract the most attention from government officials and nature-society scholars are those that affect negatively ecosystem services—by which we mean the "ecosystem-based processes" that provide benefits to people, such as clean water, pollination, or recreational opportunities (Goldman 2010)—or that present some kind of environmental hazard. They tend to be cross-boundary phenomena, transcending political and ecological delineations. Issues related to wildfire (Eriksen and Prior 2011), forest and water resource management (Howard 2008; Meadows et al. 2012), wildlife conservation (Haggerty and Travis 2006), and invasives (e.g., Gosnell et al. 2006; Epanchin-Niell et al. 2010) are noteworthy examples.

The actions of individual landowners *vis-à-vis* these cross-boundary issues affect not only their own properties but also the broader landscape (Abrams et al. 2012). If managing invasives to control their spread or intensification is the goal—although differing values may mean that it is not (Hall 2003)—then some degree of overlap in stewardship ideology and corresponding collaboration among landowners is necessary. For example, the invasive grass serrated tussock (*N. trichotoma*) is on the Australian federal government's list of the country's weeds of national significance due to its degradation of pasture and negative effects on native flora (Department of the Environment 2013). Individual serrated tussock plants can live up to 50 years and produce 100,000 seeds per year (Michelmore 2003). Dispersed by wind up to 15 km seeds easily cross property boundaries. If one landowner controls the grass

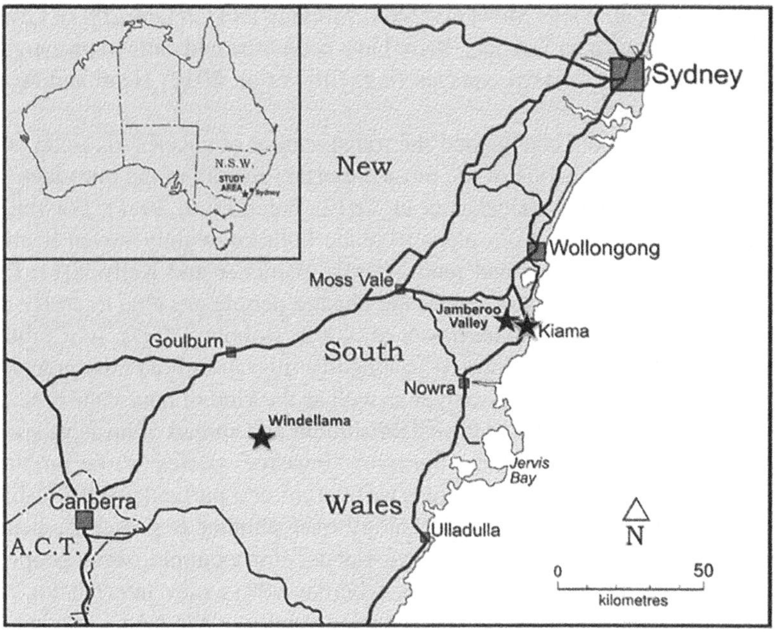

Fig. 10.1 Locations of study areas in New South Wales—inland (Windellama) and along the coast (Jamberoo Valley and the community of Kiama)—both part of the Kiama Local Government Area (Modified from Gill et al. 2010, p. 320)

but her neighbor does not then the attempt, along with a considerable investment in time, money, and effort, is likely to be ineffective (Graham 2013; Klepeis et al. 2009). If land productivity and biodiversity are not of concern, however, managing serrated tussock may be deemed unimportant.

Australia provides ample opportunity to study the intersection between invasives and rural to exurban transitions. Invasives are present in peri-urban (formerly) agricultural and rangeland areas throughout the country (Agtrans 2005; Alam 2012; Argent et al. 2007; Burnley and Murphy 2004; Williams and West 2000). Control of invasives is a top federal, state, and local government priority due to concerns about environmental biosecurity, which the Australian Department of the Environment (2014) defines as "the protection of the environment and social amenity from the negative effects associated with invasive species" (DPI 2014; Kiama Municipal Council 2014).

In our research on the towns of Windellama and Jamberoo Valley—representing inland and coastal amenity landscapes (respectively, about 234 km southwest and 114 km south of Sydney)—we interviewed both exurbanites and ranchers (Fig. 10.1) (Gill et al. 2010; Klepeis et al. 2009). The research explores qualitatively the land management activities and perceptions rural landowners have of nature, invasives, and land use. A third case study focuses on the Kiama Local Government area (LGA), and explores similar themes quantitatively using a mail survey (Gill and

Klepeis 2011; Ikutegbe et al. 2014). Collectively, the three cases show the degree to which there is cross-fertilization between exurbanite and rancher ideologies concerning land-use management and measures to control invasives.

Consistent with scholarship on rural to exurban transitions elsewhere (e.g., Abrams et al. 2012), we show that exurbanites are stimulating socio-ecological shifts at landscape and property scales (e.g., fire regimes, water management, species composition of home gardens). In particular, we demonstrate changes in landowner types, stewardship attitudes and behaviors, and weeds management (see Sect. 10.6). Ranchers and exurbanites have diverse attitudes about nature and society. For example, there are a plethora of reasons why individual landowners may disagree about the importance of invasives and willingness to enact control measures (e.g., Mendham et al. 2012; Moss 2006; Seidl and Klepeis 2011; Tidwell and Brunson 2008). Some of these reasons constitute pragmatic constraints, such as absenteeism (e.g., time devoted to removal), expense (e.g., cost of herbicides), access to information and trust of the source (e.g., the capacity to identify which species are invasive), perceived efficacy of the control measures (e.g., some rural landowners see the spread of invasives as inevitable), and willingness or capacity to work collaboratively with community members. Other reasons are linked to livelihood and the nature of a person's engagement (actual and idealized) with nature, including whether or not the landowner wishes to use the land for primary production and attitudes about land stewardship (e.g., the value placed on native versus non-native plants).

Fundamentally, the Australia case studies explore how growth in the number of exurbanites affects land stewardship and the capacity to manage plants that are identified as invasives. How do exurbanite and rancher notions of what constitutes strong land stewardship differ? Do differences in how people perceive and value nature have consequences for local ecology? Are there opportunities for cooperation between groups that derive different value from the landscape? From whom do landowners seek advice? How does information flow among community members? In other words, how do rural landowners, collectively, build knowledge about local ecologies?

10.3 Stewardship, Social Capital, and Knowledge Building

Sorice et al. (2014) ask whether exurbanites and ranchers are likely to have different motivations regarding land stewardship, and which group is most likely to put their ideas into practice. Defining "sound land management," in essence, as behaviors that maintain or enhance ecosystem services, they find that exurbanites in the rangelands of central Texas are less likely to use best practices than longer-term landowners. Other case studies show a less pronounced difference in land stewardship practices between exurbanites and ranchers (e.g., Mendham et al. 2012). Our case studies show that generalizations about rural landowners and stewardship attitudes are hard to make.

Notions of land stewardship evolve (Cooke and Lane 2015; Gill 2014), and community discourse in amenity landscapes creates and re-crafts knowledge about socio-ecological systems (Larsen and Hutton 2012). Emerging amenity landscapes experience dynamic change. It is reasonable to expect networks of information flow and levels of social cohesion between increasingly heterogeneous types of landowners to be in a constant state of flux. Halseth (1998) describes amenity landscapes as sites of ongoing contestation where different environmental commitments are prioritized depending on the stakeholder group in question. Similarly, Larsen and Hutton's (2012) idea of "co-opetition" underscores ways in which stakeholders within communities experiencing amenity migration both cooperate and compete over landscape meaning (e.g., landscape of production versus consumption) and control. They note that "it is essential to understand not only how and why residents compete in the struggle over the landscape's form and function, but also the circumstances under which they cooperate with one another as well as with long-term residents and management professionals" (Larsen and Hutton 2012, p. 664).

An international review of research on the environmental consequences of amenity migration highlights the importance of access to information and cooperation: "the particular land-uses and activities of amenity migrants carry unique implications based, at least in part, on a lack of practical, local knowledge regarding resources and their management" (Abrams et al. 2012, p. 276). In their study of the conditions leading to cooperative forest management in Queensland Australia, Meadows et al. (2012, p. 12) note that exurbanites "require information and on-ground assistance if widespread sustainable environmental management is to be achieved." The study demonstrates that cooperative cross-boundary forest management and control of invasives are critical in the reduction of forest fragmentation and the long-term health of forests (Meadows et al. 2012, p. 13).

Eriksen and Prior (2011, p. 617) find that both exurbanites and longtime landowners agree about the importance of knowledge sharing and community, and there are many cases of cooperation between the two groups, although fundamental challenges exist:

> "The high turnover of property owners in amenity-led communities can lead to an erosion of intra-community familiarity and trust, where the local knowledge concerning, for example, wildfire and wildfire preparedness is slowly lost, or held by long-term residents with little reason to share their knowledge with the 'blow-ins'" (Eriksen and Prior 2011, p. 617)

In addition to different "values, attitudes, knowledge, land-use, and management practices," Mendham et al. (2012) find that absenteeism inhibits the degree to which exurbanites engage with natural resource management issues. They also show that exurbanites are more likely to use the Internet, whereas longtime landowners tend to interact with people via trade associations or groups such as Landcare, a government funded community-based organization that focuses on natural resource management issues.

Eriksen and Prior (2011) identify the importance of both agency-community and intra-community efforts in building local environmental knowledge about wildfire. Their conclusion that information sharing and community problem solving are essential in mitigating wildfire points to the critical role strong social capital plays when addressing socio-ecological challenges. A community with strong social capital has networks of civic engagement that enhance social cohesion and collective problem-solving capacity. Putnam (1995, p. 67) describes social capital as "social organization such as networks, norms and social trust that facilitate coordination and cooperation for mutual benefit." Wilson (2012, pp. 5, 23) expands on the description, defining strong social capital as consisting of "networks of interaction between individuals and stakeholder groups that form a community," which are rooted in participatory and inclusive decision-making processes. He identifies social capital as one of a constellation of critical factors in a community's resilience to environmental, political, and socio-economic challenges. Often applied to social issues, such as public health, the concept of social capital has been invoked regularly in a host of nature-society studies, including investigations of society's capacity to adapt to climate change (Adger 2003; Goulden et al. 2013; Klinenberg 2002), natural resource development potential (Bebbington and Perreault 1999; Sobels et al. 2001), and the strength of cooperative natural resource management arrangements (Hoogesteger 2013; Kasperson et al. 1999).

The strength of social capital is relative to the ease with which knowledge flows through particular networks, a flow that social change can often disrupt. Among the factors leading to strong social capital are knowing your neighbors, the availability of skills training and education, good communication between stakeholder groups, and the ability of communities to accept change (Wilson 2012). In contrast, weak social capital is often represented by the outmigration of young people in rural areas, a lack of leadership, mistrust of neighbors, lack of control over the destiny of the community, and poor communication between stakeholder groups (Wilson 2012, pp. 28–29).

In evaluating the level of knowledge sharing between exurbanites and ranchers we assume that if social capital is strengthened there is a greater potential for rural landowners of all types to get to know each other, build trust, identify others who are knowledgeable about regional ecology and land management, develop new norms of behavior, and develop a coordinated response to invasive weeds (Fiege 2005; Meadows et al. 2012; Sobels et al. 2001). With differing socio-economic and cultural backgrounds, high rates of property turnover, often distinct attitudes about rural life and nature, and a high degree of absenteeism among exurbanites, however, social capital and the opportunity for knowledge sharing it provides may be weak in areas undergoing rural to exurban transitions (Klepeis et al. 2009).

10.4 Study Areas in New South Wales

10.4.1 Inland "Tree Change": The Town of Windellama

Located in the hills of the Southern Tablelands, a region celebrated since the late nineteenth century for fine wool production, the Windellama area's population grew from some 500 in 1976 to over 1500 in 2001. Amenity migration to the area

Fig. 10.2 A mosaic of rangeland and woodland in the Town of Windellama, NSW. *Photo credit*: P. Klepeis

is creating a diversity of owner types and a complex mix of land uses, including livestock production (sheep, cattle, and alpaca) and vineyards, but also a host of consumptive uses, such as recreation, rural retreat, and land investment. Large-scale subdivisions of grazing properties in the town started in the 1970s as real prices for wool declined, farm families began to seek alternative career paths, and property values began rising due to easy access to both Sydney (2-h drive to the north) and the country's capital city, Canberra (1-h drive to south) (Barr et al. 2005; Klepeis et al. 2009). In contrast to the stereotypical amenity landscape in Australia—"sea change" along the coast—Windellama is a mosaic of rangeland and woodland, representing an area of "tree change" (Fig. 10.2). Six key cross-boundary socio-ecological challenges facing Windellama are wildfire, feral animals, soil erosion, dryland salinity, water scarcity, and invasive weeds.

10.4.2 Coastal "Sea Change": Jamberoo Valley and the Kiama Local Government Area (LGA)

The Kiama LGA has an area of 259 km^2 and was established in 1857. On its western edge it includes a stark sandstone escarpment that runs along the coast of NSW (Fig. 10.3). The central portion of the LGA contains the Jamberoo Valley, home to

Fig. 10.3 Jamberoo Valley, NSW, and the steep escarpment along its western edge. *Photo credit*: P. Klepeis

some 900 people. Most of the 20,906 inhabitants of the LGA live in urban centers along the Tasman Sea coast (Kiama Municipal Council 2013). The LGA is a 30-min drive to the city of Wollongong and a 2-h commute to Sydney, both to the north. The area is celebrated for its many amenities: ready access to beaches, scenic views, plentiful recreation opportunities, and proximity to popular natural features, including rainforest and waterfalls; much of the LGA presents a bucolic landscape with a mosaic of rolling hills, green pastures, forest cover, and waterways, with the escarpment as a spectacular backdrop.

In addition to its lifestyle amenities, rural parts of the Kiama LGA are valued for primary production and natural resource conservation. With high annual rainfall (1800 mm) and good soils, the LGA has long been an important center for dairy production (Hindle et al. 1987). The number of dairy cattle peaked in the 1960s with the area maintaining roughly 15,000 per year during that decade. Subsequently, there has been a steady decline, with recent estimates at less than 8000 dairy cattle (ABS 1955–2005). Explanations for the reduction center on diminishing industry protection, low returns, and dramatically increasing land values as well as the oft-identified mix of factors lead to declines in farming elsewhere, such as shifting livelihood choices within farming families, a growth in alternative employment opportunities, and rising production costs (Edwards 2003; Gillespie 2003; Hindle et al. 1987; Houston 2005; Klepeis et al. 2013). Farmers have strong incentives to subdivide and sell their land.[1] In Jamberoo Valley, the percentage of the workforce in primary industry (excluding mining, but including agriculture) fell from 66 % in 1976 to roughly 12 % in 2001 (Gill et al. 2008).

Assessments of ecosystem services in the Kiama LGA identify many endangered ecological communities, such as wetlands, subtropical rain forest, and Red Gum/Stringybark forest (*Eucalyptus tereticornis*, *Eucalyptus eugenioides*, *Eucalyptus bosistoana*), the latter of which is listed under Australia's Threatened Species Conservation Act of 1995 (Mills and Associates 2006). Other concerns are endangered flora and fauna, soil erosion, invasive species, and watershed management (Kiama Municipal Council 2013). Among the invasive species in the LGA are 24 of those listed among Australia's 32 Weeds of National Significance—the control of which is required by law—including lantana (*Lantana camara*), prickly pear (*Opuntia* spp.), blackberry (*Rubus fruticosus aggregate*), and serrated tussock (Australian Department of the Environment 2014).

10.5 Study Designs

The approaches used in the Windellama and Jamberoo Valley case studies are described in detail elsewhere (Gill et al. 2010; Klepeis et al. 2009). In brief, both projects rely primarily on semi-structured, in-depth interviews with a mix of exurbanites and ranchers (totaling 36 interviews in Windellama and 25 in Jamberoo) as well as state-level regulators at the Department of Primary Industries and regional weeds officers (charged with inspecting properties and ensuring management of particular invasives as required by law). In addition, respondents owning land led walking tours of their properties during which the interview continued and observations of land management and land conditions were made. The interviews were analyzed using qualitative content analysis and "descriptive coding," an approach that facilitates the analysis of words, concepts, and responses to particular questions within the broader context of the group of respondents' collective answers (Hay 2010).

The third case study builds on the interview data to construct a survey of both ranchers and exurbanites in the Kiama LGA (Ikutegbe et al. 2014). Following standard methodology (see Dillman et al. 2009), surveys were mailed to 1000 households in the Kiama LGA's rural landholder database, 355 of which were returned. Consisting of 38 questions, and taking around 25 min to complete, the survey is split into five sections: "rural life and community"; "activities on your land"; "details

[1]One estimate of rural land-use in the Kiama LGA finds that rural residential properties cover 6711 ha (43.1 %), extensive agriculture covers 3549 ha (22.8 %), and dairy operations cover 4087 ha (26.2 %) (Sinclair, personal communication). In a country where people tend to celebrate using land "productively", the trend away from using land for dairy or agriculture and towards rural residential development contributes to debates about rural change and food security (Miller and Roots 2011).

Table 10.1 Questions indicating the strength of social capital within the Kiama Local Government Area and key results

	Question	Summary result
1.	Have you attended a local community event in the Kiama LGA in the past 6 months (such as a church or school event, craft exhibition, bushcare project)?	The large majority (77.7 %) have attended at least one community event in the last 6 months. There is not much variation between exurbanite and rancher groups
2.	Have you visited a neighbor's home in the past week?	57.2 % have visited a neighbor at least once in the past week. A large portion (42.7 %) reported no visits in the past week while 25.2 % have visited two times or more. There is not much variation between exurbanite and rancher groups
3.	How often do you do volunteer work for a local group or club?	45.1 % volunteer for a local group or club once a month or more. Roughly one-third of respondents never partake in this activity while one-quarter report volunteering rarely. There is not much variation between exurbanite and rancher groups
4.	Owning rural land in the Kiama LGA has brought me into contact with types of people with whom I have not interacted before	59.9 % have had contact with types of people with whom they had not interacted before. There is not much variation between exurbanite and rancher groups
5.1	Do you work with other people in managing your land?	A small majority (53.3 %) work with other people in managing their land. A great deal more ranchers (78 %) work with people in managing their land than exurbanites (49.5 %)
5.2	If yes (to question 5.1) then with whom have you had significant contact?	Overall, neighbors, friends, and family are the largest groups landowners have contact with in managing their land, followed by contractors. Exurbanites rely much more on neighbors and contractors than ranchers

about your rural property"; "prospects for the future"; and "details about you and your household." It explores respondents' land-use and natural resource management practices, their perceptions of nature and society, indicators of different landowner types, and the strength of social capital (Table 10.1). Also, it includes questions requested by the Council and staff from associated bodies, such as the Illawarra District Noxious Weeds Authority, the Small Farms Network, and the Department of Primary Industries.

The extent to which land-based income is important to landowners—reflecting their "production or consumption orientation" (Mendham et al. 2012)—is an oft-used point of difference between exurbanites and ranchers (Abrams et al. 2012). The mail survey gathers data on whether land provides primary, significant secondary, minor, or no income. Here, to simplify the analysis, we collapse these four categories into two groups—those who derive a primary or significant secondary income from land (50 respondents who we call ranchers) and those who derive minor or no income from their land (288 respondents who we identify as exurbanites).

Overall, 21 % of respondents own 0.4 ha or less, 86 % own 40 ha or less, and 4 % own 100 ha or more: the average landholding is 22 ha and the maximum is 400 ha.

Using in-depth and semi-structured interviews in early phases of the research and then generating formal surveys to test hypotheses generated by the earlier qualitative work is highly effective at identifying both context-specific dynamics as well as general trends in the study areas. The main weakness in our approach is that we do not assess ecological conditions directly; that is, ideally, our team would measure ecological conditions at the same properties where we conducted our interviews. Such an ecological assessment would allow us to connect specific ideologies about nature with particular land management strategies and their impacts on local ecology. Instead, our research uses interview and survey data to document land management impacts.

10.6 Results

In addition to fulltime ranchers, the study areas include both fulltime and part-time exurbanites (i.e., occasional visitors or absentee landowners who have secondary residences elsewhere or who own rural real estate for investment purposes). Approximately 65 % of Windellama's exurbanites are part-time. In the Kiama LGA—a site of dynamic demographic change—there is a high level of absenteeism with some 28 % of rural landowners spending only an average of 24 weeks per year on their property. In addition, the frequency of land transfers is high: 58 % of ranchers are likely to sell their land in the next 5 years due to "council rates," "burden of maintaining the property," and agricultural viability. And 40 % of exurbanite respondents plan to sell within the same period because of age, health, and the "burden of maintaining the property."

Fulltime exurbanites in all three study areas tend to be either commuters or retirees whose main or only source of income is off-farm, although some seek to generate profit from hobby farming. Two-thirds of exurbanites in Jamberoo Valley, for example, raise livestock on either a commercial or semi-commercial basis. But even for exurbanites with significant land-based businesses, the income is usually secondary to that provided by investments or jobs elsewhere.

10.6.1 Stewardship Attitudes and Land Management

Overall, ideology about nature and land stewardship among exurbanites and ranchers is surprisingly similar when described in general terms (Table 10.2). In all three study areas, most rural landowners see themselves as stewards of the land. The overwhelming majority of both exurbanites and ranchers advocate for active land management to address fire hazards, soil erosion, water management, and invasives. That there is consensus about conserving land and water resources is not surprising

Table 10.2 Forms of stewardship among exurbanites (after Gill et al. 2010, p. 30)

Stewardship type	Description	Examples of priorities
Lifestyle agrarian	Akin to traditional rural or farming senses of stewardship: primary production and conservation are seen as compatible and a pragmatic part of land management	Focus on animal husbandry Fencing of existing native vegetation and riparian protection. Plantings of natives and non-natives, but local species not a priority
Regenerative	Interest in improving land management as a whole with improved ecological management and restoration as important goals. Production goals are often present but they may be equivalent or subservient to conservation goals	Significant efforts at replanting or restoration of native vegetation. Extensive weeding. Semi-commercial grazing, cropping, or horticulture. Limited pasture management focused on herd size and slashing
Conservationist	Primary focus is on ecological restoration and provision of habitat. Agricultural land use is perceived as having had largely negative consequences, some of which are ongoing, and exurbanite ownership is seen as an opportunity to remedy past mistakes even if in a small way	Extensive efforts at replanting or restoration of native vegetation. Extensive weeding. Local species usually preferred but "Australian natives" or non-natives may meet their aims. No stock or only small numbers of "recreational" animals

given that these issues receive a tremendous amount of attention in Australia both in the popular press and scholarly literature. Also, most ranchers and exurbanites advocate for some kind of "productive" use of the land by which they mean production of food, fiber, or timber.

Not surprisingly, ranchers identify "making a living" as the main reason they own land, followed by family connections and land as an investment: 15 % of survey respondents identify production on their land as the primary or secondary source of income. Sixty-six percent of those surveyed, in contrast, generate no income from their land. But a common value both to ranchers who use the land to earn a living (usually relying significantly on off-farm income as well) and exurbanites is a desire to reside in "rural nature." Exurbanites, in particular, value the opportunity to experience wildlife and natural amenities while learning about the importance of rural production and agriculture. They also seek seclusion and the capacity to have complete control over what they do with their land, an attitude that may inhibit involvement with community-based natural resource management groups. Exurbanites identify "lifestyle" followed by "being closer to nature" as their primary reasons for owning land.

Dairy is the most significant production activity for ranchers in the Kiama LGA (average production value in 2008–2009, AU$143,100), distantly followed by beef (average production value in 2008–2009, AU$9650). For exurbanites who dabble with hobby farming beef production is the most significant activity with an average production value of AU$1397. Exurbanites are more likely to use their land as a

Table 10.3 The number of ranchers and exurbanites who have taken particular actions on their properties in the past 3 years (note that two of the actions listed in the table—reducing native vegetation cover and planting non-native vegetation outside of the garden—represent poor management as defined by Australian government agencies, whereas the rest of the actions contribute to strong land management)

	Ranchers ($n=50$)		Exurbanites ($n=288$)	
Actions taken on property in the past 3 years	Number	%	Number	%
Soil testing for nutrient monitoring/application	20	40.0	52	18.1
Cooperative land use or management with neighbors	12	24.0	47	16.3
Weed management with a group or organization	12	24.0	45	15.6
Feral animal control with a group or organization	25	50.0	18	6.3
Reduced native vegetation cover including understory, by any means	3	6.0	18	6.3
Considered or have changed land-use/management due to climate change	9	18.0	17	5.9
Tried to control/prevent soil erosion	15	30.0	60	20.8
Water quality testing	11	22.0	24	8.3
Planted non-native vegetation outside the garden	5	10.0	31	10.8
Weed management undertaken privately	35	70.0	56	19.4
Feral animal control undertaken privately	18	36.0	56	19.4
Received/spent grant money on environmental or agricultural projects	10	20.0	23	8.0
Implementation of ideas/techniques gained from a course/field day/training program	11	22.0	34	11.8
Undertaken a flora and fauna assessment	4	8.0	28	19.7

retreat or for conservation. Beyond the obvious time difference spent on pasture maintenance, responses to a question about the number of days respondents spent on specific activities in the last year expose additional differences, including choices about recreation (e.g., hunting versus horseback riding) (Table 10.4).

What seems to be a common ideology about nature—maintaining "productive" land— breaks down when specific management options are considered. The weight placed on maintaining land productivity compared to maintaining or promoting habitat for native species varies, for example. One exurbanite felt he was helping to "win nature back" by removing invasives and planting native species, but also through grazing cattle; he sees a mix of ecological restoration and food production as the right kind of "nature." Surprisingly, given the general emphasis on improving ecological conditions, exurbanites in the Kiama LGA outperform ranchers in only 1 (undertaking a flora and fauna assessment) out of 14 actions that represent elements of sound natural resource management (Table 10.3).

With respect to invasives, ranchers place much greater emphasis on controlling pasture weeds and feral animals that might attack livestock or dig-up pastures. Exurbanites tend to support aggressive control measures, but they are, for the most part, ill-informed about invasives; in one example, an exurbanite in Windellama

Table 10.4 The number of respondents undertaking specific activities in the past year and the average number of full days they spent on the activity

Average time spent on activities in the past year (full days)				
	Number of exurbanites ($n=288$)	Days (mean)	Number of ranchers ($n=50$)	Days (mean)
Improving soil/fertilizing/ planting legumes	70	15.2	22	72.1
Weeding/weed management	166	22.4	30	52.6
Pasture maintenance	136	26.0	33	93.8
Walking/bushwalking	63	26.7	6	6.7
Planting in the garden	288	3.7	50	9.7
Planting native trees/plants outside of the garden	288	3.7	50	2.4
Planting perennial pasture	23	3.4	12	31.7
Motor biking	15	9.9	6	120.7
Horse riding	14	68.2	3	67
Hunting	6	19.0	5	74.8
Other	24	6.3	6	90.5

went so far as to *plant* the invasive grass serrated tussock in his garden because he thought erroneously that it was a native species and that it looked nice; he planted it despite the fact that he is in favor of controlling invasives. Other factors standing in the way of effective, landscape-scale management are time (exurbanites and ranchers alike do not want to, as one rancher put it, "spend their lives chasing tussocks"), money, and effort, but also social barriers to exurbanite and rancher interaction. For example, 90 % of exurbanites and 82 % of ranchers said "privacy" is one of their reasons for owning land. And many primary producers see exurbanites as interlopers or uneducated (about rural living and land management) city dwellers "invading" rural landscapes.

Ranchers might be more knowledgeable than exurbanites, but often do no better when managing invasives. Most exurbanites do not manage invasives effectively, which breeds a sense of fatalism about the spread of weeds among ranchers and they often do not implement effective weeds control measures despite the negative impact on land productivity. Around 91 % of rural landowners in the Kiama LGA survey undertake activities to control weeds, although ranchers spend 52.6 days per year weeding when compared to an average of 22.4 by exurbanites (Table 10.4). The seemingly greater emphasis on weeding by ranchers is turned on its head, however, when considering weeding intensity. Alam (2012) demonstrates that exurbanites in the study area spending much more time on weeds management per hectare (15 days/ha/year) compared to ranchers (1 day/ha/year). Seventy-five percent of ranchers answer "no" to a question about whether they spend as much time on controlling weeds as they would like compared to 56 % of exurbanites. In general, despite a shared concern about weeds, the plants of concern to exurbanites in the Kiama LGA, such as fireweed (*Senecio madagascariensis*) and lantana, appear to reflect a greater concern with plants that are problems for conservation as compared to those affecting pasture. Time constraint is the main limiting factor in weed control effort: financial resources, age, health, and a perception that the weed problem is overwhelming are secondary factors.

In sum, while there is broad consensus within Jamberoo that invasives are to be controlled, the degree to which that desire translates into action is mixed:

> Most interviewees were not actively managing for restoration or regeneration of native vegetation, although most were interested in protecting vegetation on their land. Some were effectively maintaining the property as they found it, for example by slashing at the margins of native vegetation to maintain pasture and control invasive plants. Thus in most cases, vegetation management consisted of sympathetic or benign neglect. (Gill et al. 2010, p. 326)

Benign neglect, in the case of invasives, is insufficient to address the challenge. While some exurbanites commit significant time and money to managing vegetation, their activities are neither strategic nor collaborative. Knowledge building through interacting with other stakeholders in the community might aid exurbanites in Jamberoo in realizing some of their land management goals, but a desire to be private and withdraw from society, to some degree, stands in the way.

10.6.2 The Potential for Collaboration

Exurbanites and ranchers tend to run in different social circles. It is the predominance of these social divisions that lead us to hypothesize that social capital in amenity landscapes is weak. Ranchers are more likely to participate in the fire brigade, for example, and knowledge about wildfire among exurbanites is poor (Eriksen and Gill 2010). Exurbanites are also less likely than ranchers to be willing to receive training on the application of weeds-killing chemicals and they are less likely to maintain fence lines that border property. In addition, exurbanites are more likely to be part-time residents of their rural properties and they tend to have less skills training and education about natural resource management, both factors affecting negatively the social capital of the study area.

Most people are involved with multiple community-based organizations and activities; however, there is a degree of social separation between the different landowner types. In addition, a moderate level of social capital is reflected in whom people trust for information about land management. Ranchers are more likely than exurbanites to consult friends and family regarding natural resource management. While also seeking advice from friends and family, exurbanites tend to seek more input from neighbors and contractors than ranchers do (Ikutegbe et al. 2014, p. 9).

Despite our assumption going into the research that social capital would be weak, however, its strength is higher than expected: the networks of interaction (see Wilson 2012) in which exurbanites and ranchers participate *do* overlap, although only partially. As reported in Ikutegbe et al. (2014), networks of communication and interaction seem to differ depending on whether the landowner emphasizes amenity-oriented land use or seeks significant income from the land. Due to high turnover rates among property owners and high rates of absenteeism—both of which inhibit the capacity to interact with other community members (Klepeis et al. 2009; Mendham et al. 2012)—exurbanites tend to acquire more information from the Internet, nurseries, and representatives from professional agencies than neighbors and other community members. The community members with whom they *do* interact tend to be other exurbanites and not necessarily those landowners with the greatest local environmental knowledge. The Kiama LGA survey data show that exurbanites and ranchers differ in whom they considered the most important contact in managing land. Primary or significant secondary income producers are significantly more likely to consider family and friends the most important contact in managing their land, where exurbanites were more in contact with neighbors and contractors as well as friends and family (Gill and Klepeis 2011).

While knowledge about an environmental issue is not the only or even the most important factor in explaining whether people adopt practices that mitigate a particular problem (e.g., Guagnano et al. 1995; McCaffrey 2004; MacNaghten and Urry 1998; Robbins 2007), access to information and technical advice on land management is a vital element in the mix. Ranchers are more likely than exurbanites to seek expert advice regarding their rural property, although only 34 % are likely to seek advice before planting noncommercial trees compared to 55 % of exurbanites. The most important sources of advice for tree planting by exurbanites are books, magazines, the Internet, and nurseries: ranchers consult those sources as well, but place additional emphasis on advice from experts within natural resource management groups such as Landcare and the Southern Rivers Catchment Authority. Ranchers are much more likely than exurbanites to have cooperated about land-use management with neighbors, weed management with a group organization, and feral animal control with a group organization (Table 10.3). That said, only 24 % of ranchers undertook cooperative land management, a relatively low percentage if cross-boundary socio-ecological problems are to be managed effectively. Ranchers are almost twice as likely to participate in field days and training programs. Overall, ranchers tend to be more engaged on natural resource management issues and tend to reflect stronger social capital than exurbanite groups, although cooperative behaviors are undertaken by a relatively small number of people in both categories.

10.7 Shifting Socio-ecologies

Throughout our research, we have assumed that rural to exurban transitions will stimulate change in knowledge about flora, notions of stewardship, and how people learn about nature. Our assumption has been proven correct. Despite a moderate level of social capital as reflected by attendance at events, visits to neighbors, and community volunteer work, both exurbanites and ranchers value their privacy and tend to avoid significant collaboration on natural resource management. A lack of time is the primary reason identified by landowners for the minimal collaboration between the two groups, although absenteeism and cultural differences and traditions are important as well, such as the value placed on volunteering for the fire brigade or attitudes toward hunting, motor biking, and other recreational activities.

Exurbanites and ranchers may not cooperate with one another often, but there are elements of overlap in notions of stewardship. Opportunities exist for enhanced cooperation among them (Larsen et al. 2011). Both rely on similar sources of information and advice, and both groups are worried about invasive weeds (Ikutegbe et al. 2014). Also, the desire for knowledge building and access to advice seems to be strong in both cases. A majority of rural landowners are engaged with other community members in one way or another, and it is reasonable to assume that some kind of cooperation on land management might result. But there is a paradox: despite moderate intra-community engagement on natural resource management, landowners are largely disconnected on management of invasives.

We assume that the role of government should be to reduce environmental harms and encourage environmental benefits (which we define using the concept of ecosystem services). One pathway for the government is to improve the exchange of environmental knowledge among all rural stakeholders. We acknowledge that creating or enhancing social capital through public policy is inherently challenging (Ballet et al. 2007), but given that addressing the spread of invasive weeds requires landscape-scale, multi-stakeholder collaborative action, it follows that natural resource management experts need to help strengthen social capital in amenity landscapes to the degree possible.

In their study of "co-opetition," Larsen and Hutton (2012, p. 664) conclude that "periodic and sometimes intense interdependence and mutual aid" may create opportunities for dialog that are sensitive to distinct ideologies about nature and landscape orientations. A critical issue is to identify the mechanisms by which community members or government officials can engage with those rural landowners who prefer to remain *dis*engaged with each other. These opportunities may be found in key areas of consensus, such as water management and mitigating soil erosion. And given that almost all landowners are concerned about invasives, even if for different reasons, then dialog about other issues may spill over into discussion of weeds management.

We suggest that exurbanites would benefit from greater interaction with ranchers, and more direct contact with professional natural resource managers, such as weeds inspectors. Currently, the fines for poor invasive weeds management are quite small, and do not serve as a stimulus for rural landowners to manage weeds effectively. One policy response is to make the fines larger. But our work on social capital and information flow suggests that government officials might serve as catalysts for creating dialog across rural landowner type. For example, weeds inspectors might organize neighborhood seminars and workshops regarding a number of issues, including those related to water, soil, and weeds. In other words, officials can create a host of topics that will attract a broader audience and construct a workshop format that allows for neighbors to interact directly with one another.

Graham (2013) identifies a second pathway to control the invasive weed serrated tussock, which involves enhancing information flow by building relationships among neighbors. In particular, she notes that the collaborative, community-based conservation group Landcare represents a mechanism through which different landowner types might build relationships, strengthen social capital, and work toward more effective collective weed management. The use of existing institutions, such as Landcare, is consistent with our finding that networks of information flow that cross landowner type are already present in many community landscapes.

10.8 Conclusions

The Australia case studies connect the behaviors, motivations, ideologies, and socio-economic relations among rural actors to particular types of land management. We use the example of invasive weeds as an important window into how "competing rural capitalisms" play out in amenity landscapes. The economic relation to land of different owners is important in how they know, perceive, and act on weeds (e.g., which ones they worry about, if they worry at all). And the degree of co-opetition among ranchers and exurbanites affects the landscape ecologies that develop, including patterns of land-use and weed distribution. The issue of invasives highlights tension between people celebrating collective interests, such as managing an important transboundary socio-ecological challenge, and those generated by private property rights under capitalism (reflected by some who want isolation and privacy).

The issue of invasive weeds also presents a window into the aspirations and value for nature held by landowners, perhaps most interestingly the exurbanites. The way people manage weeds and the broader activities within which weeding exists (e.g., bush restoration) reflects the kind of nature they want to create. Apart from any influence exurbanites might have, most significant changes in amenity landscapes have come about through passive land management and the retreat of agriculture—steep slopes are abandoned, woody vegetation recovers, and invasives spread (Klepeis et al. 2013). In contrast, exurbanites bring an active and intentional agency to rural land management through activities such as weeding, native plant restoration, and efforts to promote forest recovery. We have observed in Jamberoo Valley, for example, exurbanites removing the invasive *Lantana camera* from forests on a scale a rancher would never do, simply because most ranchers do not have the time and money to do so.

To the degree that there is more intense land management per hectare by exurbanites compared to ranchers—and given that exurbanites and ranchers tend to focus on different types of invasives—our results suggest the possibility of an enhanced capacity to control invasives in the future. Many exurbanites express more interest in conservation than ranchers; however, they often lack the skills, knowledge, and time to implement land management that reflects their expressed values. As the rural to exurban transition matures, the number of exurbanites goes up, and the new skills and knowledge exurbanites have about invasives and land management grows, there are likely to be even greater attempts by exurbanites to assert their agency. The result may be even more pronounced shifts in ideologies and ecologies than those observed to date.

References

Abrams, J., Gosnell, H., Gill, N., & Klepeis, P. (2012). Re-creating the rural, reconstructing nature: An international literature review of the environmental implications of amenity migration. *Conservation and Society, 10*(3), 270–284. doi:10.4103/0972-4923.101837.

ABS (Australian Bureau of Statistics). (1954/1955–2005/2006). *Agricultural censuses*. Retrieved from http://www.abs.gov.au/websitedbs/c311215.nsf/web/agriculture+-+agricultural+census.

Adger, N. W. (2003). Social capital, collective action, and adaptation to climate change. *Economic Geography, 79*(4), 387–404. doi:10.1111/j.1944-8287.2003.tb00220.x.

Agtrans. (2005). *Review of the progress on invasive species*. Final report to Department of Environment and Heritage. Agtrans research in conjunction with Noel Dawson, Brisbane, 142 pp. Retrieved from http://deh.gov.au/biodiversity/invasive/publications/review/index.html.

Alam, M. J. (2012). *Invasive plant management in complex social landscapes: A case study in coastal New South Wales in Australia*. Unpublished master's thesis, School of Earth and Environmental Science, University of Wollongong.

Argent, N., Smailes, P., & Griffin, T. (2007). The amenity complex: Towards a framework for analyzing and predicting the emergence of a multifunctional countryside in Australia. *Geographical Research, 45*(3), 217–232. doi:10.1111/j.1745-5871.2007.00456.x.

Australian Department of the Environment. (2014). *Invasive species*. Retrieved from http://www.environment.gov.au/biodiversity/invasive-species.

Ballet, J., Sirven, N., & Requiers-Desjardins, M. (2007). Social Capital and Natural Resource Management: A Critical Perspective. *The Journal of Environmental Development, 16*, 355-374. doi:10.1177/1070496507310740.

Barr, N., Karunaratne, K., & Wilkinson, R. (2005). *Australia's farmers: Past, present and future*. Canberra, Australia: Land & Water Australia.

Bebbington, A., & Perreault, T. (1999). Social capital, development, and access to resources in highland Ecuador. *Economic Geography, 75*(4), 395–418. doi:10.1111/j.1944-8287.1999.tb00127.x.

Breen, J., Hurley, P. T., & Taylor, L. E. (2016). No (back)sliding: Amenity migration, viewsheds, and contesting steep slope ordinances in Western North Carolina. In L. E. Taylor & P. T. Hurley (Eds.), *A comparative political ecology of exurbia: Planning, environmental management, and landscape change*. New York: Springer.

Burnley, I., & Murphy, P. (2004). *Sea change: Movement from metropolitan to arcadian*. Sydney, Australia: University of New South Wales Press.

Butcher, E. R., & Kelly, D. (2011). Physical and anthropogenic factors predict distribution of the invasive weed Tradescantia fluminensis. *Austral Ecology, 36*, 621–627. doi:10.1111/j.1442-9993.2010.02196.x.

Cadieux, K. V., & Hurley, P. (2011). Amenity migration, exurbia, and emerging rural landscapes: Global natural amenity as place and as process. *GeoJournal, 76*(4), 297–302. doi:10.1007/s10708-009-9335-0.

Cadieux, K. V., & Taylor, L. (Eds.). (2012). *Landscape and the ideology of nature: Green sprawl*. London: Routledge.

Cooke, B., & Lane, R. (2015). How do amenity migrants learn to be environmental stewards of rural landscapes? *Landscape and Urban Planning, 134*, 43–52. doi:10.1016/j.landurbplan.2014.10.006.

Crosby, A. W. (1986). *Biological imperialism: The biological expansion of Europe, 900–1900*. New York: Cambridge University Press.

Decker, K. L., Allen, C. R., Acosta, L., Hellman, M. L., Jorgensen, C. F., Stutzman, R. J., et al. (2012). Land use, landscapes, and biological invasions. *Invasive Plant Science and Management, 5*(1), 108–116. doi:10.1614/IPSM-D-11-00007.1.

Department of the Environment. (2013). *Weeds of national significance*. Australian Government. Retrieved from http://www.environment.gov.au/biodiversity/invasive/weeds/weeds/lists/wons.html.

Dillman, D. A., Smyht, J. D., & Christian, L. M. (2009). *Internet, mail and mixed-mode surveys: The tailored design method*. New York: Wiley.

DPI (Department of Primary Industries). (2014). *Biosecurity*. Retrieved from http://www.dpi.nsw.gov.au/agriculture/pests-weeds/weeds/definition.

Dunlap, T. R. (1997). Australian nature, European culture: Anglo settlers in Australia. In C. Miller & H. Rothman (Eds.), *Out of the woods: Essays in environmental history* (pp. 273–289). Pittsburgh, PA: University of Pittsburgh Press.

Edwards, G. (2003). The story of deregulation in the dairy industry. *The Australian Journal of Agricultural and Resource Economics, 47*, 75–98.

Ellis, E. C., Antill, E. C., & Kreft, H. (2012). All is not loss: Plant biodiversity in the anthropocene. *PLoS ONE, 7*(1), e30535. doi:10.1371/journal.pone.0030535.

Epanchin-Niell, R. S., Hufford, M. B., Aslan, C. E., Sexton, J. P., Port, J. D., & Waring, T. M. (2010). Controlling invasive species in complex social landscapes. *Frontiers in Ecology and the Environment, 8*(4), 210–216. doi:10.1890/090029.

Eriksen, C., & Gill, N. (2010). Bushfire and everyday life: Examining the awareness-action 'gap' in changing rural landscapes. *Geoforum, 41*, 814–825. doi:10.1016/j.geoforum.2010.05.004.

Eriksen, C., & Prior, T. (2011). The art of learning: Wildfire, amenity migration and local environmental knowledge. *International Journal of Wildland Fire, 20*(4), 612–624. doi:10.1071/WF10018.

Fiege, M. (2005). The weedy West: Mobile nature, boundaries, and common space in the Montana landscape. *The Western Historical Quarterly, 36*(1), 22–47. doi:10.2307/25443100.

Flannery, T. (1994). *The future eaters: An ecological history of the Australasian lands and People*. New York: Grove.

Gill, N. (2014). Making country good: Environmental change and stewardship in central Australian pastoral culture. *Transactions of the Institute of British Geographers, 39*(2), 265–277. doi:10.1111/tran.12025.

Gill, N., Chisholm, L., Klepeis, P., Wickramasuriya, D. R., & Marthick, J. (2008). *Land management and land cover on land owned by amenity oriented rural landowners in Jamberoo Valley*. Retrieved from http://ro.uow.edu.au/scipapers/999/.

Gill, N., & Klepeis, P. (2011). *Data summary: Living, working, and playing on the land survey*. Retrieved from http://socialsciences.uow.edu.au/ausccer/UOW002998.html.

Gill, N., Klepeis, P., & Chisholm, L. (2010). Stewardship among lifestyle oriented rural landowners. *Journal of Environmental Planning and Management, 53*(3), 317–334. doi:10.1080/09640561003612890.

Gillespie, P. D. (2003). *Agricultural trends in the Sydney region 1996–2001 census comparisons*. Sydney, Australia: NSW Agriculture.

Goldman, R. (2010). Ecosystem services: How people benefit from nature. *Environment, 52*(5), 15–23. doi:10.1080/00139157.2010.507140.

Gosnell, H., & Abrams, J. (2010). Amenity migration: Diverse conceptualizations of drivers, socioeconomic dimensions, and emerging challenges. *GeoJournal, 76*(4), 303–322. doi:10.1007/s10708-009-9295-4.

Gosnell, H., Haggerty, J. H., & Travis, W. R. (2006). Ranchland ownership change in the Greater Yellowstone Ecosystem, 1990–2001: Implications for conservation. *Society and Natural Resources, 19*(8), 743–758. doi:10.1080/08941920600801181.

Goulden, M., Adger, W. N., Allison, E. H., & Conway, D. (2013). Limits to resilience from livelihood diversification and social capital in lake social-ecological systems. *Annals of the Association of American Geographers, 103*(4), 906–924. doi:10.1080/00045608.2013.765771.

Graham, S. (2013). Three cooperative pathways to solving a collective weed management problem. *Australasian Journal of Environmental Management, 20*(2), 116–129. doi:10.1080/14486563.2013.774681.
Guagnano, G. A., Stern, P. C., & Dietz, T. (1995). Influences on attitude-behavior relationships: A natural experiment with curbside recycling. *Environment and Behavior, 27*, 699–718. doi:10.1177/0013916595275005.
Haggerty, J. H., & Travis, W. R. (2006). Out of administrative control: Absentee owners, resident elk and the shifting nature of wildlife management in southwestern Montana. *Geoforum, 37*(5), 816–830. doi:10.1016/j.geoforum.2005.12.004.
Hall, M. (2003). Editorial: The native, naturalized, and exotic—Plants and animals in Human History. *Landscape Research, 28*(1), 5–9. doi:10.1080/01426390306534.
Halseth, G. (1998). *Cottage country in transition: A social geography of change and contention in the rural recreational countryside*. Montréal, Quebec, Canada: McGill-Queen's University Press.
Hay, I. (Ed.). (2010). *Qualitative research methods in human geography*. Oxford: Oxford University Press.
Head, L. (2000). *Cultural landscapes and environmental change*. London: Oxford University Press.
Head, L., & Atchison, J. (2009). Cultural ecology: Emerging human-plant geographies. *Progress in Human Geography, 33*(2), 236–245. doi:10.1177/0309132508094075.
Hindle, J. P., Nott, M. J., & Crichton, J. R. (1987). *Illawarra region—Agricultural land classification study*. Sydney, Australia: Department of Agriculture New South Wales.
Holmes, J. (2005). Impulses towards a multifunctional transition in rural Australia: Gaps in the research agenda. *Journal of Rural Studies, 22*, 142–160. doi:10.1016/j.jrurstud.2005.08.006.
Hoogesteger, J. (2013). Trans-forming social capital around water: Water user organizations, water rights, and nongovernmental organizations in Cangahua, the Ecuadorian Andes. *Society and Natural Resources: An International Journal, 26*, 60–74. doi:10.1080/08941920.2012.689933.
Houston, P. (2005). Re-valuing the fringe: Some findings on the value of agricultural production in Australia's peri-urban regions. *Geographical Research, 43*(2), 209–223. doi:10.1111/j.1745-5871.2005.00314.x.
Howard, J. (2008). The future of the Murray River: Amenity re-considered? *Geographical Research, 46*(3), 291–302. doi:10.1111/j.1745-5871.2008.00524.x.
Ikutegbe, V., Gill, N., & Klepeis, P. (2014). Same but different: Sources of natural resource management advice for lifestyle oriented rural landholders. *Journal of Environment Planning and Management, 58*(9), 1530–1543. doi:10.1080/09640568.2014.936551.
Kasperson, R. E., Golding, D., & Kasperson, J. X. (1999). Trust, risk, and democratic theory. In G. Cvetkovich & R. Lofstedt (Eds.), *Social trust and the management of risk* (pp. 22–44). London: Earthscan.
Kiama Municipal Council. (2013). *Kiama Municipal Council—Annual report 2011–2012*. Retrieved from http://www.kiama.nsw.gov.au/Your-Council/Major-Publications---Reports.
Kiama Municipal Council. (2014). *Flora and fauna*. Retrieved from http://www.kiama.nsw.gov.au/environment/our-environment/biodiversity/flora---fauna.
Klepeis, P., Gill, N., & Chisholm, L. A. (2009). Emerging amenity landscapes: Invasive weeds and land subdivision in rural Australia. *Land Use Policy, 26*(2), 380–392. doi:10.1016/j.landusepol.2008.04.006.
Klepeis, P., Scull, P., Lalonde, T., Svajlenka, N., & Gill, N. (2013). Changing forest recovery dynamics in the Northeastern United States. *Area, 45*(2), 239–248. doi:10.1111/area.12016.
Klinenberg, E. (2002). *Heat wave: A social autopsy of disaster in Chicago*. Chicago: University of Chicago Press.
Larsen, S. C., Foulkes, M., Sorenson, C., & Thompson, A. (2011). Environmental learning and the social construction of an exurban landscape in Fremont County, Colorado. *Geoforum, 42*(1), 83–93. doi:10.1016/j.geoforum.2010.10.003.
Larsen, S., & Hutton, C. (2012). Community discourse and the emerging amenity landscapes of the rural American West. *GeoJournal, 77*(5), 651–665. doi:10.1007/s10708-011-9410-1.

MacNaghten, P., & Urry, J. (1998). *Contested natures*. London: Sage.
Maestas, J. D., Knight, R. L., & Gilgert, W. C. (2003). Biodiversity across a rural land-use gradient. *Conservation Biology, 17*(5), 1425–1434. doi:10.1046/j.1523-1739.2003.02371.x.
Matarrita-Cascante, D., & Stocks, G. (2013). Amenity migration to the global south: Implications for community development. *Geoforum, 49*, 91–102. doi:10.1016/j.geoforum.2013.06.004.
McCaffrey, S. (2004). Thinking of Wildfire as a Natural Hazard. *Society and Natural Resources, 17*, 509–516. doi:10.1080/08941920490452445.
McNeely, J. A. (2004). Strangers in our midst: The problem of invasive species. *Environment, 46*(6), 17–30. doi:10.1080/00139157.2004.10545159.
Meadows, J., Herbohn, J., & Emtage, N. (2012). Supporting cooperative forest management among small-acreage lifestyle landowners in Southeast Queensland, Australia. *Society and Natural Resources: An International Journal, 26*(7), 745–761. doi:10.1080/08941920.2012.719586.
Mendham, E., Curtis, A., & Millar, J. (2012). The natural resource management implications of rural property turnover. *Ecology and Society, 17*(4), 5. doi:10.5751/ES-05071-170405.
Michelmore, M. (2003). *The serrated tussock managers' factpack*. The State of New South Wales, Australia: NSW Agriculture.
Mills, K., & Associates. (2006). *The natural vegetation in the municipality of Kiama New South Wales—Final report*. Kiama, New South Wales, Australia: Report prepared for Kiama Municipal Council.
Mooney, H. A., & Cleland, E. E. (2001). The evolutionary impact of invasive species. *Proceedings of the National Academy of Sciences of the United States of America, 98*(10), 5446–5451. doi:10.1073/pnas.091093398.
Moss, L. A. G. (Ed.). (2006). *The amenity migrants: Seeking and sustaining mountains and their cultures*. Cambridge, MA: CABI Publishing.
Polce, C., Kunin, W. E., Biesmeijer, J. C., Dauber, J., Phillips, O. L., & the ALARM Field Site Network. (2011). Alien and native plants show contrasting responses to climate and land use in Europe. *Global Ecology and Biogeography, 20*(3), 367–379. doi:10.1111/j.1466-8238.2010.00619.x.
Putnam, R. (1995). Bowling alone: America's declining social capital. *Journal of Democracy, 6*(1), 65–78.
Robbins, P. (2004). Comparing invasive networks: Cultural and political biographies of invasive species. *The Geographical Review, 94*(2), 139–156. doi:10.1111/j.19310846.2004.tb00164.x.
Robbins, P. (2007). *Lawn people: How grasses, weeds, and chemicals make us who we are*. Philadelphia, PA: Temple University Press.
Seidl, D., & Klepeis, P. (2011). Human dimensions of earthworm invasion in the adirondack state park. *Human Ecology, 39*(5), 641–655. doi:10.1007/s10745-011-9422-y.
Sobels, J., Curtis, A., & Lockie, S. (2001). The role of landcare networks in rural Australia: Exploring the contribution of social capital. *Journal of Rural Studies, 17*(3), 265–276. doi:10.1016/S0743-0167(01)00003-1.
Sorice, M. G., Kreuter, U. P., Wilcox, B. P., & Fox, W. E., III. (2014). Changing landowners, changing ecosystem? Land-ownership motivations as drivers of land management practices. *Journal of Environmental Management, 133*, 144–152. doi:10.1016/j.jenvman.2013.11.029.
Tidwell, L. S., & Brunson, M. W. (2008). Volunteering to manage rangeland weeds. *Rangelands, 30*(4), 19–24. doi:10.2111/1551-501X(2008)30[19:VTMRW]2.0.CO;2.
Vitousek, P. M., Dantonio, C. M., Loope, L. L., Rejmanek, M., & Westbrooks, R. (1997). Introduced species: A significant component of human caused global change. *New Zealand Journal of Ecology, 21*(1), 1–16.
Williams, J. A., & West, C. J. (2000). Environmental weeds in Australia and New Zealand: Issues and approaches to management. *Austral Ecology, 25*(5), 425–444. doi:10.1046/j.1442-9993.2000.01081.x.
Wilson, G. A. (2007). *Multifunctional agriculture: A transition theory perspective*. Cambridge, MA: CABI Publishing.
Wilson, G. A. (2012). *Community resilience and environmental transitions*. New York: Routledge.

Chapter 11
Politics of Landscape Transformation in Exurban King County, Washington

Jenna H. Tilt and Lee K. Cerveny

11.1 Introduction

Over the last two decades, many cities in the western United States have experienced rapid exurban growth particularly along corridors leading to areas rich in natural and rural amenities (e.g., mountains, seaside, lakes, farms) or that offer an abundance of parks and open spaces. King County (Seattle), Washington represents an ideal setting to explore the complex dynamics and competing interests that shape the physical and social landscapes being cultivated in exurbia. Large land tracts in the Cascade foothills once owned by timber corporations and agricultural producers are being sold to land developers who navigate complex land-use planning policies to create a variety of residential forms, including master-planned communities, low-density residential lots, and gated communities. The dynamics of these shifting landscapes from productivist to post-productivist regimes have underlying causal mechanisms that come into focus when viewed through a political ecology lens.

In this chapter, we examine the interests and negotiated approaches of various actors (i.e., municipal and county planners, developers, real estate professionals, and residents) in southeastern King County who are involved in creating and consuming exurban landscapes. We focus on three neighboring exurbs situated along State Highway 169, which radiates from Seattle's suburbs to Mt. Rainier. The largest community, Maple Valley (pop. 22,684), is one of the fastest growing communities

J.H. Tilt (✉)
Geography, Environmental Sciences, and Marine Resource Management,
College of Earth, Ocean, and Atmospheric Sciences, Oregon State University,
104 CEOAS Admin, Corvallis, OR 97331, USA
e-mail: tiltj@onid.orst.edu

L.K. Cerveny
Pacific Northwest Research Station, U.S. Forest Service,
400 N. 34th St. Suite 201, Seattle, WA 98103, USA
e-mail: lcerveny@fs.fed.us

L.E. Taylor, P.T. Hurley (eds.), *A Comparative Political Ecology of Exurbia*,
DOI 10.1007/978-3-319-29462-9_11

in Washington, emphasizing dense residential development that caters to young families and outdoor enthusiasts. Ravensdale (pop. 1101) is a dispersed, wooded settlement that attracts landowners seeking privacy, large lots, and facilities for livestock and horses. Black Diamond (pop. 4141) is an historic mining community surrounded by forests and farms. Black Diamond is the future home of two master-planned communities with 6050 proposed residential units, which has ignited an intense debate over competing visions of "rural" lifestyle. Over time, these three exurbs have embodied residential forms that challenge our notions of "urban" and "rural."

Exurban areas are shaped and reshaped by evolving discourses from planning institutions, regional market economies, and cultural norms which in turn influence residential preferences and perceptions of access to urban and rural amenities. The pace and pattern of exurban development is continually shaped by private landowners, local governments, and nongovernmental actors who operate within the complex arena of growth management policies. The construction of new communities marketed by developers as, "rural living close to the city" and by planners as "rural by design" has fragmented traditional landscapes and fractured local identities. While designed to bring the consumer closer to nature and rural life, these developments have, in fact, transformed rural areas into manufactured exurbs that bring the city closer to the rural.

11.2 Political Ecology of Exurbia

Exurban discourses often become tangled by competing notions of exurbia and exurban landscapes. Three elements receiving treatment in the exurban literature help to frame our discussion: (1) cultural landscape of exurban studies (exurbia); (2) physical or spatial characteristics of exurban landscapes; and (3) political ecology of exurban residential growth.

Previous explorations of the cultural landscape of exurbia have emphasized the pull of natural amenities, rural lifestyles, and cultural identity that is located well beyond the urban core. Taylor (2011) describes exurbia as "a cultural landscape where the wild natural or pastoral rural landscapes of the imagination motivate exurbanites' residential choices and individual land management decisions" (p. 331). Within the same vein, Johnson (2008) defines the exurban individual as someone who (a) resides in sparsely populated areas beyond the suburbs; (b) resided previously in an urban or suburban area; and (c) derives the majority of household income from sources other than production from their land. These exurban relationships are best understood from a political ecology perspective that explores the margins between urban, suburban, and rural, where political actors negotiate for control of a contested physical landscape. The results can be highly variable, with the emergent exurban landscape form often conforming to the will of the most persuasive and persistent political forces.

The physical transformations occurring in exurban areas are most commonly characterized as low-density residential growth (Theobald 2001) with post-productivist economies (Taylor 2011) by scholars seeking to understand environmental and social impacts of rural transformation (see Abrams et al. 2012 for review). Using this operational definition of exurbia, exurban research has included studies examining the effects of second-home ownership in exurban "gateway" communities that served as rural recreational centers (Stedman and Hammer 2006) as well as amenity landowners engaged in "hobby farming" (Busck 2002; Holloway 2002). Often these studies revolve around familiar themes of conflict and contested spaces between long-time residents and new-comers (Crump 2003; Hurley and Walker 2004; Walker and Fortmann 2003) or polarized "pro-growth" and "anti-growth" rhetoric. This work has given much breadth and depth in our current understanding of the environmental, social, and economic impacts of low-density exurban landscapes.

Yet, as Lichter and Brown (2011) point out, physical landscape transformations from rural to urban can occur through processes other than amenity migration, such as annexation or incorporation of a rural area into a municipal government structure. These processes can lead to radically different exurban residential landscape patterns that could include traditional suburban-like developments or large master-planned communities with built-in retail, schools, pathways, and parks, and a mixture of residential homes from condominiums to townhouses, and single family homes of all sizes and densities. What results is a patchwork of residential forms in the exurban zone. These land-use patterns call into question whether these areas are truly exurban or just suburbs near natural resource production zones. This "definitional knot" (Taylor 2011, p. 334) frustrates scholars, particularly demographers, who try to pinpoint clear demarcations between urban, rural, suburban, and exurban. But, rather than squaring the circle by overlooking rural transformative landscapes that do not conform to a pattern of low-density residential development, researchers should examine the underlying causal mechanisms arising in these distinct and diverse landscape patterns.

Political ecology presents a useful framework for considering these causal mechanisms, particularly the role of power and politics in environmental events by acknowledging that actors have differential access to resources and often craft alliances and strategic partnerships to achieve their goals. In this chapter, we explore how a diverse set of actors negotiated for the use and control of spaces resulting in diverse land-use patterns that transformed a formerly rural landscape into a dynamic and complex exurban zone. We highlight the interactions among actors operating at different scales and how these relations affect the manifestation of exurbia in the far-reaches of King County (Bryant 1992). Specifically, we seek to understand the decisions various community actors have made about development in their community within the context of their political, economic, and social environment. These communities are not isolated and bounded spatial entities, but are tied to the global economy where negotiations among actors, local and global, have implications for the shape of exurban neighborhoods and resident lifestyles (Blaikie and Brookfield 1987). We learn how alliances between local municipal officials and corporate developers can serve to finagle regulations and creatively interpret legislation. We see how bargains are made, where exurban residents are given assurances that their rural lifestyles will be protected, but where new developments can also take hold. We also see evidence of citizen action to protect communities from urbanizing forces and retain a desired lifestyle using creative planning mechanisms.

Our study extends the conceptualization of exurban transformation to include a diverse set of land-use patterns created by global, regional, and local power regimes, economic conditions, and cultural preferences. We find that planning policies, corporate development strategies, and grassroots organizing each have the power to transform an exurban landscape. Further, we find that the emergent shape of these exurban landscapes attracts a particular set of residents whose preferences conform to the one being idealized by these powerful actors. By understanding these causal mechanisms behind the creation of this exurban landscape, we gain a deeper understanding of the cultural production of exurbia.

11.3 Regional Context

Our study takes place in the exurbs of King County, Washington (pop. 2 million, 2013), which encompasses Seattle, Bellevue, and many large suburban cities. King County occupies a vast landscape (2307 square miles) that extends from the Puget Sound to the crest of the Cascade Mountains (Fig. 11.1). The county contains 850,000 acres of forestlands including large portions of the Mt. Baker-Snoqualmie National Forest, state and county forests, commercial forests, and 50,000 acres of farmland (King County 2010). King County is rich in natural beauty with territorial views of Mt. Rainier (elevation: 14,409 ft) just 54 miles southeast of Seattle, as well as the Cascade Range to the east, the Olympic Mountains to the west, Lake Washington, separating Seattle and Bellevue, and the Puget Sound connecting to the Strait of Juan de Fuca and the Pacific coast.

King County is home to Native American tribes including the Snoqualmie, Duwamish, Muckleshoot, Puyallup, and Skykomish. The county has had a long history of natural resource production and extraction, primarily timber, farming, commercial fishing, and mining. World Wars I and II brought significant population growth, as Seattle's emerging industries built ships and airplanes for the war effort. In the post-war era, the county grew due to the baby boom and an abundance of manufacturing jobs. Interstate highways brought growth out to the suburbs. Since 1970, the county has grown steadily with first-ring suburbs Bellevue (pop. 132,100) and Renton (pop. 95,448) growing into large metropolitan areas. King County has transitioned from its natural resource extraction origins to become the home to high-tech, manufacturing, and retail powerhouses, with vast corporate campuses of Microsoft, Boeing, Starbucks, Costco, and Amazon located in Seattle's urban core and first-ring suburbs. We focus our study on State Route 169 (SR-169)

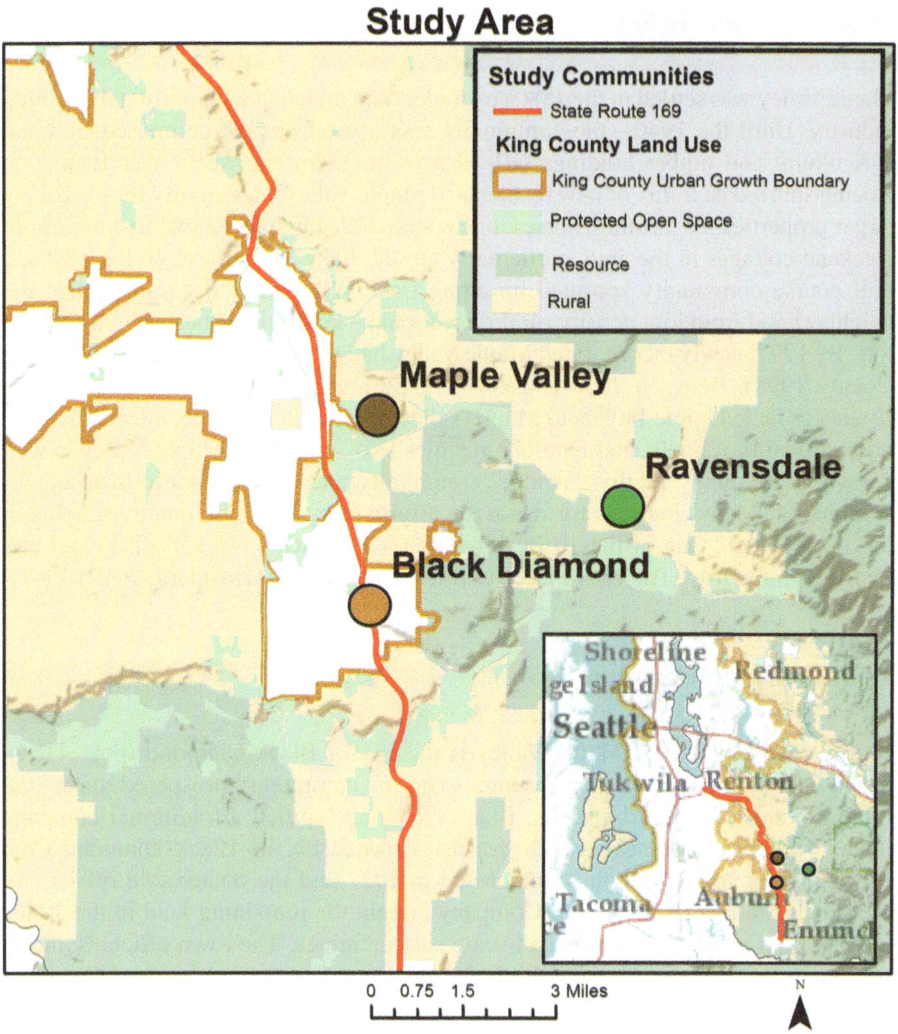

Fig. 11.1 Map of study region. Produced by J. Tilt using ESRI software

in southeast King County. SR-169 begins in Renton and follows the Cedar River east for 15 miles before turning south to cross the Green River Gorge. The route ends in Enumclaw, the gateway to Mt. Rainier National Park. Along SR-169, about 25 miles from Seattle's urban core, three adjacent communities exemplify King County's proliferating exurbs: Maple Valley, Black Diamond, and Ravensdale (Fig. 11.1). Although these communities are within close proximity to one another, their development has been dramatically different.

11.3.1 *Maple Valley*

Maple Valley was settled in the 1880s by traders and merchants supporting the mining industry. Until the 1940s, the community was a rural outpost composed of small agricultural and timber holdings (10–50 acre lots). During World War II, jobs at Boeing spurred an influx of new residents to Maple Valley, particularly those seeking larger properties for farming, horses, or livestock. Meanwhile, Seattle residents built weekend cottages in the area, particularly around Lake Wilderness. In the 1990s, a golf course community sprouted up along Lake Wilderness that transformed the neighborhood from low-density rural/recreational parcels into suburban quarter-acre lots. By 1997, newly incorporated Maple Valley had grown to 11,000 residents (King County 1998). Between 2000 and 2007, rising housing prices elsewhere in King County attracted home-buyers to Maple Valley, which offered large, affordable, new homes, quality schools, and natural amenities (Kossen 2002). Maple Valley is well known as a young "family-friendly" community and has won various accolades from national news journals for its safety, affordability, and high quality schools; it has been called "One of the 10 Best Towns for Families" (Nayyar 2011) and was included in the list, "The Best Places to Raise Kids 2013" (D'Addario 2012).

11.3.2 *Black Diamond*

Nearly 5 miles south of Maple Valley is the city of Black Diamond (pop. 4154). Black Diamond is rich in coal resources and the community prospered during the mining boom of the early 1900s. Black Diamond typified a traditional company town, with many businesses and properties owned by the Black Diamond Coal Mining Company. The main mine closed in 1927 and the owner sold off its city properties. Palmer Coking Coal Company bought the remaining land in the 1940s and has continued to operate small-scale surface mines. The town officially incorporated in 1959, having lost one-third of its peak population (City of Black Diamond 2009). Since 2000, Black Diamond has seen steady population growth due to spill-over pressure from Maple Valley. While still predominantly rural in character, new subdivisions have taken root on former forests and farmlands. In 2006, Palmer Coking and Plum Creek Timber sold 1567 acres to the Yarrow Bay Development Company, which designed two large master-planned communities. The social conflict between residents supporting and opposing the development has deeply divided the community.

11.3.3 *Ravensdale*

East of Maple Valley is Ravensdale (pop. 1101), an unincorporated community with a rich history of natural resource extraction, including coal mining, agriculture, and forestry. The Northern Pacific Company mined in the area through most of the

Table 11.1 Community demographics* and characteristics

Community	Population (2010)	Median sale price (2012)[a]	Percent White (2010) (%)	Land area (square miles)	Median age (years)	Density (population per square mile)	Percent of residents commuting 10–24 miles (2011) (%)
Maple Valley	22,684	$250,000	85.8	5.9	34.2	3965	66.5
Black Diamond	4154	$200,000	92.0	6.7	40.4	690	61.4
Ravensdale	1101	$350,000	95.7	5.1	41.0	162	72.2
Seattle	608,660	$400,000	69.5	142.7	36.1	7251	16.9
King County	1,931,249	–	68.7	2307.0	37.1	908	30.3

[a]Trulia.com
*U.S. Census 2010

twentieth century, eventually selling the land and claim to Palmer Coking Coal Company. Today, the major employer of the area is the Sila Sand Works, an aggregate extraction company, but most Ravensdale residents commute toward Seattle and its suburbs. Ravensdale is part of the Rock Creek Valley, a 32 square mile area rich in natural resources and habitat diversity and includes several county parks and open spaces. These natural areas are connected by trail systems used by hikers, mountain bikers, and equestrians. Ravensdale and the Rock Creek Valley experienced growth during the height of the housing boom in the early 2000s with approximately 1300 acres approved for development (FRCV 2004). New homes were developed on lots from 2.5 acres and greater. The low-density, wooded properties attract residents seeking privacy and space for large gardens, horses, and farm animals.

Table 11.1 compares community demographics and residential trends among the study communities as well as Seattle and King County. The study communities are similar in many respects in terms of land area, housing prices, and commuting rates. In addition, all three communities indicate a higher percentage of non-minority residents than found in King County. Maple Valley is the largest and most densely developed of the communities and has the lowest median age, suggesting a preponderance of families. Black Diamond represents an older demographic and is less dense and populated than Maple Valley. Ravensdale residents also are older than the county average and experience lower housing density. All study communities have a high proportion of residents commuting 10–24 miles. Sales prices of homes in the study area are much less than neighboring Seattle, making the area an affordable choice for home owners.

11.4 Methodology

To study the political factors producing the exurban landscapes, we focused on the actors involved in planning and development. Data collection took place between 2011 and 2013. We relied on three major data sources for this case study comparison. First, we conducted 15 open-ended interviews to explore changes in the social and political landscape and the emergence of exurban communities along the SR-169 corridor. Four categories of exurban actors were interviewed: real estate developers, real estate agents, municipal and county planners, and area residents. Real estate developers were identified by researching current development projects in the area and contacting the developer agencies; we also consulted local planners and real estate professionals about our interviewee selection. Similarly, real estate agents were identified through research of local home sales in the study communities and consultation with local planners and residents. Local planners were identified by contacting local community planning offices. In 2013, seven focus groups were conducted in communities along the SR-169 corridor, including the communities featured in this analysis. Focus groups included typically 8–10 residents who were guided through a structured series of questions related to residential choices and exurban experiences. Notices of focus group meetings were posted throughout the community and local organizations (e.g., Parent-Teacher Associations, Elks club, Home-Owner Associations) were directly contacted and invited to participate in the focus group interviews.

All interviews utilized a semi-structured format with flexibility to pursue relevant questions that reflected the interviewee's expertise or lived experiences. Community focus group questions focused on understanding community growth, the forces shaping particular developments, and on contemporary community needs, issues, and challenges. Interviews with developers explored factors that shape community design and negotiations with local government officials to create communities consistent with city objectives. Interviews with real estate professionals emphasized marketing and selling of homes in these neighborhoods, with insights in shifting consumer trends. City planner interviews provided historical context for city growth and insights into political aspects of community development. Qualitative analysis of interview data was undertaken using the NVivo software (Nvivo 2012).

We supplement the individual interviews with analysis of additional information gathered in local newspapers, city websites, planning documents, developer packages, and promotional materials. A content analysis of these materials was undertaken to explore the underlying causal factors in shaping the environmental, social, and economic landscapes of these exurban areas (Neuman 2003). Taken together, these three sources of data (individual interviews, focus groups, and document analysis) allow us to understand the complex dynamics among actors that lead to exurban development.

11.5 Politics of Exurban Growth

Development in the exurban communities along the SR-169 corridor has been shaped by the interaction of global economic forces, regional governing institutions, and the area's physical geography. The communities of Maple Valley, Black Diamond, and Ravensdale have not only been influenced by their history of dependence on natural resource extraction, but they are continually shaped and reshaped by regional and global economies which in recent years have stymied resource-based industries, while fueling growth in manufacturing and technology. Concerned about the implications of rapid population growth, the Washington State legislature passed the Growth Management Act (GMA) in 1990–1991 to guide future growth and protect natural resources. This was accomplished by mandating comprehensive planning at the city and county level, including the identification of urban containment areas (e.g., urban growth areas) and allowable land uses for areas within and outside those areas. Designated urban growth areas cover 22 % (460 square miles) and agricultural production zones cover 65 % (1380 square mile) (King County 2007).

Squeezed between the urban and production lands, rural designated lands are often contested exurban areas experiencing pressure both from residential developers as well as open space conservation interests. These rural areas have been called the "left-over meat in the GMA refrigerator" (Western Washington Growth Management Hearings Board 1995), and currently cover approximately 13 % of King County (290 square miles). We examine how these three contested communities negotiate the growth and development process vis-à-vis comprehensive planning between urban, resource, and rural designated lands to reimagine and transform the social, economic, and environmental exurban landscape in dramatically different ways.

11.5.1 Maple Valley

In King County's first comprehensive plan after the passage of the Growth Management Act, the western half of Maple Valley was included inside this original urban growth boundary and identified as an "unincorporated urban area" or a "potential annexation area." This meant that King County's long-term vision for the community was either annexation into an existing city or incorporation into its own community. Fearing loss of local control with annexation and frustration over reduced county services (Clutter 1996b), community leaders advocated for incorporation in 1995. A Maple Valley planner reflected on the incorporation this way:

> The city incorporated in 1997 because the local group here, … the town power brokers, said we can do it cheaper and better than the county. 'We're tired of listening to the county. We want our control.' … . Their mantra at the time is 'we can do it cheaper than the county.' Well, the county services were already extremely limited out here because this was the end of the trail.

Fig. 11.2 Maple Ridge Highlands in Maple Valley. *Photo credit*: J. Tilt

With incorporation, Maple Valley created its first comprehensive plan that paved the way for more intense residential development. Plum Creek Timber, which owned property near the Maple Valley Urban Growth Boundary, sold 750 acres to Polygon Northwest, a national home-building corporation. Polygon planned to develop the property into 20-acre residential parcels—a residential density permissible in King County's designated rural lands that exist outside urban growth areas. However, King County and the City of Maple Valley negotiated with Polygon to take an alternative development approach using a land-use set-aside policy known as "4-to-1." This land-use policy allows for high-density residential development to occur outside an urban growth area, in exchange for permanent protection of open space by requiring that for each acre developed outside of the urban growth area, four additional acres are retained as open space (King County 2015). Using this policy, Polygon was able to build 575 homes on 150 acres while 600 additional acres were deeded as permanent open space to King County. The two high-density residential subdivisions built by Polygon, Maple Woods and Maple Ridge Highlands, were annexed into the city in 2006 (Fig. 11.2).

New development throughout Maple Valley led to significant population growth. Between 1990 and 2000, the population soared from 1211 to 14,209 (U.S. Census 2000). Over the next 10 years, development in the newly annexed areas led to even more residents, with an estimated 24,804 in 2013. Currently, 70 % of Maple Valley is dedicated to residential use with densities ranging from four dwelling units per acre or higher. Commenting on these dramatic changes, one long-time resident of the area says:

> Maple Valley is not what it was. Having been in Maple Valley since 1972, we've seen huge changes. It was quite rural. I think most people twenty years ago would come to Maple Valley for the farm.... There's not that much farming community left.

The transformation of Maple Valley was enabled by the very same set of planning tools used to contain and limit growth throughout King County. Rather than limit growth in Maple Valley, King County encouraged high-density development of this exurb to alleviate growth pressures elsewhere by granting incorporation. The decision to encourage developers to adopt a 4-to-1 policy in rural areas adjacent to Maple Valley, rather than implement the allowed low-density rural residential zoning, facilitated dense urban growth even closer to the forest production zone. This shift in planning policies has since created new development pressures on the remaining rural lands squeezed in between "urban" Maple Valley and the natural resource lands. Moreover, a high-density community at the forest edge left little space for anything else, including active parks and open spaces within the community boundaries. As one city official explained:

> They [the city leaders] didn't forecast what was going to happen in this 5.6 square miles. It's completely built out. The city forefathers thought they were land banking at a high enough rate for park and recreational facilities. Think again.

The need for active park space within a community with an exploding population of school children has now become the surrogate issue for defining Maple Valley's future. This city official continued:

> Now, the council is struggling: "Do we go into debt? Do we not? Are we willing to invest in our own community or not?" They're hanging onto these rural connections with the past thinking that undeveloped property is attractive.... And, certainly in some of their minds it's this, "hey we're a small town." It's like really we're not a town anymore. This is now a city.

The rapid development of dense residential housing in an effort to keep tax bases low and keep to the original promise of "doing business cheaper than the county" has left community residents without some of the services they desire, while the old political leadership comes to grips with their new identity as an exurban city rather than a rural town.

However, there is an abundance of "undeveloped," natural open space adjacent to and within close proximity to Maple Valley, due to land set-asides and other preserved public lands and easements. However, access and use of these spaces is primarily for equestrians or mountain bikers. Maple Ridge Highlands has 600 acres of open space surrounding the development that was preserved through the 4-to-1 set aside, yet the open space is not heavily utilized by local residents. While a few residents comment that they "enjoy the ability to be closer to mountain biking," or they have "discovered mountain biking" once they moved to the community, residents more often mentioned features such as active parks, basketball courts, and cul-de-sacs of the Maple Ridge Highlands community as primary recreational areas, rather than the ample open space around them.

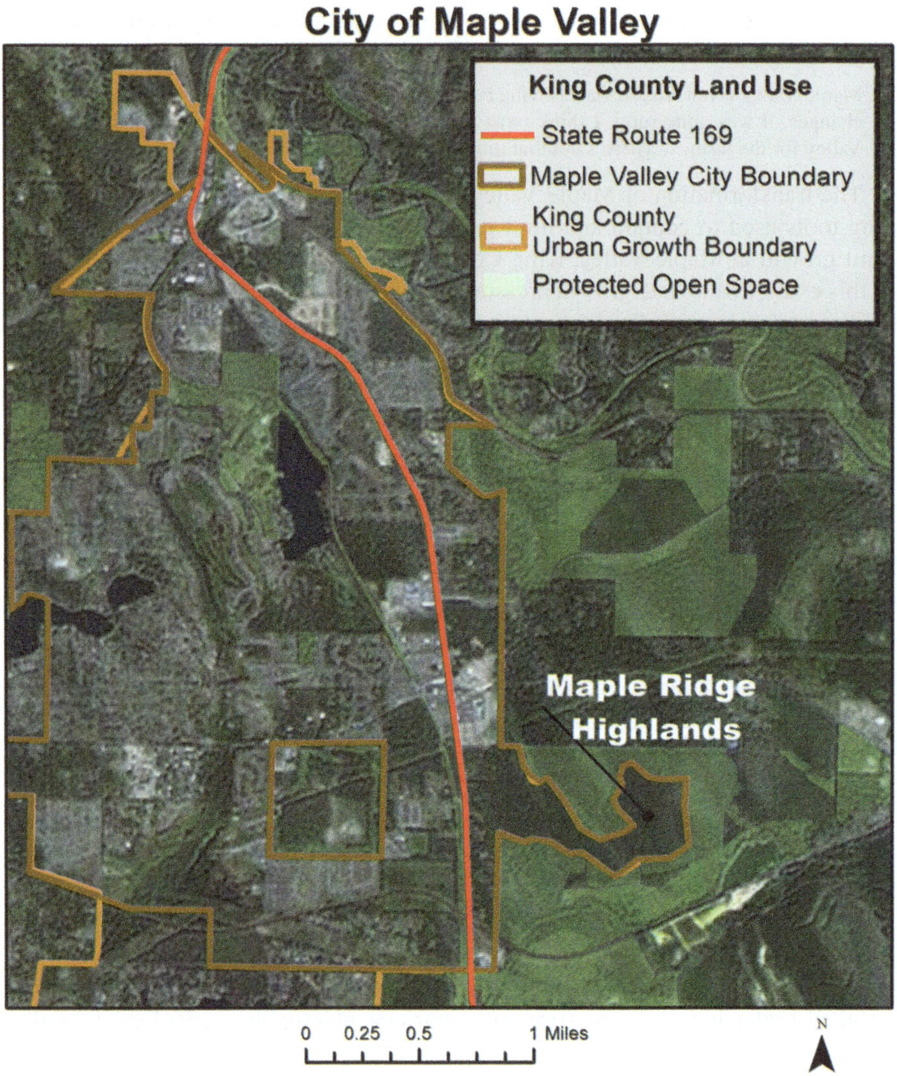

Fig. 11.3 Map of Maple Valley land use. Produced by J. Tilt using ESRI software

In a relatively short time, Maple Valley underwent a dramatic transformation from a rural community to an exurban city (Fig. 11.3). The incorporation of the 5.6-mile area allowed community leaders to set the breakneck pace of development and subsequently keep their low tax promise to local constituents. However, this vision did not fully realize that incoming residents would be primarily young families seeking affordable housing, successful schools, and active parks. While the exurban community was able to successfully negotiate with King County and Plum Creek to secure access to 600 acres of undeveloped open space through the 4-to-1 policy, access to the open space is difficult and use of it is limited mostly to equestrians and mountain bikers; the type of outdoor space desired most by Maple Valley (parks, basketball courts, etc.) is severely lacking. Although envisioned as a family-focused community with parks and open space, the choices made by developers and city officials ultimately restricted access to these spaces. The growth and development of Maple Valley is a cautionary tale of how powerful local actors can transform a landscape dramatically even under the purview of growth management.

11.5.1.1 Black Diamond

While causal factors of a hot economic and housing market allowed King County to justify its use of the 4-to-1 policy to alleviate some of the growth pressures in Maple Valley, no such conclusion can be made about Black Diamond. The City of Black Diamond and two large corporate landowners, Palmer Coking Coal and Plum Creek Timber, successfully negotiated a similar land exchange long before the city had experienced any regional growth pressure. Here, the pressure of seeking Black Diamond's compliance with the new growth management legislation led King County to accept a land exchange that significantly increased the urban growth area for Black Diamond.

Palmer Coking and Plum Creek Timber privately negotiated with King County over the size of the urban growth area in Black Diamond. Both companies held large land holdings surrounding Black Diamond and feared being left out of possible future development opportunities with the designation of urban growth areas. King County was pressured to consider the prospect of 5-acre residential lots scattered across Plum Creek and Palmer Coking land in unincorporated King County versus allowing Black Diamond to expand its urban growth area (UGA) and permit high-density growth. In 1996, an agreement was struck based on the county 4-to-1 policy: providing more acreage for urban growth inside the UGA while also setting aside large acreage for open space preservation outside the boundary. The final deal, referred to as the Black Diamond Urban Growth Area Agreement (BDUGAA), stipulated 782 acres of land could be designated as urban within the new UGA and an additional 1670 acres would be given to the city or county and be protected as permanent open space—the total area would be annexed into Black Diamond when the city was ready to pursue development (Clutter 1996a). As the media reported at the time, the final deal was little more than a 2-to-1, "but based on the principle of the 4-to-1"; nevertheless, it was reported as an "agreement to be proud of" (Clutter 1996a). However, opposing the deal, King County Councilman Brian Derdowski stated: "This is, in fact, a rural town; yet we are now targeting Black Diamond with massive amounts of urban growth" (Clutter 1996a). Reflecting on the land-use policies and deals that set the stage for the development in Black Diamond, the planner for the community stated:

> I think it [BDUGAA] was seen as a unique opportunity to do something unique out here. And get a lot for it at the same time as far as open space goes. And, to prevent what some people were seeing as the swirl of this southeast King County area which [is] a lot of five acre lots with one house on it out here.

Similar to Maple Valley, local residents worried growth management policies would lead to a loss of property tax revenue. Working with large corporate landowners, they were successful at using King County's own land-use policies to gain a greater foothold for urban development. Even though King County's zoning policies allowed for low-density resident development, some residents feared that "rural sprawl" could take over wide swaths of exurban land currently held by logging and mining companies if these lands remained outside of the urban growth area. Thus, the deal brokered between King County, Black Diamond, Palmer Coking, and Plum Creek Timber enabled a reimagined Black Diamond with high residential density and permanent open space woven throughout the community. Still, at the time of the agreement (and for many years after), Black Diamond faced no growth pressure to justify an expansive urban growth area.

Population in Black Diamond began to grow steadily in the 1990s—particularly around the shores of Lake Sawyer and other small developments near the historic downtown—rising from 1422 in 1990 to 3970 in 2000 (Fig. 11.4). Current population stands now at 4300. With this steady rise in population growth, Black Diamond decided in 2005 that the time was ripe to annex about 2000 acres stipulated under the BDUGAA. The following year, the Yarrow Bay Development Company purchased all of the newly annexed property. Three years later, the city adopted a new comprehensive land-use plan and master-plan community development code to guide the Yarrow Bay developments. The final master plan, approved in 2011, included two separate residential developments that would build a total of 6050 residential units (75 % will be single family residential with an average density of four single family homes per acre), 643 acres of open space, a 102-acre commercial/retail center, and two schools. The development has been controversial and appeals delayed ground-breaking. The original city council members who once advocated for the development all lost their political positions in subsequent elections. Clearing and site preparation finally began in December 2013 and a final oppositional appeal failed in January 2014 (Fig. 11.5).

Many residents remain skeptical of how the development can retain the rural character of the area. Bob Edelman, a founding member of Toward Responsible Development stated, "Clearing 7000 trees off of 95 acres and flattening the landscape doesn't fit their promise of a 'rural by design' development that blends with the environment" (Box 2013). One resident puts it this way:

> We're living, and we've lived it for a few years now, and obviously the huge challenge of managing this significant growth that we've got in front of us.... There's a lot of very significant emotional and important trade-offs that the city is going to keep running into all along the way, and so that's going to be a real challenge for our community.

Fig. 11.4 Historic Black Diamond. *Photo credit*: J. Tilt, 2014

Like Maple Valley, the social impacts of this development may not be fully realized by the local political actors, developers, and others. When discussing the new master-plan development with one developer, it became clear that the developer plans to reinvent Black Diamond to facilitate home sales:

> In Black Diamond we need to create a place and sense of community. Right now no one knows where Black Diamond is. Ask someone and they'll say, "where's that?" We have to create awareness and create a story which will be paramount to getting things sold. In order to do this, it is upon *us* [emphasis] to be on the forefront and provide services—shops, schools, fire station because there is nothing there.

Fig. 11.5 Yarrow Bay Development, Black Diamond. *Photo credit*: J. Tilt, 2014

This perception of "nothing there" was also echoed by a local real estate professional who repeatedly stated that "no one is buying in Black Diamond"

However, there *is* something in Black Diamond: the community has experienced steady population growth and has a small historic downtown area with an esteemed bakery and shops. Community leaders may be justified in their concern about a local tax base; however, the historic community center of Black Diamond is decidedly "there." Discourse focused on the "nothingness" of Black Diamond allows the developer, with implicit approval of the planning commission, to create a new town center spatially and socially separate from the existing community center. As the developer stated while pointing to a map:

We will move the city center from here [points to old town] to here [new development]. There is nothing there in Old Town …. We will be bringing in a new population and commercial [activity]. We have to create a sense of place, a sense of cool and unique.(Interview with Black Diamond developer 2013)

Moving the town center can have negative economic and social impacts on the existing, "unseen" community, as demonstrated in a previous study of master-planned developments (Tilt and Cerveny 2013). The creation of a parallel town center and the shifting of city services, schools, retail opportunities, and youth and community programs to the new town center can create rifts between existing and new residents.

Another impact associated with the Yarrow Bay development is the closure of popular trails used "unofficially" by area residents for many years. Many residents were concerned about the effects of the proposed development on access to trails and open spaces. As one Black Diamond resident explained:

> On the side of the road all the way out [is], posting, owned by Palmer or it's been bought up by Yarrow Bay. But, it's still got the sign on it. You know, "Don't come in." You've gotta drive up towards the mountain, I guess, Crystal Mountain, Mount Rainier. There's no place out east you can get to. You can't get into nature.

Black Diamond is being re-packaged by local and regional political and economic actors who seek to shape the landscape for high-density growth surrounded by open space (Fig. 11.6). Using heavy-handed negotiating strategies, the stage was set to transform the rural community of Black Diamond long before growth pressures arrived. Although the new master-planned community has barely broken ground, the social and economic impacts of this transformation are already emerging as local residents experience reduced access to natural places and loss of community. Concerns about growth and how the city of 4300 residents will handle a potential influx of 20,000 new Yarrow Bay residents are paramount in local conversations.

11.5.2 Ravensdale

Being on the "other" side of the urban growth boundary, Ravensdale has not had the high-density growth and development experienced by nearby Maple Valley and soon to be felt by Black Diamond (Fig. 11.7). Yet, Ravensdale has not been immune to transition either. Three large private natural resource commercial industry landowners: Plum Creek, Weyerhaeuser, and Palmer Coking Coal are located within and around Ravensdale. With urban services extending into Maple Valley in 1997, the economic benefit of divesting land for real estate revenue rather than remaining in natural resource production became increasingly attractive for these major landowners. A senior analyst in the county's Office of Regional Policy and Planning described the situation this way:

> More homes mean more development, which leads to more traffic, the more you allow this to happen, then the lines become blurred: What's the difference between 'rural' and 'urban'? What's the difference between 'forest' and 'rural'? (Solomon 2000).

A rural real estate agent and developer described growth in rural King County during the high housing/economic bubble of the 1990s and early 2000s: "Before 2008, people were able to get loans for land and a lot of people got caught up in the land speculation and bought land."

In response to this continued growth in rural areas, then King County Executive, Ron Sims, attempted to slow down growth outside the urban growth boundaries by proposing downzoning rural lands from one dwelling unit per 5 or 10 acres to

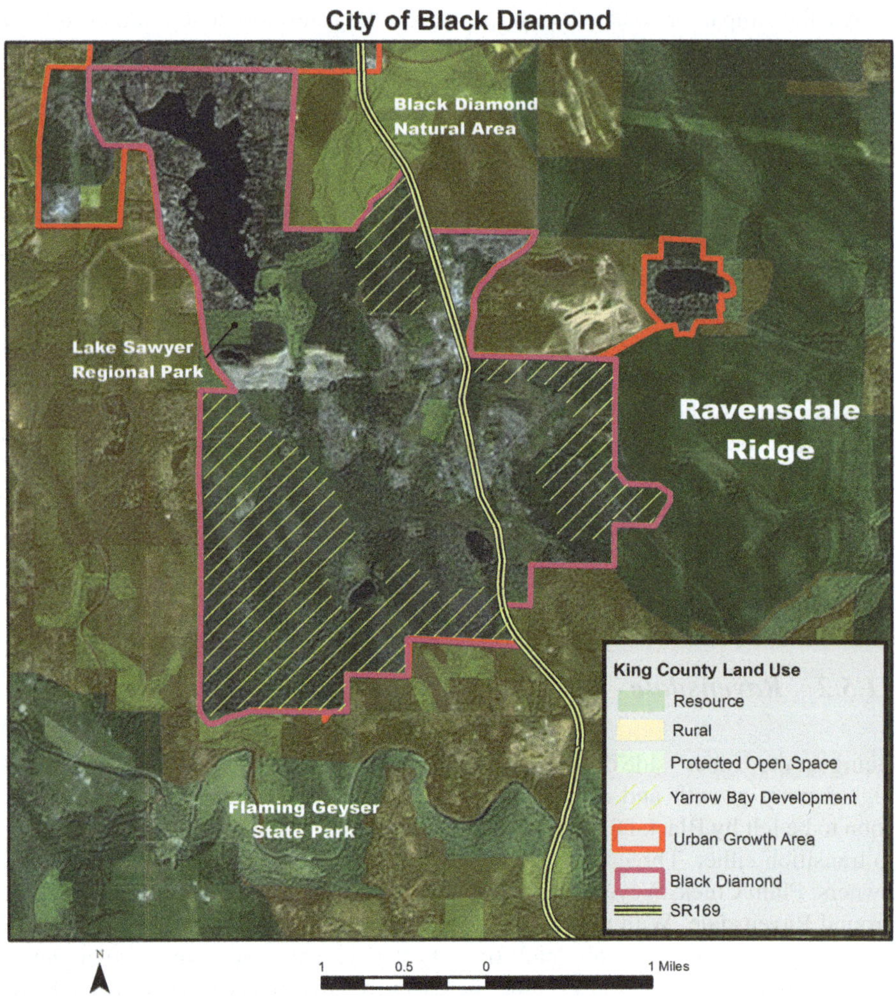

Fig. 11.6 Map of Black Diamond land use. Produced by J. Tilt using ESRI software

one dwelling unit per 10 or 20 acres. However, the County council saw this as an economically and politically unpopular move and did not approve the change in the 2000 update to the King County Comprehensive Plan (Dudley 2000). Taking matters into their own hands, residents of Ravensdale and other nearby hamlets joined forces in 1999 to create the grassroots organization, Friends of Rock Creek Valley (FRCV). The group's purpose was to conserve and protect natural and recreational resources in the valley. The Friends of Rock Creek Valley emphasized

Fig. 11.7 Ravensdale residence. *Photo credit*: J. Tilt, 2010

non-regulatory and voluntary strategies to retain forest cover and a rural lifestyle in the area (FRCV 2004; personal communication with FRCV members).

Meanwhile, King County council shifted its priorities for its use of the Washington State Conservation Futures tax, which previously was used to acquire open spaces for active and passive uses. Now, the funds would be used to prioritize lands with high ecological integrity, particularly for Chinook Salmon (*oncorhynchus tshawytscha*), listed as a threatened species under the Endangered Species Act. Friends of Rock Creek Valley was successful in obtaining grants and matching funds from local and federal sources to purchase land, easements, and create a conservation plan for the area. Their vision was endorsed by King County Executive Ron Sims in 2000: "We must encourage voluntary land-use incentives and pursue existing resources, like Conservation Futures funds, to ensure this area retains its rural character" (King County Executive Office 2000).

By 2004, the Friends of Rock Creek Valley had created land-use and conservation plans outside the purview of King County's institutionalized planning framework. For the first time in King County history, a local community plan was formally endorsed and adopted by the King County council and executive. This plan calls for the protection of critical natural, scenic, and recreation areas using easements, clustered developments, and other non-regulatory strategies. Using this plan, several large land deals occurred in the Rock Creek Valley between industrial landowners, King County, and a regional land trust, the Cascade Land Conservancy. One of the largest land deals included the purchase of development rights from 1600 acres of an area known as Ravensdale Ridge, owned by Plum Creek Timber, plus another 645 acres near Black Diamond deeded for permanent open space managed by King County. The deal also included an additional 329 acres of Plum Creek land near Black Diamond, which were annexed into the city.

By using new economic capital provided in part by the Chinook Salmon endangered species listing, Ravensdale, along with land conservation interests and King County officials, negotiated with large corporate landowners to identify which lands would be removed from active resource production and whether these lands would be used for restoration (allowing passive recreation use), or exchanged for other parcels for development. Utilizing this approach, Ravensdale was able to negotiate the terms of growth and development in their community. The Friends of Rock Creek was able to successfully transform their valley and their communities by leveraging regional economic growth to convince King County to provide market-based incentives to establish areas "off-limits" to growth development where county planning policies fell short.

Though the conservation efforts of the Friends of Rock Creek Valley have helped stem the tide of development in the area, the area has not been immune to the growing pains felt along the SR-169 corridor. Ravensdale's population grew modestly, from 816 residents in 2000 to 1100 residents in 2010. But dramatic growth in neighboring Maple Valley and Black Diamond brought mixed responses from Ravensdale residents:

> It's really convenient to have some of the new stores.... At the same token, the traffic is getting worse. And it will continue to get worse as things start building out more towards Ravensdale.... Land is being cleared here in Black Diamond. It's coming to us. And unfortunately, that's the way it is.

Ravensdale residents also complained bitterly about the lack of access to services that are easily provided just across the urban growth boundary: "We lost our sheriffs precinct. How many bus routes did we lose? I can't tell you how many bus routes...."

Most significantly, growth and development of Maple Valley next door brought more users to the open spaces that act as a de facto border between urban and rural zoned King County. As a King County planner explained:

> It's pretty high-density in a lot of those [4-to-1s] and that's encouraged use [of surrounding open space]. From a management side, it's very difficult to maintain open space lands when you have super high-density, so we've got lots of issues with that.

With more users in these urban/rural buffer green spaces, conflict between user groups is intensifying particularly between equestrians and mountain bikers. Historically, open space along the SR-169 corridor was used primarily by equestrians. Many Ravensdale residents moved to that area specifically because they could keep horses and mules on their property, while also having access to an extensive trail network that was protected, in part, through the efforts of the Friends of Rock Creek Valley. However, this trail network was increasingly being used by mountain bikers from throughout King County, whose bike routes were being hampered by development elsewhere. Meanwhile, popular bike trails were closing in on Black

Diamond due to the Yarrow Bay development. Thus, pressure for mountain bikers to lay claim on the primarily equestrian trails in the Ravensdale/Rock Creek area intensified. Local Ravensdale equestrians feel that "There are less and less trails for horseback riders" (Interview with Ravensdale equestrian 2013). While mountain bikers argue that their efforts to build and maintain trails can be quickly ruined by a horse. To alleviate the conflict, King County designated a specific section of the Maple Ridge Highlands open space for equestrian use (Danville-Georgetown) and another section for mountain bike use (Henry's Ridge). Yet, loss of undeveloped space elsewhere in King County will continue to put pressure on the open spaces created by the 4-to-1 programs.

Similar to the stories of Maple Valley and Black Diamond, the actions of local and regional actors in shaping the exurban environment had unintended consequences. A long-term vision and coordinated effort by grassroots groups such as Friends of Rock Creek and Cascade Lands Conservancy helped to protect open spaces from development in Ravensdale. The timing of these efforts coincided with the desire for the King County Commissioner to slow the pace of change in lands beyond the rural growth boundary. Moreover, the large private landowners were brought on board. Ravensdale was able to avoid some rural sprawl by creating a network of conserved open spaces and easements. However, the community's proximity to urban growth areas of Maple Valley and Black Diamond has put pressure on Ravensdale's undeveloped spaces. User conflicts among recreationists are one example of this spillover pressure. In addition, residents have had to contend with increased congestion and traffic from growth in nearby Maple Valley, although they do enjoy better access to retail, restaurants, and health care services that accompany this growth.

11.5.3 Summary

Though only a few miles apart, Maple Valley, Black Diamond, and Ravensdale have experienced dramatically diverse growth trajectories (Table 11.2). Yet, each of these communities leveraged their exurban transformation by utilizing and manipulating formal regulatory land-use planning mechanisms. The execution of the state's Growth Management Act set the stage to carve up the exurban landscape space into disputed territories of urban and rural. Even though King County had the greatest financial and political resources to designate the urban and the rural in this landscape, each local community successfully maneuvered within the confines of the legislation to achieve their distinctive community vision. Mechanisms such as incorporation, the 4-to-1 policy, conservation easements, annexation, and transfer of development rights were employed to achieve this vision. The results of these political and economic negotiations have created dynamic exurban communities that are dramatically different in their spatial layout, residential density, and commercial growth, despite their geographic proximity and shared extractive natural resource economy legacy. The economic, political, and planning forces that shaped

Table 11.2 Political ecology of exurban communities along state route-169

	Maple Valley	Black Diamond	Ravensdale	Common to all
Relationship of community to Urban Growth Boundary (UGB)	• Inside UGB	• Inside UGB	• Outside UGB	N/A
Approximate percentage of protected lands in each community	• 9 % (Maple Valley owned) • 18 % (King County and other ownerships)	• 20 % (Black Diamond owned) • 30 % (King County and Yarrow Bay)	• 31 % (King County and other ownership in the Rock Creek Valley)	
Major actors involved	• Plum Creek Timber • Polygon homes • King County • City of Maple Valley	• Palmer Coking • Plum Creek • King County • Yarrow Bay Devel. Co. • City of Black Diamond	• Palmer Coking • Plum Creek • Weyerhaeuser • King County • Friends of Rock Creek Valley • Cascade Lands Conservancy	• Private landowners • King County
Planning tools shaping the community	• Comprehensive planning; • 4-to-1 • Incorporation	• Comprehensive planning; • Annexation	• Purchasing lands or development rights using Conservation Futures and other funding sources; Transfer of Development Rights	• King County comprehensive planning; • Washington State Growth Management Act
Community shape	• High density, surrounded by protected open space	• *Current:* medium to low density • *Expected:* high density with protected open space	• Low density, surrounded by protected open space	• Protected open space
Implications f community shape	• Lack of developed park space; limited retail development	• Current: Loss of trail networks • Future: Potential loss of town center	• Pressures from nearby development lead to increased user conflicts in open spaces; access to urban amenities	• Increased population growth
Notable natural amenities, outdoor recreation areas	• Henry's Ridge (mountain biking) • Lake Wilderness (swimming, boating, park)	• Black Diamond Natural Area • Lake Sawyer (swimming, boating, park)	• Danville-Georgetown Trail System (hiking, equestrian) • Henry's Ridge • Black Diamond Natural Area	• Mt. Rainier National Park • Crystal Mountain Ski Area • Flaming Geyser State Park
Population growth rates (2000–2010)	59.6 %	4.5 %	31.8 %	

the physical landscape of each exurban community also created dynamic and changing community identities that attract distinct types of exurban migrants.

Table 11.2 shows that these three communities had diverse planning strategies which influenced the shape of the exurban environment, the pace of growth, and the changing social environment. Maple Valley developed using a model of high-density growth that allowed for more affordable housing and an influx of young families. However, the community had not fully developed its parks, community, and youth programs, or retail spaces to keep pace with population growth. Black Diamond is situated in the middle of a large conservation area. Most of the new land that was slated for development previously had been used by locals for outdoor recreation. Ravensdale created a vast network of equestrian trails conserved by conservation easements. These spaces are increasingly being used by other recreational users, especially mountain bikers.

11.6 Discussion

The exurban landscape in our case study provides the stage on which different actors—local and regional planners, productivist landowners, real estate professionals, land conservation non-profits and residents themselves—negotiate the physical and social production of these three exurban communities. We return to key elements to make connections between our case studies and the emerging academic field of exurban landscapes.

11.6.1 Political Ecology of Exurban Residential Growth

A political ecology perspective highlights the roles of political and economic actors and helps us understand the deep causal factors of the emergence and transformation of exurban spaces. For example, a key concept in King County's 2030 vision document is a "*Viable Rural Area. The rural area, established in 1992, is permanently protected with a clear boundary between Rural and Urban areas*" (King County 2011). Yet, policies stipulated within the very same document undermine and blur this boundary between urban and rural by allowing urban growth areas to expand using the 4-to-1 policy with a minimum requirement of 80 acres (20 for urban growth/60 for open space) (King County 2015). The planning principle of "clustering" to the exurban area has been celebrated as a "win-win": allowing for new housing in a desirable "rural" landscape, while also preserving some of this landscape as natural open space. This planning tool, developed and promoted by Randall Arendt's book *Rural By Design*, has been implemented in Midwest agricultural communities as well as the southwest and New England (Arendt 1994) since the 1980s. However, its ability to deliver on the promise of retaining "rural character" (Ryan 2002; Tilt et al. 2007; Tilt and Cerveny 2013) or containing rural sprawl

(Partridge et al. 2011) has been called into question. A political ecology approach helps to illuminate the underlying political and economic mechanisms driving exurban landscape change.

Global and regional mechanisms also transform an exurban landscape. In our case study, exurban landscapes were shaped at the local level by land exchanges with large industrial landowners; but at another level we can see that these land swaps were exacerbated by regional, national, and global economic forces. The 1990s and the 2000s were eras of high economic growth fueled by housing developments and dubious mortgage-backed securities ("Crash course" 2013), while US natural resource production continued to decline. Increasingly, it became less economically viable to harvest or mine in exurban communities across the US, when the land could be sold for residential or commercial development at a higher rate of return. Throughout King County, agricultural producers as well as timber and mining companies divested their lands and sold them to development communities, while others reorganized their business plans and entered the development industry. In collaboration with local and county planners, large-scale developers transformed the landscapes of communities in the Cascade foothills and other areas at the fringes of King County from rural and undeveloped areas into residential communities of various densities. Understanding how regional and global economic mechanisms contributed to exurban growth and transformation is essential in identifying new areas that may experience future development.

11.6.2 Physical and Spatial Characteristics of Exurban Landscapes

The political ecology forces behind exurban growth also give it its physical manifestation. However, current conceptualizations of exurbia based on residential density (Irwin et al. 2007; Theobald 2001) often do not recognize the range of exurban residential densities created by political and economic actors: for example, high-density exurban clustering. By limiting exurban studies to a focus on the spatial analysis of low-density residences, we limit our temporal understanding of these landscapes: we fail to explore how these communities became "exurban" or what happens to them once they no longer meet this definition. In our case study, the physical exurban landscape was manipulated by the interplay of political, economic, and social forces. Exurban communities were able to broaden the urban growth area because policymakers were caught deciding between either planned urban growth or rural sprawl as a future regional (re)imagined exurban landscape. Productivist landowners, along with the assistance and support of the exurban municipalities, were able to skillfully negotiate large land deals that transitioned natural resource production lands into future high-density residential developments. As a result, the pattern of development met standard spatial definitions of exurban (Theobald 2001) in only some areas and not in others. Furthermore, we found that exurban areas are not static but can be physically transformed, shifting their spatial identities over time from "rural" to "suburban" to "urban," depending on the temporal and spatial scale of analysis. Thus, we challenge planners, demographers, and geographers to be cautious when examining exurban phenomena using only spatial constructs and encourage careful consideration of how political, economic, social, and environmental factors blur and transform the landscape beyond typical conceptions of exurbia.

11.6.3 Cultural Landscape of Exurban Studies

Previous social discourses of exurban cases have revolved around a few key themes: factors driving amenity migration (Gosnell and Abrams 2011); the divergent social, economic, and political capital and expected cultural clash between "new-comers" and "old-timers" (Crump 2003; Hurley and Walker 2004; Walker and Fortmann 2003); and the emergent use of natural resource lands (both public and private) for natural amenity use (views, recreation, "hobby farming") (Busck 2002; Holloway 2002), or ecological restoration (Mendham and Curtis 2010). While these studies have greatly expanded our understanding of life and challenges in "exurbia," they take place in a narrow band of exurban landscapes and do not exhibit the diverse set of exurban actors as shown in our case study. As a result, our collective understanding of the "exurban resident" has been limited primarily to discourses regarding wealthy, retired migrants to high-amenity remote landscapes (see Gosnell and Abrams 2011, for review). Omitted from this understanding are exurban actors like local and regional planners, developers, city councils, non-profit conservation groups, recreational clubs, and even natural resource extractive industries. Our case study illustrates that a multitude of powerful actors can influence the changes to the cultural landscape of an exurban area by negotiating physical landscape change. Moreover, in rapidly changing exurban landscapes, the dichotomy between "new" and "old" residents or "resource production" vs. "amenity production" breaks down as exurban residents adapt to the changing landscape.

11.7 Conclusion

The exurban land-use patterns that emerged from the three communities led to unintended consequences. Ravensdale and Black Diamond residents spoke about the expanded use of popular trails and growing conflicts among user groups. As with the transformation of the rural economy during the colonial land rushes in the 1880s, we found that exurban land is claimed and its territory marked, fenced off, and posted. Land is reimagined and its future use is determined by select actors and political, social, and economic processes that result in sanctioned land-use plans that determine residential, commercial, open space, natural resource production, or watershed protection uses. As a result, residents living in these communities continue to hit these land-use boundaries both physically and metaphorically as the geography of their exurban landscape remains in flux. Expectations for exurban lifestyles, (e.g., proximity to outdoor recreation opportunities, expectation for heightened community bonds, rural character, or privacy) must be reconciled with other imagined futures held by other incoming residents (including those who seek access to urban and suburban features), local government officials, private landowners, and developers.

References

Abrams, J., Gosnell, H., Gill, N., & Klepeis, P. (2012). Re-creating the rural, reconstructing nature: An international literature review of the environmental implications of amenity migration. *Conservation and Society, 10*(3), 270–284. doi:10.4103/0972-4923.101837.

Arendt, R. (1994). *Rural by design: Maintaining small town character*. Chicago: American Planning Association Press.

Blaikie, P., & Brookfield, H. (Eds.). (1987). *Land degradation and society*. London: Methuen.

Box, D. (2013, December 12). Groundbreaking work on Yarrow Bay's The Villages development begins. *Maple Valley Reporter*. Retrieved from http://www.maplevalleyreporter.com/news/235654391.html#.

Bryant, R. L. (1992). Political ecology: An emerging research agenda in Third-World studies. *Political geography, 11*(1), 12–36. doi:10.1016/0962-6298(92)90017-N.

Busck, A. G. (2002). Farmers' landscape decisions: Relationships between farmers' values and landscape practices. *Sociologia Ruralis, 42*(3), 223–249. doi:10.1111/1467-9523.00213.

City of Black Diamond. (2009). *City of Black Diamond comprehensive plan*. Department of Community Development. Retrieved from http://www.ci.blackdiamond.wa.us/Depts/CommDev/planning/2013/FinalPlan_092209with20112012amendments.pdf.

Clutter, S. (1996a, November 26). County oks Black Diamond deal—In land swap, rural area would be developed. *Seattle Times*. Retrieved from http://community.seattletimes.nwsource.com/archive/?date=19961126&slug=2361814.

Clutter, S. (1996b, November 6). Threat of being torn apart unites residents—Voters worried about being annexed. *Seattle Times*. Retrieved from http://community.seattletimes.nwsource.com/archive/?date=19961106&slug=2358474.

Crash course: The effects of the financial crisis are still being felt, five years on. (2013, September 7). *The Economist*. Retrieved from http://www.economist.com/news/schoolsbrief/21584534-effects-financial-crisis-are-still-being-felt-five-years-article.

Crump, J. R. (2003). Finding a place in the country: Exurban and suburban development in Sonoma County, California. *Environment and Behavior, 35*(2), 187–202. doi:10.1177/0013916502250207.

D'Addario, D. (2012, December 17). America's best places to raise kids 2013. *Bloomberg Businessweek*.

Dudley, B. (2000, March 1). Sims would tighten hold on rural area. *Seattle Times*. Retrieved from http://community.seattletimes.nwsource.com/archive/?date=20000301&slug=4007495.

Friends of Rock Creek Valley. (2004). *Rock Creek Valley conservation plan and priorities*. Retrieved from http://www.frcv.org/content/cpfinal/rcvcpp062004_fm.pdf?lbisphpreq=1.

Gosnell, H., & Abrams, J. (2011). Amenity migration: Diverse conceptualizations of drivers, socioeconomic dimensions, and emerging challenges. *GeoJournal, 76*(4), 303–322. doi:10.1007/s10708-009-9295-4.

Holloway, L. (2002). Smallholding, hobby-farming, and commercial farming: Ethical identities and the production of farming spaces. *Environmental and Planning A, 34*, 2055–2070. doi:10.1068/a34261.

Hurley, P. T., & Walker, P. A. (2004). Whose vision? Conspiracy theory and land-use planning in Nevada County, California. *Environment and Planning A, 36*, 1529–1547. doi:10.1068/a36186.
Irwin, E. G., Cho, H. J., & Bockstael, N. E. (2007). Measuring the amount and pattern of land development in nonurban Areas. *Review of Agricultural Economics, 29*(3), 494–501. doi:10.1111/j.1467-9353.2007.00360.x.
Johnson, B. E. (2008). Nature, affordability, and privacy as motivations for exurban living. *Urban Geography, 29*(7), 705–723. doi:10.2747/0272-3638.29.7.705.
King County. (1998). *1997 Annual growth report*. King County Budget Office. Retrieved from http://your.kingcounty.gov/budget/agr/agr97/.
King County. (2007). *King County buildable lands report: Recent growth & land capacity in King County & its cities*. King County Office of Management and Budget. Retrieved from http://your.kingcounty.gov/budget/buildland/bldlnd07.htm.
King County. (2010). *About the region: Environment*. Retrieved from http://www.kingcounty.gov/about/region/environment.aspx.
King County. (2011). Chapter 5: Vision and Framework. *2011 King County Countywide Planning Policies*. Retrieved from: http://www.kingcounty.gov/depts/permitting-environmental-review/codes/growth/GMPC/~/media/property/permits/documents/GMPC/2011/CPPs/VisionandFramework_FinalDraft.ashx.
King County. (2015). *King County code, chapter 20: Planning. Sections 20.18.170 & 20.18.180*. Retrieved from http://your.kingcounty.gov/mkcc/clerk/code/23_Title_20.pdf.
King County Executive Office. (2000). *Sims proposes Rock Creek Valley vision*. Retrieved from http://your.kingcounty.gov/exec/news/2000/103000.htm.
Kossen, B. (2002, April 23). Bargain homes turn Maple Valley, Covington into boomtowns. *Seattle Times*. Retrieved from http://community.seattletimes.nwsource.com/archive/?date=20020423&slug=boomtowns230.
Lichter, D. T., & Brown, D. L. (2011). Rural America in an urban society: Changing spatial and social boundaries. *Annual Review of Sociology, 37*, 565–592. doi:10.1146/annurev-soc-081309-150208.
Mendham, E., & Curtis, A. (2010). Taking over the reins: Trends and impacts of changes in rural property ownership. *Society and Natural Resources, 23*(7), 653–668. doi:10.5751/ES-05071-170405.
Nayyar, S. (2011, August). 10 Best towns for families: 2011. *Family Circle*.
Neuman, W. L. (2003). *Basics of social research: Qualitative and quantitative approaches*. Boston: Pearson Education.
NVivo. (2012). *Qualitative data analysis software*. QSR International Pty Ltd. Version 10.
Partridge, C., Boese, J., & Bernath, S. (2011). *2010 Forest carbon workgroup: Final report*. Water Quality Program, Washington State. Retrieved from https://fortress.wa.gov/ecy/publications/summarypages/1110006.html.
Ryan, R. L. (2002). Preserving rural character in New England: Local residents' perceptions of alternative residential development. *Landscape and Urban Planning, 61*, 19–35. doi:10.1016/S0169-2046(02)00066-X.
Solomon, C. (2000, January 19). County tries to slow sprawl into rural areas. *Seattle Times*. Retrieved from http://community.seattletimes.nwsource.com/archive/?date=20000119&slug=4000208.
Stedman, R. C., & Hammer, R. B. (2006). Environmental perception in a rapidly growing amenity-rich region: The effects of lakeshore development on perceived water quality in Vilas County, Wisconsin. *Society & Natural Resources, 19*(2), 137–151. doi:10.1080/08941920500394733.
Taylor, L. (2011). No boundaries: Exurbia and the study of contemporary urban dispersion. *GeoJournal, 76*(4), 323–339. doi:10.1007/s10708-009-9300-y.
Theobald, D. (2001). Land-use dynamics beyond the American urban fringe. *Geographical Review, 91*(3), 544–564. doi:10.1111/j.1931-0846.2001.tb00240.x.
Tilt, J. H., & Cerveny, L. (2013). Master-planned in exurbia: Examining the drivers and impacts of master-planned communities at the urban fringe. *Landscape and Urban Planning, 114*, 102–112. doi:10.1016/j.landurbplan.2013.03.003.

Tilt, J. H., Kearney, A. R., & Bradley, G. (2007). Understanding rural character: Cognitive and visual perceptions. *Landscape and Urban Planning, 81*, 14–26. doi:10.1016/j.landurbplan.2006.09.007.

U.S. Census. (2000). *Total population for cities*. Retrieved from http://censtats.census.gov/cgi-bin/pct/pctProfile.pl.

U.S. Census. (2010). *Total population for cities*. Retrieved from http://quickfacts.census.gov/qfd/states/53000.html.

Walker, P., & Fortmann, L. (2003). Whose landscape? A political ecology of the 'exurban' Sierra. *Cultural Geographies, 10*, 469–491. doi:10.1191/1474474003eu285oa.

Western Washington Growth Management Hearings Board. (1995). *Achen, et al. vs. Clark County, et al. No. 95-2-0067*. Final Decision and Order 9/20/1995.

Chapter 12
Redux: A Comparative Political Ecology of Exurbia Photo Album

Laura E. Taylor, Innisfree McKinnon, and Jacob McLean

This chapter is an album of images to illustrate the sometimes quite disparate geographies the term "exurbia" is enlisted to describe. Ideally, every reader would be able to visit all of the places discussed throughout the book. We think that only through an in-person visit can landscape observers fully appreciate the scale of the view, the color of the light and sky, and the sounds and smells unique to each place. As a poor substitute, we have gathered here in an album a Bing map at the same zoom level (500 ft/100 m) for each chapter and photographs of a "typical" exurban homes chosen by authors to be evocative of the places they discuss and to illustrate the diversity of exurban landscapes.

L.E. Taylor (✉)
Faculty of Environmental Studies, Planning Program, York University,
4700 Keele Street, Toronto, ON M3J 1P3, Canada
e-mail: taylorL9@yorku.ca

I. McKinnon
Department of Social Sciences, University of Wisconsin, Stout,
Menomineee, WI 54751, USA
e-mail: mckinnoni@uwstout.edu

J. McLean
Faculty of Environmental Studies, York University,
4700 Keele Street, Toronto, ON M3J 1P3, Canada
e-mail: mcleanja@yorku.ca

L.E. Taylor, P.T. Hurley (eds.), *A Comparative Political Ecology of Exurbia*,
DOI 10.1007/978-3-319-29462-9_12

2 From Swamp to Ridgeline: Exploring Exurbia in Southeastern Pennsylvania and the Sierra Nevada Foothills of California

Patrick T. Hurley and Laura E. Taylor

Fig. 12.1 Exurban Pennsylvania. *Source*: Bing Maps, 2015

Fig. 12.2 East Rockhill historic farmhouse with exurban owner. *Photo credit*: P. T. Hurley

Fig. 12.3 East Rockhill typical newer exurban home. *Photo credit*: P. T. Hurley

Fig. 12.4 Exurban Nevada County. *Source*: Bing Maps, 2015

Fig. 12.5 Typical exurban scene Nevada County. *Photo credit*: P. T. Hurley

3 Divergent Perspectives and Contested Ecologies: Three Cases of Land-Use Change in Calaveras County, California

Colleen C. Hiner

Fig. 12.6 Exurban Calaveras County. *Source*: Bing Maps, 2015

Fig. 12.7 Typical farmhouse near Trinitas golf course. *Photo credit*: C. C. Hiner

Fig. 12.8 Exurban backyard landscape Calaveras County. *Photo credit*: C. C. Hiner

4 Rural Residential Development and its Discontents: A Political Ecology of Sprawl Containment in Wallowa County, Oregon, USA

Jesse Abrams

Fig. 12.9 Exurban Wallowa County. *Source*: Bing Maps, 2015

Fig. 12.10 Typical new exurban home Wallowa County. *Photo credit*: J. Abrams

5 Competing or Compatible Capitalisms? Exurban Sprawl and High Value Agriculture in Southwestern Oregon

Innisfree McKinnon

Fig. 12.11 Exurban Jackson County. *Source*: Bing Maps, 2015

Fig. 12.12 Typical new exurban home. *Photo credit*: I. McKinnon

Fig. 12.13 Exurban Jackson County. *Photo credit*: I. McKinnon

6 "In the Real Estate Business Whether We Admit it or Not": Timber and Exurban Development in Central Oregon

Brent Olson

Fig. 12.14 Exurban landscape near Sisters OR. *Source*: Bing Maps, 2015

Fig. 12.15 Exurban home near Bend OR. *Photo credit*: Leisl Carr Childers

Fig. 12.16 Black Butte Ranch. *Photo credit*: Leisl Carr Childers

7 Death by a Thousand Cuts? The Moral Terrain of Neoliberal Environmental Governance in the South Carolina Lowcountry

Annette Watson and Kate Skaggs

Fig. 12.17 Exurban South Carolina Lowcountry. *Source*: Bing Maps, 2015

Fig. 12.18 Typical new exurban home. *Photo credit*: A. Watson

Fig. 12.19 East Edisto development area. *Photo credit*: Nick Rubin Bowman Consulting

8 "If That Would Have Happened": The Moral Imperative of Environmental History

Michael H. Finewood and Lou Martin

Fig. 12.20 Exurbia along the Colleton River SC. *Source*: Bing Maps, 2015

Fig. 12.21 Coastal exurbia South Carolina. *Photo credit*: M. Finewood

Fig. 12.22 Coast exurbia South Carolina. *Photo credit*: M. Finewood

9 No (Back)Sliding: Amenity Migration, Viewsheds, and Contesting Steep Slope Ordinances in Western North Carolina

Jessica McCallum Breen, Patrick T. Hurley, and Laura E. Taylor

Fig. 12.23 Exurban Appalachia. *Source*: Bing Maps, 2015

Fig. 12.24 Exurban house under construction Jackson County. *Photo credit*: The Sylva Herald

Fig. 12.25 Exurban house under construction Jackson County. *Photo credit*: The Sylva Herald

10 The Paradox of Engagement: Land Stewardship and Invasive Weeds in Amenity Landscapes

Peter Klepeis and Nicholas Gill

Fig. 12.26 Exurban Windellama. *Source*: Bing Maps, 2015

Fig. 12.27 Exurban home Windellama. *Photo credit*: N. Gill

11 Politics of Landscape Transformation in Exurban King County, Washington

Jenna H. Tilt and Lee K. Cerveny

Fig. 12.28 Exurban Washington State. *Source*: Bing Maps, 2015

Fig. 12.29 New amenity development Snoqualmie Ridge. *Photo credit*: J. Tilt

Fig. 12.30 New amenity development with farm. *Photo credit*: J. Tilt

Chapter 13
Conclusion: Moving Beyond Competing Rural Capitalisms and Uneven Environment Management in Exurbia

Patrick T. Hurley and Laura E. Taylor

13.1 Toward a Truly Comparative Political Ecology of Exurbia

This book has provided a comparative political ecological examination of exurbia. In doing so, we have connected case studies of local exurban landscape change in particular places to the larger-scale regional, national, and global processes producing these changes. The collective analytical focus of this volume draws solidly on the work of Walker and Fortmann (2003) and the ways competing rural capitalisms produce dynamics that shape uneven environmental governance regimes (Reed 2007). The case studies discussed in the previous chapters describe contemporary nature–society relationships in the transitional landscapes of not only the United States but also in other contexts, both in the developed and developing worlds (see e.g., Hurley and Arı 2011). As the cases here illustrate, the rural to exurban transition appears at different times in different places, ranging from the South Carolina Lowcountry to the Cascade Mountains of western Washington to the tablelands of southeastern Australia.

Discussing the social, political, and ecological challenges arising in the case studies here can provide food for thought for those who seek to better understand exurbia and its social dimensions from a scholarly perspective as well as those who see exurban change on the horizon in their own communities. By following local communities—and the constellation of regional, national, and global actors that

P.T. Hurley (✉)
Department of Environmental Studies, Ursinus College,
P.O. Box 1000, Collegeville, PA 19426, USA
e-mail: phurley@ursinus.edu

L.E. Taylor
Faculty of Environmental Studies, Planning Program, York University,
4700 Keele Street, Toronto, ON M3J 1P3, Canada
e-mail: taylorL9@yorku.ca

L.E. Taylor, P.T. Hurley (eds.), *A Comparative Political Ecology of Exurbia*,
DOI 10.1007/978-3-319-29462-9_13

align in these places—through a critical moment in their environmental and social histories, the extensive case studies in the book provide excellent examples of the diverse local experiences had when dealing with the exurban transition. Indeed, a key goal of this book has been to identify the similarities and differences in this very common, yet complex and dynamic process of landscape-scale change, which we have identified as the exurban transition. In doing so, we have sought to more critically and fully examine the black box that is the land-use planning process, given what we see as its particular relevance for viewing the social dynamics that shape exurbanization.

Exurbanization has been the source of scholarly attention for more than two decades. This attention has often focused on the ecological and social shifts that occur in transitional rural areas in the U.S., western Europe, Australia, and beyond. Scholarly attention in the United States has frequently sought to explore what these transitions mean for local ecologies and communities. With this scholarly attention, it has become evident that exurban change often results in new patterns of land use, associated vegetation changes, and concerns for aesthetic features across relatively large areas of the United States. However, much of the research has focused on identifying exurban areas, quantifying their extent, and characterizing the ecological patterns that have resulted (Berube et al. 2006; Brown et al. 2005). By contrast, fewer systematic and comparative approaches—by which we mean those studies that collectively pick case study areas for examination using the same methods, question frameworks, and analysis—have been employed in describing the types of conflicts exurbanization has produced. Moreover, not all aspects of the exurban transition fit neatly into the binary of urban/rural so often deployed by scholars as an analytical tool for examining exurbia. This book, then, has been an attempt to address these shortcomings within exurban scholarship; that is, in order to consider more than just ecological and social shifts we have undertaken a comparative approach to examining exurbia and have throughout attempted to theorize the uniquely rural/urban/suburban characteristics of exurbia.

While this volume has considered exurban conflicts, landscape change, and land-use outcomes, and the "urban" dynamics that produce them, many of the chapters have attempted to bring a more ethnographically grounded perspective to understanding the complex social, political, economic, and ecological dynamics that shape exurban landscape transformations in places more recently thought of as "rural." In doing so, contributors eschew a frame that sees only newcomers in conflict with longtime locals. Instead, these chapters have focused on the roles that governance and land-use decision-making processes, influenced by competing forms of capitalism and diverse ideologies of nature, have played in creating new landscapes. As such, these case studies also highlight the key aspects of the "rural" that animate decision-making processes, the ways that natural resource users (both newcomers and longtime locals) engage in these processes, how different types of urban development come to be embraced (or not), and the ways that rural landscapes are maintained into the future.

By focusing on the ways these social–cultural–political dynamics of decision-making influence the exurban transition, particularly within land-use governance and decision-making processes, this book seeks to provide comparative insight into the forces that shape the emergence of particular exurban places. In doing so, we hope to bring greater focus to the role that power and land-use negotiations of different types among economic and cultural actors play in the production of new exurban landscapes. To this end, the chapters have, on the one hand, sought to highlight the distinctive dynamics of exurban change in particular places, often focusing on details and dynamics seemingly distinctive to a particular case. On the other hand, our contributing authors have sought to place their case studies within the framework of existing political ecology examinations of exurbia. In taking this analytical approach, this volume provides new insights into exurbia as place and exurbanization as process. Among these insights are the myriad ways in which diverse actors—often thought of as holding divergent and competing perspectives—are collectively engaged in producing emergent landscape outcomes.

13.2 Focusing on Emergent Landscape Outcomes in Exurbia

Focusing on emergent outcomes in exurbia helps to reveal the diversity of responses by exurban actors and the variability in landscape outcomes that characterizes exurbia. Seeing exurbia as a series of similar and divergent outcomes, as illustrated by the works gathered here, will help to open up new analyses of the drivers of this type of landscape change. But to do so, we suggest three particular ways in which a focus on emergent outcomes can be systematically pursued. First, developing a better understanding of the exurban transformation means paying closer attention to the ways that land-use policies and plans get created, how they influence landowner decisions, and the extent to which particular strategies or planning tools are implemented (or not) at different stages of decision-making. Second, developing this better understanding also means examining the ideologies of nature that shape proposed and realized parcel-by-parcel changes as part of development, both by landowners and other actors within the land-use planning process. Third, examination of exurbia must focus on the ways that landscape ideologies play out in time and space, often by detailing the way individual and aggregate parcels are transformed by changes in the types of land-use regulations, management approaches, and stewardship styles applied to these places. Importantly, eventual landscape outcomes are best understood by keying in on the "defining moments" when parcels are subjected to new visions associated with the destabilization of ideologies and their subsequent recoding (Hiner forthcoming), including how these ideologies engage with and work through the technical aspects of the land-use decision-making process.

The above three broad areas of analysis should be used in combination with the key markers of exurban change proposed in the Introduction (Chap. 1, Table 1.1) to this book. The seven markers were proposed to identify common characteristics of emergent outcomes in order to promote more comparative research in the future (i.e., rural landscape character, accessibility to urban centers, nature ideology, land management, amenity-driven increase in land values, persistence of resource-based activities, increased participation of coalitions of people in community politics, and the emergence or expansion of the role of land-use planning). The wide diversity of exurbanization experiences means that each of the seven common characteristics will tend to look very different across case studies but they likely emerge through very similar processes. The three approaches to analysis suggested in the previous paragraph encourage readers to see how their experience with local exurban change is produced through the interaction of land-use regulations and landscape ideologies, each of which are often tied to regional, national, or even global dynamics, but also how the enactment of those regulations and ideologies is accomplished by very specific constellations of individuals and/or communities. Research that incorporates those three approaches to exurban analysis described above could include the following: documenting the different landscape forms and types that propel economic valorization within the amenity economy (i.e., prioritizing particular landscape elements or material aspects of nature through, primarily, real estate markets); documenting the diverse types of economic changes experienced by natural resource producers within markets ranging from the local to the global that precipitate this new type of landscape valorization; studying the uneven and unexpected ways in which other types of changes in the culture or attitudes of residents in a particular area engage with or respond to exurbanization at different points in time and affect different parts of the landscape; and observing the convergent and divergent groups that seek to influence land-use planning and its control over landscape change trajectories.

Comparative research, such as that presented in this book, provides a first step toward identifying the ways that particular ideologies are inscribed into specific landscapes, the circumstances and dynamics that seem to enable or constrain these types of landscape inscriptions, and the ease or difficulty with which alternative forms are rejected. At the same time, we hasten to add that further efforts are needed in this regard. Indeed, while the work presented here is intended to provide comparative insights on these dimensions, we see a need for future political ecology perspectives of exurbia that are more systematically comparative in their project formulations and methodological approaches (i.e., where cases are chosen deliberately and concurrently, and parallel sets of methods are employed). Such efforts might begin to trace the ways that particular ideologies of nature are at work in and through key actors or "defining moments" in the development and land-use decision-making process more widely.

If more systematically comparative research is to be undertaken, it is also appropriate that we further reflect on the specific insights that our cases have for the existing political ecology research on exurbia. Thus, we now turn to thinking about what these cases mean for the current thinking in political ecology about transitional rural landscapes. In the first of the following sections, we suggest ways that the cases presented in this book challenge the idea of competing rural capitalisms and we suggest how this concept might be restructured and expanded. In the next section, we turn our attention to a rethinking of the idea of "uneven environmental management," and we specifically consider how the focus throughout this book on the types of conservation territories, commons, and other environmental management areas helps us to better understand the dynamic and emergent outcomes produced by rural capitalisms in "co-opetition" (see below).

13.3 Competing or Compatible Rural Capitalisms?[1] Emergent Coalitions and the Transformation of Rural Landscapes

A key aspect of political ecology's insights about exurbia has been that different forms of rural capitalism are *competing* with one another to extract value from the material resources and landscapes in particular places (Chap. 1; see also Walker and Fortmann 2003). These struggles are tied to sociodemographic shifts and associated political economic transformations, including changes in who controls access to an area's resources and the forms of governance that shape their development. But these struggles are also clearly shaped by the ways that new ideologies of nature seek to extract particular values from the landscapes and biophysical features that characterize areas in transition. This aspect of competing rural capitalisms in exurbanization has been conceptualized in some studies as a duality wherein landscapes of production are replaced by landscapes of consumption. Yet the chapters here, like earlier work in political ecology (see e.g., Abrams and Bliss 2013; Hurley 2013), point to the ways in which the binary between productive and consumptive economies dissolves under further scrutiny. Indeed, the case studies in this book show that coalitions of actors characterized by different political economic interests, often reflecting histories of livelihood relationships to particular resources (e.g., hunting/fishing/gathering, mining/farming/forestry, land development/real estate), emerge to forge distinctive trajectories of development and landscape transformation. These trajectories are clearly influenced by particular ways of envisioning area landscapes and their future ability to extract value (or not). In the process of breaking down this binary, however, the relational dimensions between resource production and landscape consumption can be hidden in fundamental ways.

Landscapes of production often create the very landscape aesthetics that are consumed by exurbanites. For example, in Quakertown Swamp, as discussed in Chap. 2, small-scale farming and the fields that typify this resource form help to create the pastoral characteristics of a seemingly "rural" place. In South Carolina, not just tidal marshes and wetlands but expanses of pine forest (including plantation forests) add to the rural feel of the Lowcountry (see Chaps. 7 and 8). In the Sierra Nevada, open areas of oak woodlands signal the region's long history of ranching, while fruit orchards and vineyards remind observers of the region's historic success in producing fruit (Chap. 5). In northeastern Oregon, forestlands (both publicly and privately owned), ranchlands, and the rugged mountain landscape contribute to the area's rural beauty (Chap. 6). In the southern Appalachian mountains of western North Carolina, the hardwood forests of the region long used for timber harvests, but that also support commons-style harvesting of nontimber resources, comprise the spectacular viewsheds and rural character (Chap. 9). In each of these cases, as with others in the book, landscape features and qualities are the amenities being sought out by new migrants and being capitalized on by housing developers. But, as these cases show, the continued use of these landscapes for resource production

[1] See McKinnon (Chap. 5).

may be perfectly compatible with—and may, in fact, be essential for—emerging real estate values.

The links between these exurbanizing landscapes, moreover, may be more complicated than sometimes recognized, with landscapes of production and consumption economically and politically intertwined. Indeed, as McKinnon specifically argues in Chap. 5, the exurban transition in some contexts raises questions about whether some types of rural production landscapes are economically dependent on the consumption dimensions of the emerging real estate economy. The Jackson County, Oregon case demonstrates that, rather than conflicting with or disrupting the agricultural economy, amenity migrants have historically provided capital and labor for agriculture in the area. Moreover, recent land-use regulations proposed by the state, which are intended to restrict urban sprawl in support of agricultural land preservation, instead threaten this historic cooperative economic relationship. In so doing, land-use governance intended to maintain agricultural landscapes may, in fact, be destabilizing the *complementary* nature of rural capitalism in the area, where farmers' livelihoods are dependent upon real estate capital investment and related markets. The land-use changes indicate an assumption on the part of the state of Oregon that production and consumption are land uses in conflict rather than in symbiosis. The role for planners and this process remains a challenging one, but political resistance may result if these tensions are not addressed (see also Walker and Hurley 2011).

The political dynamics of rural capitalisms are also examined by Abrams (Chap. 4), who shows how actors in Wallowa County, Oregon associated with diverse livelihood interests are collaborating, even if not through coordinated action, to ensure the maintenance of particular landscape features. These include but also extend beyond agricultural landscapes. Specifically, Abrams documents how amenity migrants are working together with longtime residents and different kinds of resource users to ensure the maintenance of a landscape aesthetic that both suit their economic interests and match their particular environmental imaginaries. In doing so, these efforts to manage particular landscape elements highlight the tensions between the exurban real estate economy and the ways that certain types of "working landscapes" contribute to the maintenance of the landscape upon which real estate values rest. These collaborations point to the shifting commitments of specific resource users during the exurban transition.

The case of exurban development in King County, Washington (Chap. 11) reminds us that area governments may actively shepherd the relatively rapid emergence of new exurban transformations, as opposed to reactively mitigating ongoing in-migration and its impacts that create new settlements in an area. Developers may respond in ways that produce different housing patterns, including projects that offer high-density buildings and increased open space in the form of common areas or houses with large private yards, which are intended to meet different kinds of housing markets. As Tilt and Cerveny conclude, the resulting relatively high-density developments, albeit in a rural context, may confound existing quantitative characterizations of exurbia that focus on density patterns. In the South Carolina Lowcounty (Chap. 7), a similar process is underway, in which new developments are part of the process of maintaining the rural character and the natural amenities that contribute to the area's sense of place. The insights of these cases remind us that political and economic conditions create new livelihood opportunities for both the actors that already inhabit or manage exurban areas and new ones recently arrived on scene.

Indeed, actors long associated with natural resource production activities may transform themselves into promoters of exurbanization and its associated landscape aesthetics and amenity-driven real estate economy. Such was the case with Deer Creek Park 2 and Wolf Creek Ranch Estates in Nevada County, California, as discussed in Chap. 2. Further, within the context of unlikely exurban boosters, Olson's examination of Central Oregon (Chap. 6) is noteworthy for highlighting how existing natural resource companies shift their means of extracting value from the natural resources they own, thereby engaging with and participating in the transformation of exurban places. For example, Brooks Resources actively worked to recreate itself as a development entity, benefiting from the landscape aesthetics created by forests that they once had planned on harvesting for timber. A similar story is at play as part of the examination of the South Carolina Lowcountry and the development undertaken by the company MeadWestVaco as described by Watson and Skaggs in Chap. 7, where the company is engaged in a comparable transformation in the valuation of its lands from timber resource to real estate. Similar dynamics are also at play in the western Cascades of Washington, where new residential communities are emerging on lands formerly envisioned for timber harvest (see Chap. 11 by Tilt and Cerveny). Meanwhile, Hiner, in Chap. 3, further demonstrates how in Calaveras County, California individual, non-corporate landowners can seek to capitalize on the emergence of an amenity-focused real estate economy. For example, through idiosyncratic attempts some exurbanite developers (e.g., a golf course and a vineyard) seek to reconfigure landscapes in ways that simultaneously extract value from agricultural production and meet the recreational demands of new and longtime residents but without compromising the scenic landscape.

Despite there being many examples of productive and consumptive economies existing "symbiotically," the developments (and development proposals) that both produce and reinforce exurban patterns of land use and landscape change often involve moments of conflict, some more intense than others. In their case study examining Beaufort County in South Carolina (Chap. 8), Finewood and Martin provide a historical perspective on the ways the contemporary landscape is the product of intense struggle between a proposed industrial use and those who defended the area's tidal beauty. The contemporary Lowcountry landscape of the Bluffton area hides the historic rejection of the industrial use, but the same landscape is still a place of production based on a fishing industry supported by the area's natural amenity values. Similar struggles are still in full view in Jackson County, North Carolina (Chap. 9), where concerns over exurban sprawl motivated a diverse set of actors to defend new land-use regulations based on the logic of protecting the region's steep slopes. Here, though, the overlap between the interests of amenity migrants and other actors demonstrates how particular landscape features may, for very different reasons, motivate their defense. In Beaufort, the success of the amenity valuation of the landscape appears complete, while in North Carolina the debate over which land uses are acceptable is ongoing.

Thinking about future research on exurbia, we suggest that the competing rural capitalisms frame might be expanded and reinterpreted to offer ways to explore emergent dynamics in exurbia. As such, the concept of competing rural capitalisms involves a wide diversity of actors and political coalitions, which sometimes compete and sometimes cooperate to extract economic value from landscapes and their associated natural resources. This type of "co-opetition" has been documented by Larsen and Hutton (2012), who focus on the nature of community interaction and acceptance in the valuation of particular heritage landscapes in an exurban area of Colorado. Larsen and Hutton provide a counterpoint to the literature on exurban conflict by demonstrating how competing actors in the community found ways to cooperate in situations where reliance on neighbors is necessary to particular types of land maintenance or in dealing with specific emergencies. Here, in our work, we note that these forms of co-opetition extend beyond discourses on community and have implications specifically for changes to material landscapes. Further examination of how ideologies of nature play a role in co-opetition would contribute greatly to our understandings of exurbia. At the same time, we are reminded that even though these efforts to find points of agreement and collaboration take place within the context of shifting economic conditions at regional, national, and global scales, they remain highly contingent upon local and regional histories of land ownership, social–cultural interactions with these landscapes and resources, and the ongoing influence of in-migration.

Understanding the ways that exurban transitions result in new forms of landscape and natural resource valorization requires paying attention to these multiscalar dynamics. Doing so means disentangling the myriad ways that actors are seeking to extract economic value from landscapes and resources, and exploring how these ways correspond to specific sets of actors and the political coalitions that appear to support their efforts. In some cases, sets of similar types of actors, such as developers, amenity migrants, or longtime residents, may cooperate in one case study of a time and place, but the same types of actors may compete with one another in other cases. We suggest that these instances of cooperation and competition are likely to be best viewed through engagements within the land-use decision-making process as the mechanism for fixing the dominant ideology of nature in material landscapes and thus regulating the future use of those landscapes. Exploring how, when, and where particular co-opetition dynamics couple in specific ways should be a key goal for future exurban scholarship.

13.4 Beyond "Uneven" Environmental Management: Land-Use Governance Regimes and Acceptable Forms of Land Use, Open Space, and Stewardship

Besides the concept of competing rural capitalisms, an organizing concept for this book's analytical framework was the idea of uneven environmental management. To some extent, the exurban dynamics described above are shaped by dimensions not specifically or directly economic, in that exurban actors draw on specific visions and ideologies about what constitutes appropriate land management. In doing so, they are making claims on resources or parts of the landscape that extend beyond their direct ownership or control, particularly in terms of protecting landscape qualities and aesthetics. This was an explicit element of consideration in Walker and Fortmann's (2003) discussion of competing rural capitalisms in Nevada County, California (discussed in Chap. 1) and these dynamics clearly emerge in the cases from Quakertown Swamp and the contests over new subdivisions in Nevada County from Chap. 2. The dynamics associated with competing rural capitalisms point to the ways in which ideologies of nature and the outcomes that result from political conflict over those ideologies come to reflect exactly whose ideas of appropriate land use persist or emerge through new forms of stewardship and in the types of open space maintained and created. We hope that increased understanding of how the exurban transition results in uneven outcomes in nature conservation and sustainable uses of the land can help communities involved move beyond or avoid common problems in dealing with their own exurban dynamics.

As conceptualized by Reed (2007, pp. 321–322), the concept of uneven environmental management refers to the "social relations, cultural practices, and ecological conditions" that determine "the character and direction of both ecosystem change and social outcomes" in a given place (see Chap. 1). Even with similar social relations, cultural practices, and ecological conditions, outcomes vary from place to place—the success of conservation efforts or of sustainable design efforts are "uneven"—more successful in some places than others. Reed's work is intended to "explore the roots of and effects of uneven environmental management," especially on the forms of governance and associated specifics of management. Reed's examination of protected areas in Canada focuses primarily on the role of different civil society actors in "setting aside" lands from future development—whether as possible sites of natural resource extraction or residential development. In her article, she discusses the outcome of two cases with similar decision-making processes, where in one case a public regime conservation area was created and, in the other, a private regime conservation area was created. Her analysis is critical both in pushing toward a comparative approach to understanding environmental governance processes and demonstrating important differences among protected area design and the ways these differences shape ongoing constructions of and interactions with nature in specific places.

Indeed, the maintenance and creation of open spaces "set aside" from future resource or residential development is a key marker of exurban landscapes (see key marker number 3, "Sect. 1.3.3" in Chap. 1). The emergence of differential approaches to open space creation through planning and zoning strategies says much about the ways particular communities and their governments prioritize degrees and types of intervention, potentially share assumed landscape aesthetics, and/or leave resource protection and/or development up to chance (i.e., the market). The focus on the types of and ways that open spaces are created through exurban land-use decision-making processes can reveal a lot about the role that ideologies play in shaping different types of landscapes of production and/or the maintenance of landscapes of consumption, as discussed in the previous section. Understanding these land-use processes and practices also helps to inform researchers about the ways that both political and cultural commitments align with particular economic considerations. In both the Quakertown Swamp and Nevada County examples in Chap. 2, different communities and constituencies clearly set out to restrict residential development, while ensuring that aesthetic landscape qualities were retained. In one area of the Swamp, this meant the use of large-lot zoning and limited public land acquisition focused on protecting pastoral and working landscapes, while in another it meant strategically working with a local land trust to ensure the core of the Swamp ecosystem and adjacent lands would be conserved in perpetuity. In yet another part of the Swamp, a wider set of forest, fields, and "representative" landscape features would be incorporated into new conservation territories. All of these actions were facilitated by county-level and state-level technical assistance and funding, but individual township-level interventions were designed and administered by local governing bodies responsive to the political dynamics of their respective jurisdictions. These are examples of how uneven outcomes occur within similar regulatory and governance regimes. They also represent critical examples of how similar types of actors in the planning process negotiate diverse public–private partnerships. The variability in open space conservation efforts results from and produces particular aesthetics and ecologies. The fact that different outcomes emerge in the same place and within the same general economic and ecological context further suggests the roles that environmental imaginaries and particular planning tools play in reimagining and shaping the emerging landscape.

The resulting conservation territories may also reassign access rights to new groups of individuals (also described as reterritorialization; see Brogden and Greenberg 2003), including exurbanites, or reinforce longstanding patterns of access associated with particular coalitions. Outside of Charleston, South Carolina, the case of East Edisto (Chap. 7) shows how the advocacy of local residents and engagement with those residents by large-scale timber landowners (turned developers) is leading to alterations to the residential housing layouts. Moreover, this engagement helped in the production of an exurban landscape featuring significant conservation features through both the for-profit East Edisto development and the not-for-profit Savannah River Preserve. But these efforts include land trust and developer-initiated conservation areas (two different forms of private conservation regimes) committed to ensuring the persistence of the community's traditional forms of access to the area's forests. This case shows that concern over local ecologies and their amenity values often leads to permanent open space conservation and to continued community uses of that landscape, where those uses are in conformity with the ideology of nature conservation decided for those lands. Such cases challenge conclusions that exurbanization cannot accommodate longstanding land-use traditions. But they also raise further questions about how this accommodation comes about and the durability of these forms of cooperation over time, as new residents continue moving into the resulting exurban landscape.

One of the best examples of the interdependence of conservation interventions and development is Tilt and Cerveny's case study of King County in Washington State in Chap. 11. There, timber and mining companies, seeking to divest themselves of resource lands no longer profitable due to changes in global markets, are very successful in navigating the land-use planning process to create areas for exurban residential development. In the process, large areas are set aside for ecological conservation, which provide conservation areas for ecological restoration, open space recreation, and further secure real estate investments for those who purchase properties within the emerging residential communities. A similar story emerges in the South Carolina Lowcountry described in Chap. 7, as MeadWestVaco responded to community concerns by ensuring that their development project maintained large areas of open space. To ensure vast areas of the wetland and forested landscape would be kept from being transformed by residential development, an assemblage of landowners and conservation organizations created the nearby Savannah River Preserve using conservation easements and land trusts. In both cases, once again, a coalition of actors—both longtime locals and newcomers—embraces the creation of new conservation territories.

At the same time, Klepeis and Gill in Chap. 10 remind us of the ecological challenges that exurbanization creates. Their research on invasive species in New South Wales, Australia describes emerging environmental management regimes in exurbia, where ranchers and exurbanites are faced with the necessity of collaborating to manage ecological change. In collaborating, their collective efforts support the ideology of nature that drew amenity migrants by ensuring invasive weeds do not threaten the livelihoods of ranchers who maintain the working landscape. Yet this intervention also pushes a process of considering alternative ways of approaching this ecological management challenge. The authors focus on an alternative approach that includes forming alliances to improve knowledge exchange as well as regulation. Their work suggests that future studies of exurbia and conservation need to take into consideration the ways in which ideologies of nature specifically shape the logics and ethics of stewardship in exurbia.

Public entities, including local government, are actively involved in the creation of open space, both in ways that draw on private ownership and that transfer land to public ownership. In the Appalachian Mountains of North Carolina, the case of Jackson County (Chap. 9) illustrates the ways that exurban counties may embrace new regulations that are simultaneously about reducing levels of development and addressing risks associated with natural hazards. Key to the politics of land-use planning in this part of southern Appalachia appears to be the combination of low-density development and steep slope regulation in protecting forests and forest access. That these land-use interventions are supported by discourse coalitions spanning the newcomer-longtime local divide will not be entirely surprising to many who study political ecology. But this finding is a reminder that longstanding community traditions associated with resource commons (see e.g., Newfont 2012), such as hunting and foraging, may align with the ideology of nature underpinning the real estate economy, lead to regulatory outcomes (in this case steep slope ordinances) that may confound assumptions that local peoples will resist government intervention in exurban areas (Hurley and Walker 2004; Nesbitt and Weiner 2001; Walker and Fortmann 2003).

By contrast, Hiner's exploration of land-use and landscape conservation in the Sierra Nevada foothills in Chap. 3 suggests that cooperation does not always occur, nor do these types of cooperative coalitions always emerge. Indeed, in the absence of land-use planning and government intervention to ensure the conservation of large areas of open space, other types of environmental management processes will unfold. Hiner describes the success of a winery in meeting local nature ideologies, despite the fact that another attempt at economic expansion (specifically, the creation of a golf course) was rejected by the community as it failed to meet their local landscape vision. With or without state intervention, exurban landscapes exist in an often-uneasy balance between conservation and development.

The exurban transition is a dynamic socio-ecological transformation with uneven environmental management outcomes in different times and places. "Setting aside" lands for conservation, either through public or private means—or in partnership—seems to be necessary to secure the ideology of nature in the exurban landscape for the future. As discussed in this chapter, representing the rural to exurban change as a purely urban/rural or newcomer/longtime resident culture clash and power shift does not tell the whole story. Walker and Fortmann (2003; see also Hurley and Walker 2004) are clear to point out that amenity migrants often find political supporters among longtime locals, who object to particular types of natural resource extraction and wish to take control of the growth and land management agenda. Together this coalition of interests around landscape values seeks to influence decision-making about regulatory practices and priorities, including challenging traditional *and* emerging land-use practices through land-use planning processes. Further, real estate developers may influence the creation of more standardized forms of development approval within land-use decision-making processes, in part because they seek to protect natural amenity in the landscape as a form of resource commons (Robbins et al. 2012). In so doing, new types of land-use planning approaches or decisions in exurban contexts are not just intended to shape the built environment, but the natural environment as well; that is, exurban planning frameworks—often the first time strong land-use regulations are imposed in a particular place—secure the natural landscape value dimensions of newly emerging economies and influence how socio-cultural and ecological values develop. Understanding these dynamics is not only important for scholarly explorations of exurbia, but also for understanding how and whether interventions intended to achieve goals of conservation and sustainability can be achieved.

The transition from rural to exurban often feels cataclysmic for local communities and their relationships to particular places, natural resources, and community relations. Exurbanization threatens the loss of a sense of ruralness on what may be perceived as an inevitable march toward becoming the city. Cherished traditions of rural land use, like hunting, often come under scrutiny and may be challenged, thereby leading local customs to radically change or even die out. Forms of social interaction that maintained community cohesion and even household subsistence may be severely threatened or altered in a process often referred to as enclosure of the commons, whereby longstanding resource users lose access to these resources (Robbins et al. 2009). Some landowners benefit handsomely from exurban land sales, including land that was previously relatively worthless, while others struggle to retain their property and maintain their longstanding resource-based forms of income. Yet these exurban transitions are not always inevitable, nor are they straightforward, and the exurban turn may offer new opportunities for local livelihoods, albeit on new terms whose power dimensions are uneven at best (see e.g., Grabbatin et al. 2011).

13.5 Creating a More Integrated Urban Political Ecology: Drawing on Exurban Political Ecology Insights

Recent developments within political ecology have seen a divergence between approaches derived from urban studies (Heynen 2013) and approaches derived from rural studies, especially those in the developing world (Blaikie and Brookfield 1987; Peet and Watts 1996). We intend that our approach provide a model for integrating urban and rural theoretical approaches. Within urban political ecology, an increased interest in the urban/rural divide has emerged with (re)new(ed) attention to the countryside outside of major cities (Gustafson et al. 2014). With this turn, urban political ecologists have taken notice of the exurbs and other peri-urban areas. Urban political ecology scholars are seeking to examine the logics of capital in exurban areas, where the transformation of rural places is described using the metaphor of the nature being "metabolized" by both individual residents and investment strategies tied to urban centers (see e.g., Heynen et al. 2006; Gustafson et al. 2014). This work offers many fascinating possibilities, but unfortunately (and quite strangely from our perspective), urban political ecologists have almost entirely ignored existing political ecological scholarship on exurbia. While both theoretical traditions focus in part on the role of capital, we argue that the existing exurban political ecology scholarship has a stronger history of examining key aspects of land-use governance, including the role of land-use planning and decision-making specifically in producing development outcomes. Existing scholarship on the political ecology of exurbia, drawing from critical cultural geography, also provides important insights into the particular ways that nature is constructed through ideological struggles not always easily aligned with race, class, or gender during the exurban process. Moreover, exurban political ecology, having developed out of the application of "rural political ecology" and its roots in the so-called developing world (Walker 2003), may be more attentive than urban political ecology scholarship to the issue of how non-economic aspects of development (e.g., specific land uses associated with subsistence interactions with nature) as well as dimensions of land use that are not associated with residential development shape this peculiar transition (Walker and Fortmann 2003).

Moreover, representing the rural to exurban change as a purely urban transformation may risk recreating the newcomer/longtime resident culture clash and power shift binary. As illustrated in the chapters of this book, exurbia is a place where diverse values are fought over in particular times and places, embraced and supported through competitive strategies in other times and places, and where new forms of residential and natural resource development emerge as a result. While these dynamics are shaped by flows of capital and people, some from within the global urban, they are also shaped by ideas, uses, needs, and engagement by capital and individuals much less directly tied to urban centers, or capital and individuals acting in rejection of urbanism. So, too, these dynamics take place within the context of shifting global economic, social, and ecological dimensions.

Also, in overcoming the newcomer/longtime resident binary, public and private conservation area creation is a regular feature of exurban landscape change, with conflicts over the creation of these spaces tied to the different land-use and amenity needs of both competitive and cooperative capitalist groups and entities. Thus, the chapters in this book demonstrate that the concepts of competing rural capitalisms and uneven environmental management can be fruitfully expanded to think not just about the creation of new protected areas in transitional rural places, but also the constellation of emerging land-use interventions that shape exurban spaces. As we wrap up this book, we are considering the potential of "uneven land-use governance regimes"—the "formal and information institutional arrangements" (Reed 2007, p. 321) within which political economic power is wielded—as an operational concept for studying the myriad ways that new environmental management configurations emerge from within the land-use planning process, where the regulation of land is the result of the negotiation of competing ideologies of nature, and where those configurations set the rules (open up opportunities) for the type and extent of capital investment in and profit-taking from the landscape and nature.

Competing (and complementary) rural capitalisms institutionalize particular rules, the visions of future land development these codify, and the patterns of development that emerge in exurbia. By examining how these rules reflect the ideologies of different constellations of actors associated with diverse efforts to extract value from particular types of landscapes, we further suggest an ability in future research to tease apart the politics and processes of exurbanization. In the process, scholars might develop a better understanding of nature's role in politics and broad socioeconomic processes, including by revealing how material entities of nature enable, constrain, and/or marginalize particular actors and uses within exurban spaces. Indeed, wider and more systematically comparative examinations of these social-ecological dimensions are needed in the study of exurbia. Such an approach would also provide a way to begin systematically exploring the conditions under and through which particular types of land-use interventions seem to be pursued and how and why they are successful (or not).

The hybridity of exurban landscapes, where rural character and economic processes persist alongside the so-called urban ideologies of nature and real estate valuation, calls for more nuanced scholarship on the part of political ecologists, moving beyond a strong focus on capital and/or its effect on material flows. Clearly, exurbia is a rejection of the urban by many living in these transitional places, but the very process of exurbanization threatens the persistence of rural ways of interacting with, benefiting from, and using landscapes. Neither urban political ecology nor rural political ecology has satisfactorily described what is fully going on in these dynamic spaces. Exurban political ecology needs to continue comparative research into the nature and dynamics of these fascinating landscapes, albeit seeking to further detail what is distinctive about particular cases and what is generalizable to exurbia as a whole. We therefore call for a more integrated Exurban Political Ecology that is systematically comparative in its approach.

13.6 Exurbia Beyond North America and Australia

Rural landscapes across the globe have undergone and continue to undergo dynamic transitions, many of which share similarities with the exurban transitions discussed in this book. We acknowledge that this book has been about the experience in the (primarily) English-speaking, industrial, and post-industrial landscapes of the United States (although the book also contains one chapter about Australia and insights are drawn from the experiences of a Canadian editor). And indeed, as we have learned during presentations of the material in this book at conferences outside the United States, the term exurbia does not necessarily travel all that well in some global contexts. For some, the term raises notion of "peri-urban" spaces or evokes contexts of low-density development that may not be present in a particular country. Nevertheless, in our discussions with colleagues studying the types of urban and global pressures placed on transitional rural spaces, we have often found distinct parallels. Thus, we feel strongly that many of the insights presented here may offer clear guidance for a better understanding of rural transitions elsewhere, especially for developing more critically engaged studies of the social dynamics shaping those transitions.

A few points are key in thinking about the application of the insights provided in this volume to other regions. First, widespread acceptance of land-use planning, as it is generally understood in the European and British-colonial context, is largely absent within the United States. Only a few jurisdictions have anything close to paralleling the strict control over urban expansion and regulation of land uses that is a critical feature elsewhere in the developed world. This fact, however, should not lead scholars in more highly regulated environments to easily dismiss the insights provided by the U.S. case studies included here. Indeed, as several case studies reveal, the power dynamics and ideological work of landscape and conservation science are often at play outside, within, and through planning processes. Planning contexts constrain and shape negotiations over ideological differences, and, at times, layers of planning regulations have the effect of smoothing over ideological differences by leaving no room for their negotiation within political processes. Second, the lessons for thinking about low-density development within the U.S. context—specifically because planning controls are so weak—may indeed provide useful parallels for thinking about rural change in other parts of the developing world. Indeed, the socio-economic transformations associated with the exurban turn appear to be well underway in many parts of the developing world, with the same kinds of planning and conservation approaches circulating and being tested and applied within these contexts. Third, we are reminded of recent work by Nelson and

Nelson (2011), where exurbia—conceptually limited in this book to a consequence of amenity migration—also includes the movement and presence of ethnic minorities, through the existence of the so-called "linked migration" (i.e., low-wage workers, often immigrants or members of minority communities, that take up work in exurbia). So, too, ideas about what constitutes the rural or the urban—and the exurban—may differ along ethnic or racial lines (Hanlon et al. 2006; Pfeiffer 2012). Thus, the work contained in this volume—to the extent that it focuses principally on the experiences of a dominant, white majority (frequently the demographic reality in exurbia)—overlooks significant social–political and ecological dynamics and anticipates future study of these dynamics.

We hope that readers have been inspired to proceed with their own exurban political ecologies. In the book, we have offered a systematic and comparative approach to the study of the impacts of amenity migration, the processes of exurbanization, and the idea of exurbia as a fusion of urban and rural ideas worthy of study in its own right. We have compared case studies where exurban landscapes have emerged from an identified cluster of processes. We have focused on political ecological analyses of material landscapes emerging in the United States and Australia from social and political processes in which the shift from rural to exurban is negotiated. We have been especially interested in land-use planning, ideologies of nature, and the material landscape changes these processes bring about, from individual homes and properties to large areas of open space conservation. We have used the concepts of competing rural capitalisms and uneven environmental management to provide frameworks for exploring the intersection of ecologies and economics, suggesting that exurbia is a place where productive and consumptive land uses and livelihoods co-exist in dynamic tension over the long term. But we continue to see exurbia as a fusion of urban and rural with energy of its own. That is, exurbia is a phenomenon in its own right and worthy of theoretical consideration and grounded research, distinct from urban and rural analyses. We hope the analytical framework of this book and the resulting increased understanding of exurban change will provide a basis for more sustainable and just outcomes for communities and natural spaces undergoing exurbanization.

References

Abrams, J. B., & Bliss, J. C. (2013). Amenity landownership, land use change, and the re-creation of "working landscapes". *Society & Natural Resources, 26*(7), 845–859. doi:10.1080/08941920.2012.719587.

Berube, A., Singer, A., Wilson, J. H., & Frey, W. H. (2006). *Finding Exurbia: America's fast-growing communities at the metropolitan fringe*. Washington, DC: The Brookings Institution. Retrieved from http://www.brookings.edu/~/media/research/files/reports/2006/10/metropolitanpolicy-berube/20061017_exurbia.pdf.

Blaikie, P., & Brookfield, H. (1987). *Land degradation and society*. New York: Methuen.

Brogden, M. J., & Greenberg, J. B. (2003). The fight for the west: A political ecology of land use conflicts in Arizona. *Human Organization, 62*(3), 289–298.

Brown, D. G., Johnson, K. M., Loveland, T. R., & Theobald, D. M. (2005). Rural land-use trends in the coterminous United States, 1950–2000. *Ecological Applications, 15*(6), 1851–1863.

Grabbatin, B., Hurley, P. T., & Halfacre, A. (2011). "I still have the old tradition": The co-production of sweetgrass basketry and coastal development. *Geoforum, 42*(6), 638–649. doi:10.1016/j.geoforum.2011.06.007.

Gustafson, S., Heynen, N., Rice, J. L., Gragson, T., Shepherd, J. M., & Strother, C. (2014). Megapolitan political ecology and urban metabolism in southern Appalachia. *The Professional Geographer, 66*(4), 664–675. doi:10.1080/00330124.2014.905158.

Hanlon, B., Vicino, T., & Short, J. (2006). The new metropolitan reality in the US: Rethinking the traditional model. *Urban Studies, 43*(12), 2129–2143. doi:10.1080/00420980600936525.

Heynen, N. (2013). Urban political ecology I: The urban century. *Progress in Human Geography, 38*(4), 598–604. doi:10.1177/0309132513500443.

Heynen, N., Kaika, M., & Swyngedouw, E. (2006). *In the nature of cities: Urban political ecology and the politics of urban metabolism*. London: Routledge.

Hiner, C. C. (forthcoming). Beyond the edge and in between: (Re)conceptualizing the rural–urban interface as meaning-model-metaphor. *The Professional Geographer*.

Hurley, P. T. (2013). Whose sense of place? A political ecology of amenity development. In W. P. Stewart, D. R. Williams, & L. E. Kruger (Eds.), *Place-based conservation: Perspectives from the social sciences* (pp. 165–180). Dordrecht, The Netherlands: Springer Netherlands. doi:10.1007/978-94-007-5802-5.

Hurley, P. T., & Arı, Y. (2011). Mining (dis)amenity: The political ecology of mining opposition in the Kaz (Ida) Mountain Region of Western Turkey. *Development and Change, 42*(6), 1393–1415. doi:10.1111/j.1467-7660.2011.01737.x.

Hurley, P. T., & Walker, P. A. (2004). Whose vision? Conspiracy theory and land-use planning in Nevada County, California. *Environment and Planning A, 36*(9), 1529–1547. doi:10.1068/a36186.

Larsen, S., & Hutton, C. (2012). Community discourse and the emerging amenity landscapes of the rural American West. *GeoJournal, 77*(5), 651–665. doi:10.1007/s10708-011-9410-1.

Nelson, L., & Nelson, P. B. (2011). The global rural: Gentrification and linked migration in the rural USA. *Progress in Human Geography, 35*(4), 441–459. doi:10.1177/0309132510380487.

Nesbitt, J. T., & Weiner, D. (2001). Conflicting environmental imaginaries and the politics of nature in Central Appalachia. *Geoforum, 32*(3), 333–349. doi:10.1016/S0016-7185(00)00047-6.

Newfont, K. (2012). *Blue Ridge commons: Environmental activism and forest history*. Athens, GA: University of Georgia Press.

Peet, R., & Watts, M. (1996). Liberation ecology: Development, sustainability, and environment in an age of market triumphalism. In R. Peet & M. Watts (Eds.), *Liberation ecologies: Environment, development, social movements* (pp. 1–45). London: Routledge.

Pfeiffer, D. (2012). African Americans' search for "more for less" and "peace of mind" on the exurban frontier. *Urban Geography, 33*(1), 64–90. doi:10.2747/0272-3638.33.1.64.

Reed, M. G. (2007). Uneven environmental management: A Canadian comparative political ecology. *Environment and Planning A, 39*(2), 320–338. doi:10.1068/a38217.

Robbins, P., Martin, S., & Gilbertz, S. (2012). Developing the commons: The contradictions of growth in exurban Montana. *The Professional Geographer, 64*(3), 317–331. doi:10.1080/00330124.2011.601193.

Robbins, P., McSweeney, K., Chhangani, A. K., & Rice, J. L. (2009). Conservation as it is: Illicit resource use in a wildlife reserve in India. *Human Ecology, 37*(5), 559–575. doi:10.1007/s10745-009-9233-6.

Walker, P. A. (2003). Reconsidering "regional" political ecologies: Toward a political ecology of the rural American West. *Progress in Human Geography, 27*(1), 7–24. doi:10.1191/0309132503ph410oa.

Walker, P. A., & Fortmann, L. (2003). Whose landscape ? A political ecology of the "exurban" Sierra. *Cultural Geographies, 10*(4), 469–491. doi:10.1191/1474474003eu285oa.

Walker, P. A., & Hurley, P. (2011). *Planning paradise: Politics and visioning of land use in Oregon*. Tucson, AZ: University of Arizona Press.

Index

L.E. Taylor, P.T. Hurley (eds.), *A Comparative Political Ecology of Exurbia*,
DOI 10.1007/978-3-319-29462-9

U

The manufacturer's authorised representative in the EU is Springer Nature Customer Service Centre GmbH, Europaplatz 3, 69115 Heidelberg, Germany. If you have any concerns regarding our products, please contact ProductSafety@springernature.com

Printed and bound by CPI Group (UK) Ltd, Croydon, CR0 4YY
30/09/2026
02225229-0010